SERIES ON
OPTIMIZATION
VOL. 2

DIFFERENTIAL GAMES OF PURSUIT

Leon A. Petrosjan

Faculty of Applied Mathematics
St. Petersburg State University
RUSSIA

World Scientific
Singapore • New Jersey • London • Hong Kong

Published by

World Scientific Publishing Co. Pte. Ltd.

P O Box 128, Farrer Road, Singapore 9128

USA office: Suite 1B, 1060 Main Street, River Edge, NJ 07661

UK office: 73 Lynton Mead, Totteridge, London N20 8DH

Library of Congress Cataloging-in-Publication Data

Petrosjan, L. A. (Leon Aganesovich)
Differential games of pursuit / Leon A. Petrosjan.
p. cm. -- (Series on optimization; vol. 2)
Includes bibliographical references and index.
ISBN 9810209797 : $82.00 (U.S.)
1. Differential games. I. Title. II. Series.
QA272.P46 1993
519.3--dc20 93-24420
CIP

Printed in Singapore.

Contents

Preface

The process of pursuit represents a typical conflict situation and it is, therefore, not surprising that such a situation is one of the oldest topics within the field of game theory. This area was pioneered by H. Steinhaus [62], who was the first to formulate in 1925 the problem of pursuit as the differential game of pursuit. After a prolonged silence, in the 1950's the mathematicians resumed their researches in differential games. They developed a method based on the "main" first order partial differential equation, which seems to be first obtained by R. Isaacs. Among the literatures of that period are the papers by W. Fleming [52, 53] dealing with the convergency of values of discrete (discrete–time) games to a solution of the "main" equation, the papers by L. Bercovitz [50, 51], where the necessary and sufficient conditions are derived for the existence of the equilibrium situation in terms of the "main" equation, and, finally, the monograph by R. Isaacs [1], where numerous examples are provided for the illustration of the whole method of finding a solution, which is based on the consideration of the "main" equation.

The first home works on differential games appeared in the 1960's (see [11, 31, 40]). L.S. Pontryagin and his followers considered the problems of pursuit by solving it for Pursuer P, and the problem of evasion for Evader E. The problem of evasion is to describe a set of initial states from which Evader can ensure avoidance of capture. Here, for the linear case, an exhaustive solutions are obtained to both problems (see [18–21, 36–41, 45, 46]).

N.N. Krasovsky and his followers assess the quality of pursuit by the time span from the initial instant of the process up to the ℓ-capture instant ($\ell > 0$). This method is based on the extremal sighting rule which gives in a number of cases the equilibrium point. The method was conclusively formulated in the monograph by N.N. Krasovsky [10]. Important developments are also observed in the theory of differential games with depended motions, for which a fine machinery of mixed strategies has been developed, leading to the equilibrium point. All the findings in this area are described in [12]. Of special interest are the works [14, 22, 24, 60].

The proposed monograph develops an approach to the problem of pursuit from the point the view of the general theory of zero–sum two–person games. Consideration is given to the problems featuring various payoff functions (the distance between players when the game terminates, the minimum distance between players as the game proceeds, the escape from a given set before the capture instant, etc.). For each problem, the strategies of pursuit and evasion are constructed which, under some conditions, prove to be optimal in the sense of "saddle point" (equilibrium situation). The existence theorems are proved for equilibrium points, and optimal strategies are explicitly expressed for a large number of specific games of pursuit. Our considerations are not confined to the perfect information games only. The chapter deals with some of the above problems for the case with delay of information of the process state. The last chapters are devoted to the n-person differential games, where the main problem of time consistency [66] and strongly time consistency [67] of the optimal decisions is investigated.

The monograph contains mostly the results obtained by the author, with the exception of some theorems from the general theory of games (see Ch. 1) and some results given in other chapters of the book. In particular, the results presented in 3, Ch. 2, were obtained by O.A. Malafeev, and those in 4, 5, Ch. 6 by R.V. Khachaturian. Examples 11, 12 in Ch. 1 were solved by N.I. Satimov. The proof of lemma in 3, Ch. 5 was done by J. Dutkevitsch, second method of successive approximates in 6, Ch. 3 was derived by S.V. Chistiakov, and the main results in 8, Ch. 7 are prooved by N.M. Sloboshanin.

The author wishes to express his indebtedness to Prof. V.I. Zubov and Prof. N.N. Vorobyev for their involvement in the discussions during the monograph preparations. His sincere thanks also go to Prof. J.S. Osipov and V.N. Ushakov from the Ural Research Center of the Russian Acad. of Sc. for their invaluable assistance in preparing the manuscript.

The author specially thank A.A. Ovseenko and S.S. Tolstobrov for the preparation of the English publication in PC.

English translation
by
J.M.Donetz and L.A. Petrosjan.

Chapter 1

Preliminaries

1.1 Zero–sum two–person games in normal form

In order to formalize a conflict situation of decision– making, it is essential to provide mathematical description to the parties involved in the conflict and the results of their actions. In compliance with the adopter terminology, such actions are called strategies and their results are assessed by a numerical function which is called a payoff function. The strategy and payoff may vary in nature to suit the task. The notions of the strategy and payoff will be illustrated by referring to examples. We shall give a formal definition of the zero–sum two–person game in normal form [2].

Example 1 (simultaneous game of pursuit on the plane). Let S_1 and S_2 be two sets on the plane. The game proceeds as follows. Player 1 chooses a point $x \in S_1$, and Player II a point $y \in S_2$. In making their choices the Player I and II have no information on the opponents actions and, therefore, such a choices can be conveniently interpreted as simultaneous. In this case, the points x and y are the strategies for the Player I and II. Thus the sets of all strategies coincide with the sets S_1 and S_2 on the plane.

Player I aims to minimize the distance between himself and Player II. Player II pursues the opposite aim. For this reason, by the payoff of Player II in this game is meant the Euclidean distance $\rho(x, y)$ between the points $x \in S_1$ and $y \in S_2$. Players I's payoff is equal to be Player II's payoff of opposite sign (Fig.1).

Example 2 (multistage game of pursuit). Let each point z in the plane be associated with the sets U_z and V_z. Consider the game that follows. At the initial time instant, the Player I and II are respectively at the points x_0 and y_0 and make simultaneous choice of the points $x_1 \in U_{x_0}$ and $y_1 \in V_{y_0}$,

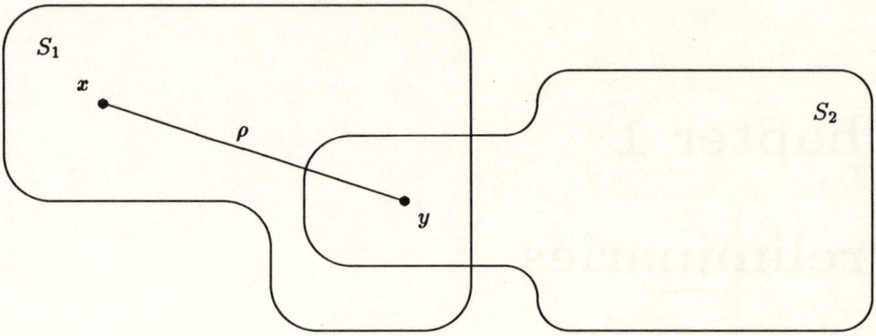

Fig. 1.

whereupon they make a transition to the states x_1 and y_1. At the points x_1 and y_1 the players choose respectively the points $x_2 \in U_{x_1}$ and $y_2 \in V_{y_1}$ and make a transitions to the states x_2 and y_2, etc. In the states x_{k-1}, y_{k-1}, the players choose the points $x_k \in U_{x_{k-1}}$ and $y_k \in V_{y_{k-1}}$ and make a transitions to the states x_k, y_k. The process terminates at the N–th stage by choosing the points x_n, y_n and making a transition to the states x_N, y_N (Fig.2).

As a result such a sequential decision making procedure realizes trajectories: $x_0, x_1, \ldots, x_k, \ldots, x_N$ being the trajectory of Player I and $y_0, y_1, \ldots, y_k, \ldots, y_N$ the trajectory of Player II. The aim of Player I is to minimize the distance between himself and Player II throughout the duration of the game, while the aim of Player II is the opposite one.

Definition of strategies in the game involves more complexities than in the earlier example. In this case, the strategy determines at each step x_{k-1} (y_{k-1}) the rule for the player to choose the next state (an element of the set $U_{x_{k-1}}$, ($V_{y_{k-1}}$)). It can be easily shown that the player's choice must be dependent on the state information he has on each step.

Assume that Player I at each k-th step has the knowledge of k, the state x_k, has the memory of all the previous states and knows the states $y_0, y_1, \ldots, y_k$ previously chosen by the opponent and his choice of $y_{k+1} \in V_{y_k}$ in the state y_k. Player II at each k–th state has complete information on k, the state y_k, his previous states $y_0, y_1, \ldots, y_{k-1}$ and the states $x_0, x_1, \ldots, x_k$ in which his opponent has been up to the instant k. By this definition of information state, the discrimination of Player II occurs at each step of the game.

Of course the formulation of the problem, where at each k–th stage of the game, in making his choice, the Player I has a knowledge of Player II's choice at this stage and Player II in making his choice has a knowledge of Player I's

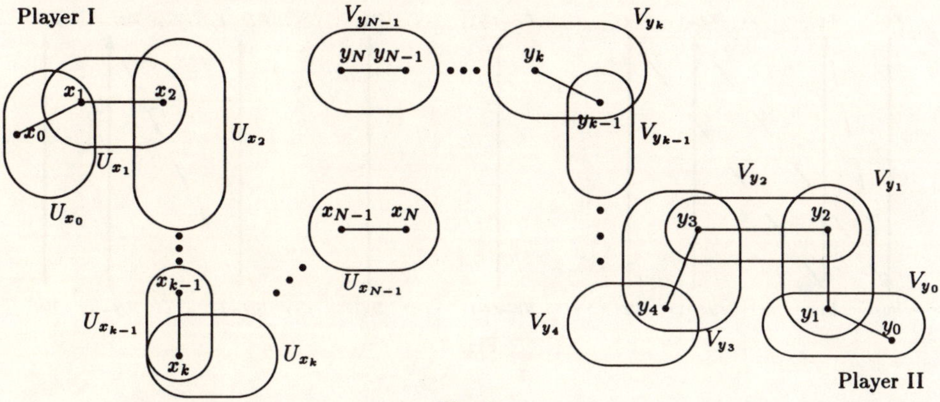

Fig. 2.

choice, would be absurd. The implication of the above information state is that Player II is the "leader". Some slight modifications in the definition enables formulation of the game where Player I is the leader.

The players have the knowledge of the game at each decision stage. For this reason this game will be called the perfect information game. The case of the perfect information game lends itself most readily for analysis, although the solution of specific games involves problems because of the large amount of information to be treated when the players' strategies are formed.

The player's strategies must enable them to make a unique choice in any possible information state. With this understanding of players strategies, the strategies for Player I must be the functions which place every possible information state of the player, i.e. every sequence $x_0, x_1, \ldots, x_k$, y_0, $y_1, \ldots, y_k, y_{k+1}, k$ in correspondence with the choice of a point, which is possible in the given information state, i.e. the choice of the point x_{k+1} from the set U_{x_k}. Thus, the strategy for Player I is a single–valued function of variables $x_0, x_1, \ldots, x_k, y_0, y_1, \ldots, y_k, y_{k+1}, k$ with values in U_{x_k}:

$$u(\cdot) = u(x_0, x_1, \ldots, x_k, y_0, y_1, \ldots, y_k, y_{k+1}, k)$$

Player II's strategy is defined in the same way except that the latter is the function only of $x_0, x_1, \ldots, x_k$, $y_0, y_1, \ldots, y_k, k$ since the choice of $x_{k+1} \in U_{x_k}$ is unknown to Player II at the k–th step. The strategy of Player II will be denoted by

$$v(\cdot) = v(x_0, \ldots, x_k, y_0, \ldots, y_k, k)$$

The values of the function $v(\cdot)$ belong to the set V_{y_k}. It is easily comprehended that each pair of initial conditions x_0, y_0 and the pair of chosen strategies uniquely determine the trajectories of the players I and II $x_0, x_1, \ldots, x_N$,

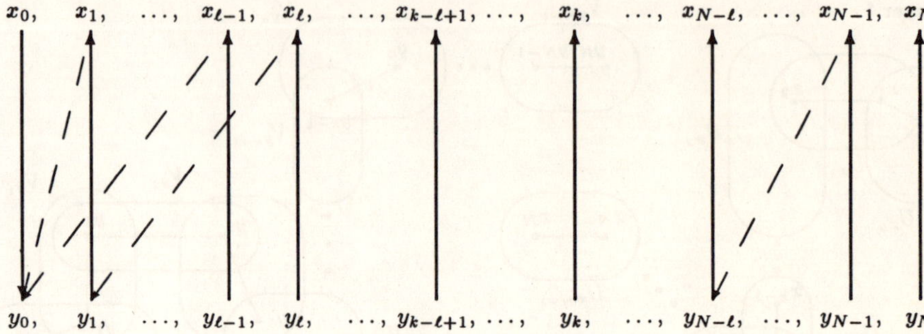

Fig. 3.

$y_0, y_1, \ldots, y_N$ that are constructed recurrently by the rule.

$$x_{k+1} = u(x_0, \ldots, x_k, y_0, \ldots, y_k, y_{k+1}, k),$$
$$y_{k+1} = v(x_0, \ldots, x_k, y_0, \ldots y_k, k),$$
$$k = 1, \ldots, N-1.$$

Consequently, the initial conditions x_0, y_0 and the pair of the strategies define the player's payoff uniquely. Considering the previously stated objective of the game,the payoff may be defined as follows:

$$K(u(\cdot), v(\cdot)) = \min_k \rho(x_k, y_k)$$

where $x_0, x_1, \ldots, x_N, y_0, y_1, \ldots, y_N$ are the pair of trajectories corresponding uniquely to the chosen strategies $u(\cdot), v(\cdot)$ and the initial conditions x_0,y_0. Unlike Example 1, the sets of strategies here are the sets of functions and have a complex structure (see Fig.2).

Example 3 (multistage perfect information game of pursuit). Kinematics and payoff in this game are defined as in Example 2, except that Player I is less informed at each stage of the game.

Let $0 \leq \ell < N$ be a given natural number. At each step $k \geq \ell$ Player I knows his choice sequence $x_0, \ldots, x_k$ and the opponent's choice sequence at the first $k - \ell$ steps of the game, i.e. $y_0, y_1, \ldots, y_{k-\ell}, y_{k-\ell+1}$, at the steps $k < \ell$ only $x_0, \ldots, x_k$ and y_0. This means that in the case involved he receives the ℓ –delay information about Player II. The information state of Player II is the same as in Example 2. Schematically, the players' informations state can be represented as follows (Fig.3) [47].

Proceeding from the notion of strategy as a rule, which places each information state of the player in correspondence with the choice of a particular decision from the set of decisions that are possible in the given information state, we may say that the strategies in this game for Player I are functions

$$u(\cdot) = u(x_0, \dots, x_k, y_0, \dots, y_{k-\ell+1}, k)$$

with values in the set U_{x_k} for $k \geq \ell$ and the functions

$$u(\cdot) = u(x_0, \dots, x_k, y_0, k)$$

with values in the set U_{x_k} for $k < \ell$. The strategies of Player II remains the same as in Example 2.

This game is called the game with delayed information for Player I and, for $\ell > 0$, refers to the zero-sum game with imperfect information. In a special case, for $\ell = 0$ we have the game with perfect information from Example 2.

Let S_1 and S_2 be the bounded sets on the plane, and C a sphere containing the set $S_1 \bigcup S_2$. We choose the initial conditions x_0, y_0 in such a manner that $\rho(x_0, y_0) \geq d(C)$ ($d(C)$ being a diameter of sphere C). Now set $N = \ell, \ell = 1$, $U_{x_0} = S_1$, $V_{y_0} = S_2$. The proposed game becomes the game from Example 1.

Indeed, since x_1 and y_1 are constrained by $x_1 \in U_{x_0} = S_1, y_1 \in V_{y_0} = S_2$, we have

$$\min_k \rho(x_k, y_k) = \min[\rho(x_0, y_0), \rho(x_1, y_1)] = \rho(x_1, y_1)$$

Example 4.(multistage time optimal pursuit game). In this example kinematics and information states of players are the same as in Example 2. At the same time, the payoff is determined here differently.

Assume that a real number $R > 0$ called the capture radius, is given. The duration of the game (the number of steps N) is not specified. Let N_n be the number of the step at which $\rho(x_{N_n}, y_{N_n}) \leq R$ for the first time. The play terminates at the step N_n and the payoff of Player I is assumed to be $(-N_n)$. The payoff of Player II is said to be $+N_n$.

By this definition, Player I is aiming to approach Player II at a distance not exceeding R in a minimal time (for a minimum number of steps). Player II seeks to delay the capture moment. Also, possible are such realizations (trajectories) that for all N's there is $\rho(x_N, y_N) > R$. Then the payoff for Player II is assumed to be $+\infty$, and the one of Player I $-\infty$, respectively. Here the strategy sets are wider than those in Example 1 since account must be taken of possible occurrence of an arbitrary long trajectories. Therefore, the strategy set for Player I(II) is composed of all possible functions of the form

$$u(\cdot) = u(x_0, \dots, x_k, y_0, \dots, y_{k+1}), k = 0, 1, \dots, N, \dots$$

$$(v(\cdot) = v(x_0, \dots, x_k, y_0, \dots, y_k), k = 0, 1, \dots, N, \dots)$$

with values in the sets U_{x_k}, V_{y_k}. As in Example 3, we can formulate the time optimal pursuit game with information delay for Player I.

Example 5 (multistage game with complete information). The game duration is not specified. Kinematics and information states of players are the same as in Example 4. Let $\{x_k\}$, $\{y_k\}$ be the trajectories realized in the game. The payoff of Player I is assumed to be +1 if there is such N that $\rho(x_N, y_N) \leq R$, otherwise it is assumed to be −1. The payoff of Player II equals the one of opposite sign for Player I . Thus, Player I aims to approach Player II for a distance R, and Player II has the opposite objective. Here as in the earlier case, we can formulate the incomplete information problem similar to the problem in Example 3.

Example 6 (multistage "Life line" game with complete information). A convex set S is given on the plane. At the initial time instant, the Players I and II are in the states $x_0 \in S$, $y_0 \in S$. Next, at each step Player I makes a choice from the set $U_{x_k} \cap S$, and Player II from the set V_{y_k}. The game terminates in one of the following two cases:

1. There is such number N_n that $\rho\ (x_{N_n}, y_{N_n}) \leq R$, $y_{N_n} \in S$;

2. There is such number N_S that $y_{N_s} \notin S$.

In the first case, the payoff of Player I(II) is assumed to be +1(−1). If one of this possibilities is not realized, i.e. $\rho(x_N, y_N) > R, y_N \in S$ for all N's, then the payoff of both players is said to be zero. From the definition of kinematics and payoff function we may conclude that the object of Player I is to approach Player II for a distance R within the set S before Player II penetrates the boundary of S. Therefore, this game is known as the "Life line" game, since the boundary of the set S turns out to be the "Life line" for Player II. Player I must remain in S throughout the game.

With the game duration unspecified, the game is represented by the alternating plays with complete information and, therefore, the strategies are determined just as in Example 4. The incomplete information case for Player I (Fig.4) can also be formulated here.

Example 7 (game with several pursuers and one evader). Here the game is played by a detail of pursuers, $P = \{P_1, \ldots, P_m\}$, acting as one player, i.e. having a common objective, and Evader II. Detail P will be referred to as Player I . All the above examples can be generalized to this case. Our consideration will be confined to Example 2.

Let $x_0^1, \ldots, x_0^m, y_0$ be the starting position of the detail partners $P_1, \ldots, P_m$ and Player I I on the plane. $U^i = \{U_{x_i}\}$, $i = 1, \ldots, m$, $V = \{V_y\}$ are the set systems on the plane governing the kinematics of the detail partners $P_1, \ldots, P_m$ and Player II.

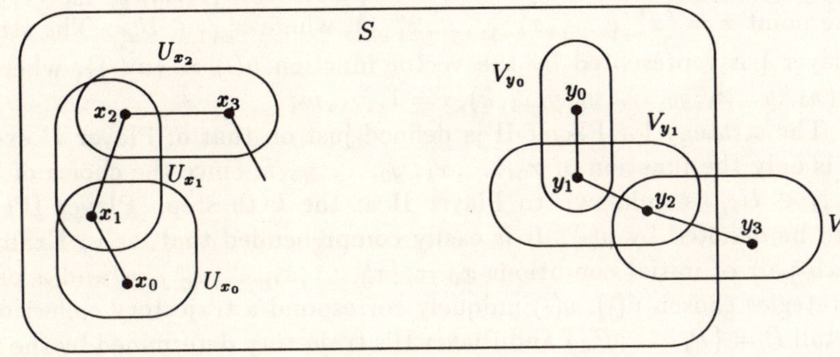

Fig. 4.

The game is as follows. At the initial time instant, the detail partners $P_1, \ldots, P_m$ and Player II make a simultaneous choice of points $x_1^1 \in U_{x_0^1}, \ldots,$ $x_1^m \in U_{x_0^m}$, $y_1 \in V_{y_0}$, and pass to the states $x_1^1, \ldots, x_1^m, y_1$. At the points $x_1^1, \ldots,$ x_1^m, y_1 the players choose the points $x_2^1 \in U_{x_1^1}, \ldots, x_2^m \in U_{x_1^m}, y_2 \in V_{y_1}$ and pass to the states $x_2^1, \ldots, x_2^m, y_2$, etc. When in the states $x_{k-1}^1, \ldots, x_{k-1}^m, y_{k-1}$ the players choose the points $x_{k_1} \in U_{x_{k-1}^1}, \ldots, y_k, x_k^m \in U_{x_{k-1}^m}, y_k \in V_{y_{k-1}}$ and pass to the states $x_k^1, \ldots, x_k^m, y_k$, etc. When in the states $x_{k-1}^1, \ldots, x_{k-1}^m, y_{k-1}$, the players choose the points $x_k^1 \in U_{x_{k-1}^1}, \ldots, x_k^m \in U_{x_{k-1}^m}, y_k \in V_{y_{k-1}}$ and pass to the states $x_k^1, \ldots, x_k^m, y_k$. The process terminates at the N–th step by choosing the points $x_N^1 \in U_{x_{N-1}^1}, \ldots, x_N^m \in U_{x_{N-1}^m}, y_N \in V_{y_{N-1}}$ and passing to the states $x_N^1, \ldots, x_N^m, y_N$. As a result of such sequential decision procedure, we obtain uniquely the $m+1$ trajectory $x_0^i, \ldots, x_N^i, i = 1, \ldots, m$, for the detail partners P_i, and the trajectory $y_0, \ldots, y_N$ for Player II.

During the game the detail P, acting as one player, aims to minimize the distance between itself and Player II at least for one of its partners. Player II has the opposite object.

The strategy for the pursuer detail P is defined as Cartesian product of strategies for every detail partner, each strategy being covered in Example 2. Assume that the information state is such as given in that example, i.e. Player I (detail P) knows k at each k–th step, his state $x_k = \{x_k^1, \ldots, x_k^i, \ldots, x_k^m\}$, remembers all his previous states, is aware of the states $y_0, \ldots, y_k$ previously chosen by his opponent, and the choice $y_{k+1} \in V_{y_k}$, by his opponent in the state y_k. At each k–th step Player II has complete information in k, his state y_k remembers his previous states $y_0, \ldots, y_{k-1}$ and the states $x_0, \ldots, x_k$, in which his opponent, has been up to the instant k.

Thus, the strategy for Player P turns out to be the rule, which correspond to each information state, i.e. each sequence $x_0, x_1, \ldots, x_k,\ y_0, \ldots, y_k, y_{k+1}, k$ the point $x = \{x^1_{k+1}, \ldots, x^i_{k+1}, \ldots, x^m_{k+1}\}$, where $x^i_{k+1} \in U_{x^i_k}$. The strategy for Player I is represented by the vector function $u(\cdot) = \{u^i(\cdot)\}$, where $u^i(\cdot) = u^i(x_0, \ldots, x_k, y_0, \ldots, y_k, y_{k+1}, k)$, $i = 1, \ldots, m$.

The strategy for Player II is defined just on that of Player P except that it is only the function of $x_0, \ldots, x_k,\ y_0, \ldots, y_k,\ k$, since the choice of the point $x_{k+1} \in U_{x_k}$ is unknown to Player II at the k-th step. Player II's strategy will be denoted by $v(\cdot)$. It is easily comprehended that, as in Example 2, to each pair of initial conditions $x_0 = \{x^1_0, \ldots, x_i, \ldots, x^m_0\}, y_0$ and a pair of the strategies chosen $u(\cdot)$, $v(\cdot)$ uniquely correspond a trajectory collection for the detail $P = \{P_1, \ldots, P_m\}$ and Player II's trajectory determined by the recurrent rule

$$x_{k+1} = u(x_0, \ldots, x_k, y_0, \ldots, y_k, y_{k+1}, k)$$

$$y_{k+1} = v(x_0, \ldots, x_k, y_0, \ldots, y_k, k),\ k = 1, \ldots, N-1$$

Consequently, the initial conditions x_0, y_0 and the pair of strategies $u(\cdot), v(\cdot)$ define the players' payoff in a unique way. Therefore, the payoff function may be defined as

$$K(u(\cdot), v(\cdot)) = \min_i\{\min_k p(x^i_k, y_k)\},$$

where $x^i_0, \ldots, x^i_N;\ y_0, \ldots, y_N,\ i = 1, \ldots, m$, are the trajectories for the partners of detail P and the Evader E corresponding uniquely to the chosen strategies $u(\cdot), v(\cdot)$ and initial conditions x_0, y_0.

Examples 3–6 can be naturally generalized to the case, where several pursuers, acting as one player, and one evader are involved.

The above examples are represent of a variety of zero–sum games even in the simple case, when they provide a qualitative model for the some physical process – pursuit. Beginning with the second example, we have been dealing with the dynamic decision processes developing in time. To obtain a correct formulation of such important notion as strategy, it is essential that the information capabilities of both parties should be defined precisely. After the strategies had been defined for both players, the games lost their dynamic character and came to bear resemblance, at least outwardly, to the game in Example 1. Indeed, having chosen strategies $u(\cdot), v(\cdot)$ at the start of the game, the players then did not make any new decisions since, by definition, the strategies envisaged a unique choice of decision at each step of the game and in any information setting. Therefore, the choice of strategies $u(\cdot), v(\cdot)$ determines uniquely the outcome of the game, i.e. the payoff. Such a transition from the dynamic game to the static state, when the entire game reduces to a simultaneous choice of strategies and subsequent computation of payoff function, is called the normalization of the game, and the game itself the game in normal form [2].

We now offer a general definition for the zero–sum game in normal form.

Definition 1. Let us assume that two arbitrary sets P, E are given. The set P is called the set of strategies for Player I , and the set E is the set of strategies for Player II. The elements of the sets P and E are denoted by $u(\cdot), v(\cdot)$, respectively. The real function K is defined on the Cartesian product $P \times E$. The triple $\Gamma =< P, E, K >$ is called the zero–sum game in normal form.

The game Γ may be interpreted as follows. The Players I and II make a simultaneous (that is, without any information on the opponent's choice) choice of elements $u(\cdot) \in P$ and $v(\cdot) \in E$. Player II receives the payoff equal to $K(u(\cdot), v(\cdot))$, and Player I the payoff equal to $-K(u(\cdot), v(\cdot))$.

In Example 1, the sets P and E coincide with the sets S_1 and S_2 on the plane, and the function $K(u(\cdot), v(\cdot))$ was the Euclidean distance between the points $x \in S_1$ and $y \in S_2$.

In the subsequent examples the strategy sets P and E had a more complex structure. They represented the sets of functions $u(\cdot), v(\cdot)$ that were defined for all possible information states governed by the rules of the game, and the payoff function was defined uniquely by the strategies chosen and was computed by the players' trajectories. Another feature of examples 2–7 is that the game in normal form depended on the initial states of players x_0, y_0. Therefore, we actually dealt with the entire family of the games in normal form Γ_{x_0,y_0} dependent on the initial states x_0, y_0 as parameters.

Definition 2. The pair of strategies $u(\cdot), v(\cdot)$,where $u(\cdot) \in P$, $v(\cdot) \in E$, is called the situation in the game Γ. The definitions of the game Γ suggests that the situation in the game Γ defines the players' payoff in a unique way.

1.2 Equilibrium Point

Let the game Γ be given in normal form $\Gamma =< P, E, K >$. For simplicity, assume that the sets P and E are compact, and the payoff function K is continuous. We shall define the minimum payoff which can be ensured by Player II in the game Γ.

If the strategy $\bar{u}(\cdot)$ chosen by the opponent were known to Player II he would always ensure the payoff

$$\max_{v(\cdot)} K(\bar{u}(\cdot), v(\cdot)) = \Phi_1(\bar{u}(\cdot))$$

With this in mind, and being "wise", Player I can choose the strategy $u_0(\cdot)$

from the condition, $\min_{u(\cdot)} \Phi_1(u(\cdot))$, i.e

$$\Phi_1(u_0(\cdot)) = \min_{u(\cdot)} \max_{v(\cdot)} K(u(\cdot), v(\cdot)) = \bar{v}. \tag{1.2.1}$$

Let $u_0(\cdot)$ be a strategy for Player I, under which

$$\min_{u(\cdot)} \max_{v(\cdot)} K(u(\cdot), v(\cdot)) = \max_{v(\cdot)} K(u_0(\cdot), v(\cdot)) = \bar{v}$$

is achieved. Then the strategy $u_0(\cdot)$ is called the minimax strategy for Player I, and $\bar{v}$ the upper value of the game. Playing the minimax strategy $u_0(\cdot)$, Player II's payoff does not exceed the quantity $\bar{v}$ under any strategy $v(\cdot) \in E$. In fact , for all $v(\cdot) \in E$ the following inequality holds

$$\bar{v} = \max_{v(\cdot)} K(u_0, v(\cdot)) \geq K(u_0(\cdot), v(\cdot))$$

The latter also means that, in playing the strategy $u_0(\cdot)$ loss of Player I (with the zero-sum game, the loss of Player I is equal to the payoff of Player II) does not exceed $\bar{v}$. Note that, to ensure the payoff $\bar{v}$. Player II has to know the choice made by Player I. In this case, choosing a strategy $v_0(\cdot)$ from the condition $\max_{v(\cdot)} K(u_0(\cdot), v(\cdot)) = K(u_0(\cdot), v_0(\cdot)) = \bar{v}$, he obtains the payoff $\bar{v}$. In choosing the strategy $v_0(\cdot)$, however, he can not ensure the payoff $\bar{v}$ under any strategy $\bar{u}(\cdot)$ for Player I, since in the situation $(\bar{u}(\cdot), v_0(\cdot))$ it may turn out that $K(\bar{u}(\cdot), v_0(\cdot)) < K(u_0(\cdot), v_0(\cdot))$.

Let us now assume that Player I acquired somehow knowledge of the strategy $\bar{v}(\cdot)$ chosen by Player II in the game Γ. Then he can always ensure that the payoff of Player II does not exceed the quantity

$$\min_{u(\cdot)} K(u(\cdot), \bar{v}(\cdot)) = \Phi_2(\bar{v}(\cdot)).$$

With this in mind, and being "wise", Player II can choose a strategy $v_0(\cdot)$ from the condition $\Phi_2(v_0(\cdot)) = \max_{v(\cdot)} \Phi_2(v(\cdot))$ i.e.

$$\Phi_2(v_0(\cdot)) = \max_{v(\cdot)} \min_{u(\cdot)} K(u(\cdot), v(\cdot)) = \underline{v} \tag{1.2.2}$$

Let $v_0(\cdot)$ be a strategy for Player II, under which

$$\max_{v(\cdot)} \min_{u(\cdot)} K(u(\cdot), v(\cdot)) = \min_{u(\cdot)} K(u(\cdot), v_0(\cdot)) = \underline{v}.$$

The strategy $v_0(\cdot)$ is called the maximin strategy of Player II, and $\underline{v}$ the lower value of the game. By playing the maximin strategy $v_0(\cdot)$, Player II ensures the payoff $\underline{v}$, whatever strategy may be chosen by Player I. Indeed,

$$\underline{v} = \min_{u(\cdot)} K(u(\cdot), v_0(\cdot)) \leq K(u(\cdot), v_0(\cdot)).$$

The definition of quantities $\bar{v}$ and $\underline{v}$ suggests that there is always

$$\bar{v} \geq \underline{v}. \tag{1.2.3}$$

Thus, Player I can always ensure that the payoff of Player II does not exceed $\bar{v}$, and Player II can always ensure the payoff, that is at least equal to $\underline{v}$.

Below is given a rigorous proof of this statement for a more general case.

Lemma 1 *Let the game* $\Gamma = < P, E, K(u(\cdot), v(\cdot)) >$ *be given. Then*

$$\bar{v} = \inf_{u(\cdot)} \sup_{v(\cdot)} K(u(\cdot), v(\cdot)) \geq \sup_{v(\cdot)} \inf_{u(\cdot)} K(u(\cdot), v(\cdot)) = \underline{v} \tag{1.2.4}$$

(the compactness of sets P, E *and the continuity of function* $K(u(\cdot), v(\cdot))$ *are not assumed here).*

Proof: By definition, for any strategy $u(\cdot)$.

$$\sup_{v(\cdot)} K(u(\cdot), v(\cdot)) \geq K(u(\cdot), v(\cdot)).$$

Hence we conclude that

$$\inf_{u(\cdot)} \sup_{v(\cdot)} K(u(\cdot), v(\cdot)) \geq \inf_{u(\cdot)} K(u(\cdot), v(\cdot)). \tag{1.2.5}$$

Since (1.2.5) holds for all $v(\cdot)$, then

$$\bar{v} = \inf_{u(\cdot)} \sup_{v(\cdot)} K(u(\cdot), v(\cdot)) \geq \sup_{v(\cdot)} \inf_{u(\cdot)} K(u(\cdot), v(\cdot)) = \underline{v}$$

We have thus completed the proof of Lemma 1. ∎

If it turns and that $\bar{v} > \underline{v}$, then it is hardly reasonable to recommend the Player I and II to adopt the minimax (1.2.1) and maximin (1.2.2) strategies $u_0(\cdot), v_0(\cdot)$, respectively. Both strategies are among the too "cautious" strategies, and their choice is justified, if the player is said to acquire, somehow, knowledge of the strategy chosen by his opponent in the game Γ. At the same time, by the rules of the game, this exactly must not occur in the game Γ, since the players make their choices simultaneously and independently of one another (none of the players knows the strategy chosen by the opponent).

An important problem in the theory of zero-sum two-person games is to reveal the class of games, for which the equality

$$\bar{v} = \underline{v} = v \tag{1.2.6}$$

is satisfied. If (1.2.6) is satisfied, then by choosing properly a maximin strategy Player II ensures the payoff , $\bar{v} = \underline{v} = v$ and with the "cautious" play, Player I cannot tolerate Player II getting the payoff more than $\bar{v} = \underline{v} = v$. Thus, if Player II has no additional information on Player I's choice (he may think that Player I makes a random choice of strategy), he must play so as to ensure the payoff v. If Player II deviates from the strategy which assures him, his payoff may be less than v. It is therefore reasonable to consider v the value of the game for Player II. The strategies $u^*(\cdot), v^*(\cdot)$, for which

$$K(u^*(\cdot), v(\cdot)) \leq v \text{ for all } v(\cdot) \in E \tag{1.2.7}$$

$$K(u(\cdot), v^*(\cdot)) \geq v \text{ for all } u(\cdot) \in P \tag{1.2.8}$$

are called the optimal strategies for the Players I and II. The following theorem relates the optimal strategies to the equality (1.2.6).

Theorem 1 *Let P and E be compact sets, and the function $K(u(\cdot), v(\cdot))$ be continuous. For the existence of strategies $u^*(\cdot), v^*(\cdot)$ and the value v in the game Γ, it is necessary and sufficient that*

$$\bar{v} = \min_{u(\cdot)} \max_{v(\cdot)} K(u(\cdot), v(\cdot)) = K(u^*(\cdot), v^*(\cdot)) = v = \max_{v(\cdot)} \min_{u(\cdot)} K(u(\cdot), v(\cdot)) = \underline{v}.$$

Proof: *Necessity.* Since for all $u(\cdot)$, $K(u(\cdot), v^*(\cdot)) \geq v$, then

$$\min_{u(\cdot)} K(u(\cdot), v^*(\cdot)) \geq v,$$

$$\max_{v(\cdot)} \min_{u(\cdot)} K(u(\cdot), v(\cdot)) \geq v.$$

Similarly, the inequality $\max_{v(\cdot)} K(u^*(\cdot), v(\cdot)) \leq v$ implies

$$\min_{u(\cdot)} \max_{v(\cdot)} K(u(\cdot), v(\cdot)) \leq v.$$

Thus, we have

$$\max_{v(\cdot)} \min_{u(\cdot)} K(u(\cdot), v(\cdot)) \geq \min_{u(\cdot)} \max_{v(\cdot)} K(u(\cdot), v(\cdot)).$$

Together with Lemma 1 we have $\underline{v} = \bar{v}$.

Sufficiency. We choose $u^*(\cdot)$ from the condition

$$\max_{v(\cdot)} K(u^*(\cdot), v(\cdot)) = \min_{u(\cdot)} \max_{v(\cdot)} K(u(\cdot), v(\cdot)) = v$$

and $v^*(\cdot)$ from the condition

$$\min_{u(\cdot)} K(u(\cdot), v^*(\cdot)) = \max_{v(\cdot)} \min_{u(\cdot)} K(u(\cdot), v(\cdot)) = v.$$

Then $K(u(\cdot), v^*(\cdot)) \geq v$ and $K(u^*(\cdot), v(\cdot)) \leq v$.

We have thus completed the proof of Theorem 1. ■

It can be easily seen that if the optimal strategies $u^*(\cdot)$,$v^*(\cdot)$ exist (see conditions (1.2.7), (1.2.8)), then the equality $K(u^*(\cdot),v^*(\cdot) = v$ is satisfied.

Definition 1. The situation $(u^*(\cdot), v^*(\cdot))$ is called the equilibrium point ("saddle point") in the game $\Gamma =< P, E, K(u(\cdot), v(\cdot)) >$ if for all $u(\cdot) \in P$ and $v(\cdot) \in E$ the inequality holds.

$$K(u(\cdot), v^*(\cdot)) \geq K(u^*(\cdot)), v^*(\cdot)) = v \geq K(u^*(\cdot), v(\cdot)). \tag{1.2.9}$$

The meaning of optimal strategies and saddle point can be readily seen from (1.2.9). If the players choose the strategies $u^*(\cdot), v^*(\cdot)$ constituting the equilibrium point then none of them seeks to change his strategy. Indeed, from the left–hand side of inequality (1.2.9) it follows that if Player I deviates from the optimal strategy then the payoff of Player II only increases, and hence his own payoff only decreases (providing Player II continues to play the strategy $v^*(\cdot)$). At the same time, the right–hand side of inequality (1.2.9) implies that if Player II deviates from the optimal strategy $v^*(\cdot)$, then his payoff only decreases (providing Player I plays the optimal strategy).

Another important feature of optimal strategies is that, before the game starts, the players may inform one another on the use of optimal strategies in the game. Such information does not increase the opponent's payoff in any possible way, since the only reasonable choice here is the choice of optimal strategies which involves obtaining the value of the game.

Definition 2. Let an $\epsilon > 0$ be given. The strategies $u_\epsilon^*(\cdot), v_\epsilon^*(\cdot)$ are called the ϵ-optimal strategies for the Players I and Player II, and the situation $(u_\epsilon^*(\cdot)$, $v_\epsilon^*(\cdot))$ the ϵ–equilibrium point, if the inequality

$$K(u(\cdot), v_\epsilon^*(\cdot)) + \epsilon \geq K(u_\epsilon^*(\cdot), v_\epsilon^*(\cdot)) \geq K(u_\epsilon^*(\cdot), v(\cdot)) - \epsilon. \tag{1.2.10}$$

holds for all $u(\cdot) \in P, v(\cdot) \in E$.

Evidently, any optimal strategies are ϵ–optimal.

Lemma 2 *Suppose there is the ϵ–equilibrium point $(u_\epsilon^*(\cdot),v_\epsilon^*(\cdot))$ for any $\epsilon > 0$ in the game Γ. Then there is a limit*

$$\lim_{\epsilon \to 0} K(u_\epsilon^*(\cdot), v_\epsilon^*(\cdot)). \tag{1.2.11}$$

Proof: Let us take the sequence $\{\epsilon_n\} \to 0$ as $n \to 0$. Let $\epsilon_k \in \{\epsilon_n\}$ and $\epsilon_{k+m} \in \{\epsilon_n\}$, where m is any fixed natural number. We have

$$K(u_{\epsilon_{k+m}}^*(\cdot), v_{\epsilon_k}^*(\cdot)) + \epsilon_k \geq K(u_{\epsilon_k}^*(\cdot), v_{\epsilon_k}^*(\cdot)) \geq K(u_{\epsilon_k}^*(\cdot), v_{\epsilon_{k+m}}^*(\cdot)) - \epsilon_k. \tag{1.2.12}$$

Next

$$\begin{aligned} K(u^*_{\epsilon_k}(\cdot), v^*_{\epsilon_{k+m}}(\cdot)) + \epsilon_{k+m} \geq \\ \geq K(u^*_{\epsilon_{k+m}}(\cdot), v^*_{\epsilon_{k+m}}(\cdot)) \geq \\ \geq K(u^*_{\epsilon_{k+m}}(\cdot), v^*_{\epsilon_k}(\cdot)) - \epsilon_{k+m} \geq \\ \geq K(u^*_{\epsilon_k}(\cdot), v^*_{\epsilon_k}(\cdot)) - \epsilon_{k+m} - \epsilon_k \geq \\ \geq K(u^*_{\epsilon_k}(\cdot), v^*_{\epsilon_{k+m}}(\cdot)) - 2\epsilon_k - \epsilon_{k+m}. \end{aligned} \tag{1.2.13}$$

From (1.2.12) and (1.2.13) we obtain

$$|K(u^*_{\epsilon_k}(\cdot), v^*_{\epsilon_k}(\cdot)) - K(u^*_{\epsilon_{k+m}}(\cdot), v^*_{\epsilon_{k+m}}(\cdot))| \leq 2(\epsilon_k + \epsilon_{k+m}) = \delta_{km}$$

With any fixed m, the sequence $\delta_{km} \to 0$, $k \to 0$. Consequently, there is a limit

$$\lim_{\epsilon \to 0} K(u^*_\epsilon(\cdot), v^*_\epsilon(\cdot)). \tag{1.2.14}$$

If in the game Γ there is the equilibrium point $(u^*(\cdot), v^*(\cdot))$ then

$$\lim_{\epsilon \to 0} K(u^*_\epsilon(\cdot), v^*_\epsilon(\cdot)) = K(u^*(\cdot), v^*(\cdot)) = v \tag{1.2.15}$$

The latter equality follows from Lemma 2 and ϵ–optimality of optimal strategies $u^*(\cdot)$, $v^*(\cdot)$ for any $\epsilon > 0$. ■

Definition 3. If in the game Γ for any $\epsilon > 0$ there is the ϵ – equilibrium point, then the limit $\lim_{\epsilon \to 0} K(u^*_\epsilon(\cdot), v^*_\epsilon(\cdot))$ is called the value of the game and is denoted by v.

From the above reasonings it follows that this definition is in agreement with the definition of the game value given above. Note that the value of the game can also exist even when the optimal strategies for players I and II not exist. The following theorem contains the necessary and sufficient condition for the existence of the value of the game.

Theorem 2 *For the existence of ϵ – equilibrium point and the value of the game in the game Γ for any $\epsilon \geq 0$, it is necessary and sufficient that*

$$\max_{v(\cdot)} \inf_{u(\cdot)} K(u(\cdot), v(\cdot)) = \min_{u(\cdot)} \sup_{v(\cdot)} K(u(\cdot), v(\cdot)) = v$$

The proof of is omitted here, since it is similar to the one for Theorem 1.

In what follows by the solution of the (zero–sum) game is meant finding the saddle–point, if any, or finding the ϵ–equilibrium point for any $\epsilon > 0$. The above definition may be illustrated by referring to Example 1 from 1.

Assume that the sets S_1 and S_2 are the closed circles with radii R_1 and R_2 ($R_1 > R_2$). Let us calculated the lower value of the game $\underline{v} = \max_{y \in S_2}$

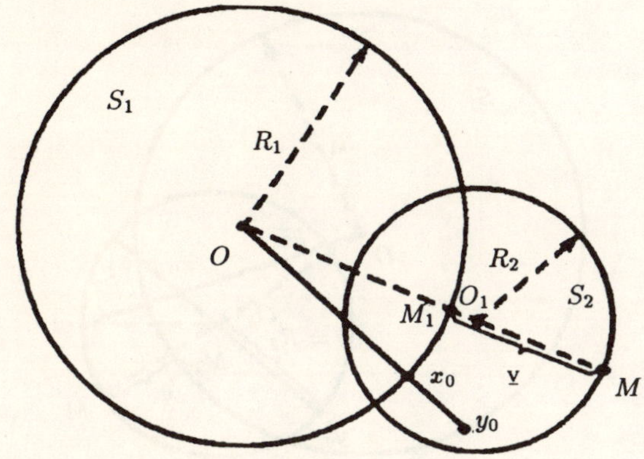

Fig. 5.

$\min_{x\in S_1} \rho(x, y)$. Let $y_0 \in S_2$. Then $\min_{x\in S_1} \rho(x, y_0)$ is achieved at the point x_0, where the straight line, passing through the center of the circle S_1 and the point y_0, intersects the boundary of the circle S_1. Evidently, the quantity $\min_{x\in S_1} \rho(x, y_0)$ attains its maximum value at the point $M \in S_2$, where the lines of centers OO_1 (Fig.5) intersect the boundary of the circle S_2 which is farthest from the point O.

Thus, $\underline{v} = \rho(O, M) - R_1$.

In order to calculate the upper value of the game $\bar{v} = \min_{x\in S_1} \max_{y\in S_2} \rho(x, y)$, consider two cases.

Case 1. The center O_1 of the circle S_2 belongs to the set S_1 (Fig.6). For each $x_0 \in S_1$ the point y_0 delivering $\max_{y\in S_2} \rho(x_0, y)$ is constructed as follows.

Let y_0^1 and y_0^2 be the points of intersection of the straight line Ox_0 with the boundary of the circle S_2, and y_0^3 be the point of intersection of the straight line O_1x_0 with the boundary of the circle S_2 which is the farthest from the point x_0. Then y_0 is determined from the condition

$$\rho(x_0, y_0) = \max_{i=1,2,3} \rho(x_0, y_0^i)$$

By construction, for all $x_0 \in S_1$.

$$\max_{y\in S_2} \rho(x_0, y) = \rho(x_0, y_0) \geq R_2$$

For $x_0 = O_1$ however we get

$$\max_{y\in S_2} \rho(O_1, y) = R_2$$

therefore

$$\min_{x\in S_1} \max_{y\in S_2} \rho(x, y) = R_2$$

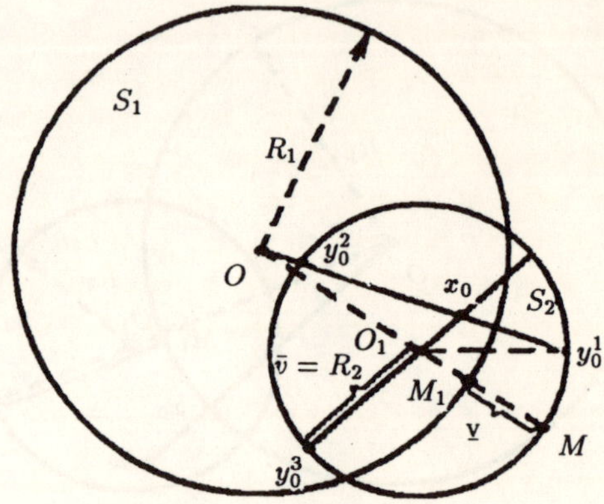

Fig. 6.

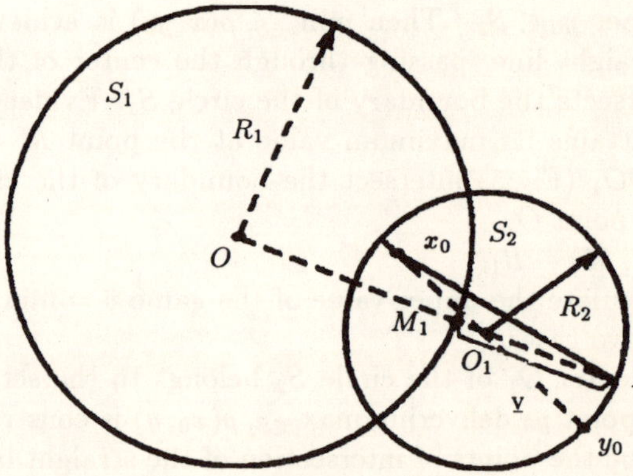

Fig. 7.

It can be directly seen that since $O_1 \in S_1$, in Case 1 $\bar{v} = R_2 \geq OM - R_1 = \underline{v}$ Moreover, the equality $\bar{v} = \underline{v}$ is possible only if O_1 belongs to the closure of the set S_1.

Thus, if in Case 1 O_1 does not belong to the boundary of the set S_1, then the equilibrium point and the value of the game does not exist. If, however, O_1 belongs to the closure of the set S_1, there is the equilibrium point. Here the optimal strategy for Player II is to choose the point M lying on the intersection of the line of the centers OO_1 with the boundary of the set S_2 and being farthest from the point O. The optimal strategy for player I is to choose the point $x \in S_1$ which is coincident with the center O_1 of the circle S_2. The value of the game here is $\bar{v} = \underline{v} = v = R_1 + R_2 - R_1 = R_2$.

Case 2. The center of the circle $O_1 \notin S_1$. This case is a version of case

1,where the center of the circle S_2 belongs to the closure of the set S_1. Let us calculate the quantity $\bar{v}$ (Fig.7). Let $x_0 \in S_1$. The point y_0 delivering $\max_{y \in S_2} \rho(x_0, y)$ then coincides with the point y_0 at which the straight line passing through x_0 and the center O_1 of the circle S_2 intersects the boundary of the circle S_2 that is farthest from the point x_0. Indeed, the circle with the radius x_0y_0, its center being at the point x_0, contains S_2 its boundary touches the boundary of the circle S_2 only at one point y_0. Evidently, the magnitude of $\max_{y \in S_2} \rho(x_0, y) = \rho(x_0, y_0)$ attains a minimum at the point M_1 where the line segment OM intersects the boundary of the circle S_1. Thus, in the this case

$$\bar{v} = \min_{x \in S_1} \max_{y \in S_2} \rho(x, y) = OM - R_1 = \underline{v}$$

The optimal strategies imply the choice of $M \in S_2$ and $M_1 \in S_1$ made by the players I and II, respectively.

If the open circles S_1 and S_2 are regarded as the strategy sets in Example 1, then in Case 2 the value of the game exist and equals

$$\underline{v} = \sup_{x \in S_2} \inf_{y \in S_1} \rho(x, y) = \inf_{y \in S_1} \sup_{x \in S_2} \rho(x, y) = \bar{v} = OM - R_1 = v.$$

The optimal strategies, however, does not exist, since $M_1 \notin S_1, M \notin S_2$. At the same time, the ϵ –optimal strategies exist that are the points from the ϵ–neighborhoods of the points M_1, belonging to the sets S_1 and S_2, respectively.

1.3 Mixed strategies and existence of equilibrium point

Example 1 in 1 shows that the equilibrium point exists not in all zero–sum two–person games. The existence of the equilibrium point is found to be exceptional father than common in nature. In this case it is not clear what may be understood by a solution of the game.

Assume that the game Γ has no equilibrium point. As stated in the preceding section, in such type of games it is very important for each player to know the opponent's intentions. Although the rules of the game do not provide such an opportunity, with a sufficient repetition of the game with the same opponent the player can statistically estimate the possibility of choosing a particular strategy and take a proper action.

What shall the player do to keep his intention secret from the opponent? The only reasonable method here is to choose a strategy at random, in compliance with a random mechanism. In this case, the player cannot guess which of the strategies has been chosen by his opponent, since the latter himself cannot foretell the result of the random choice.

Formally, such a possibility can be realized as follows. For each game Γ, we construct its mixed extension $\bar{\Gamma}$, in which the players' strategies are represented by various probability measures over the sets of strategies for the game Γ, and by the payoff is meant the mathematical expectation. We offer a rigorous definition.

Let A be some σ–algebra of the subsets of P, and B some σ–algebra of the subsets of E. Moreover, let μ,ν be probability measures on A and B respectively. Then the strategy spaces P and E convert to $(P, A, \mu), (E, B, \nu)$ spaces. Let $\bar{P}, \bar{E}$ be various probability measures on the σ –algebras A and B, and K be measurable with respect to σ– algebra $A \times B$. Set

$$M(\mu, \nu) = \int_P \int_E K(u(\cdot), v(\cdot)) d\mu d\nu,$$

where $\mu \in \bar{P}, \nu \in \bar{E}$ is the Lebesgues–Stieltjes integral representing the mathematical expectation of the payoff $K(u(\cdot), v(\cdot))$ for measures μ, ν.

Definition 1. The mixed extension of the game $\bar{\Gamma} =< P, E, K(u(\cdot), v(\cdot)) >$ is called the zero–sum two–person game in normal–form $\bar{\Gamma}$ over the set of strategies $\bar{P}, \bar{E}$ with the payoff function $M(\mu, \nu)$,i.e. the game $\Gamma =< \bar{P}, \bar{E}, M(\mu, \nu) >$.

Formally, we have the new game in normal form obtained from the game in a particular way. In the mixed extension of the game the players' behavior may be interpreted as follows.

Player II (I) chooses a probability measure $\nu \in \bar{E} (\mu \in \bar{P})$ independent of the opponent's choice, and realizes a random choice of the strategy $v(\cdot) \in E$ $(u(\cdot) \in P)$ in according with this measure (e.g., approximately with the help of the random number table). Next, Player II receives the payoff $K(u(\cdot), v(\cdot))$, and Player I the payoff $-K(u(\cdot), v(\cdot))$.

The strategies $\mu \in \bar{P}, \nu \in \bar{E}$ are called the mixed strategies in the game $\Gamma =< P, E, K(u(\cdot), v(\cdot)) >$. Thus, the mixed strategies in the game Γ coincide with the ordinary strategies in the mixed extension of this game.

We say that in the game $\Gamma =< P, E, K >$ there is the equilibrium point in mixed strategies if there is the equilibrium point in its mixed extension, i.e. the point (μ^*, ν^*) is called the equilibrium point in mixed strategies in the game Γ if fir all $\mu \in \bar{P}, \nu \in \bar{E}$ the inequality

$$M(\mu, \nu^*) \geq M(\mu^*, \nu) \geq M(\mu^*, \nu). \tag{1.3.1}$$

is satisfied.

The strategies μ^*, ν^* are called the optimal mixed strategies for the Player I and II, respectively. Evidently, ordinary strategies are a special case of the mixed strategies, i.e. the inclusion $P \subset \bar{P}, E \subset \bar{E}$ is valid. Indeed, each strategy $u(\cdot) \in P(v(\cdot) \in E)$ can be associated with the probability measure $\mu_u(\cdot) \in P(\nu_{V(\cdot)} \in \bar{E})$ concentrating the total mass on the element $u(\cdot) \in$

$P(v(\cdot) \in E)$. It can be shown that if in the game $\Gamma =< P, E, K(u(\cdot), v(\cdot)) >$ there is the equilibrium point, then the equilibrium point also exist in the game $\Gamma =< \bar{P}, \bar{E}, M(\mu, \nu) >$ (that, the equilibrium point also exists in the mixed strategies in the game Γ), then the strategies $\mu_{u^*(\cdot)}, \nu_{v^*(\cdot)}$, prescribing probability 1 to elements $u^*(\cdot) \in P$ and $v^*(\cdot) \in E$ constitute the equilibrium point in the game Γ. I.e.these are the mixed optimal strategies in the game $\Gamma =< P, E, K(u(\cdot), v(\cdot)) >$.

Indeed, let $(u^*(\cdot), v^*(\cdot))$ be the equilibrium point in the game $\Gamma =< P, E, K(u(\cdot), v(\cdot)) >$. Then

$$K(u(\cdot), v^*(\cdot)) \geq K(u^*(\cdot), v^*(\cdot)) \geq K(u^*(\cdot), v(\cdot)) \tag{1.3.2}$$

for all $u(\cdot) \in P, v(\cdot) \in E$. We shall rewrite (1.3.2) as

$$K(u(\cdot), v^*(\cdot)) \geq K(u^*(\cdot), v^*(\cdot)), \tag{1.3.3}$$

$$K(u^*(\cdot), v(\cdot)) \leq K(u^*(\cdot), v^*(\cdot)). \tag{1.3.4}$$

Integrating (1.3.3) with respect to an arbitrary measure $\nu \in \bar{E}$, and (1.3.4) with respect to an arbitrary measure, $\mu \in \bar{P}$ we obtain

$$\int K(u(\cdot), v^*(\cdot))d\mu \geq \int K(u^*(\cdot), v^*(\cdot))d\mu = K(u^*(\cdot), v^*(\cdot)), \tag{1.3.5}$$

$$\int K(u^*(\cdot), v(\cdot))d\nu \leq \int K(u^*(\cdot), v^*(\cdot))d\nu = K(u^*(\cdot), v^*(\cdot)). \tag{1.3.6}$$

We integrate (1.3.5) and (1.3.6) respectively with measures $\nu_{v^*(\cdot)}, \mu_{u^*(\cdot)}$. Then, using (1.3.5) and (1.3.6), we find

$$M(\mu, \nu_{v^*(\cdot)}) \geq M(\mu_{u^*(\cdot)}, \nu_{v^*(\cdot)}) \geq M(\mu_{u^*(\cdot)}, \nu), \tag{1.3.7}$$

i.e. the point $M(\mu_{u^*(\cdot)}, \nu_{v^*(\cdot)})$ is the equilibrium point in the mixed strategies in the game Γ, and

$$M(\mu_{u^*(\cdot)}, \nu_{v*(\cdot)}) = K(u^*(\cdot), v^*(\cdot)).$$

For simplicity, we introduce the following notation:

$$M(\mu, \nu_{v(\cdot)}) = M(\mu, \nu(\cdot))$$
$$M(\mu_{u(\cdot)}, \nu) = M(u(\cdot), \nu).$$

Theorem 3 *For the pair μ^*, ν^* to be the equilibrium point in the game Γ, it is necessary and sufficient that for all $u(\cdot) \in P, v(\cdot) \in E$ the following inequality holds.*

$$M(u(\cdot), \nu^*) \geq M(\mu^*, \nu^*) \geq M(\mu^*, v(\cdot)). \tag{1.3.8}$$

Proof: *Sufficiency.* Let $\mu \in \bar{P}, \nu \in \bar{E}$ be arbitrary mixed strategies. We integrate both parts of the left–hand inequality in (1.3.8) with respect to the measure μ, and both parts of the right–hand inequality in (1.3.8) with respect to the measure ν. As a result , we obtain

$$M(\mu,\nu^*) = \int M(u(\cdot),\nu^*)d\mu \geq$$

$$\geq \int M(\mu^*,\nu^*)d\mu = M(\mu^*,\nu^*), \tag{1.3.9}$$

$$M(\mu^*,\nu^*) = \int M(\mu^*,\nu^*)d\nu \geq$$

$$\geq \int M(\mu^*,v(\cdot))d\nu = M(\mu^*,\nu). \tag{1.3.10}$$

From (1.3.9) and (1.3.10) we find out that the pair (μ^*,ν^*) constitutes the equilibrium point in mixed strategies. The necessity is evident, since if (1.3.2) is satisfied for all $\mu \in \bar{P}, \nu \in \bar{E}$, then in particular, it also holds for all strategies $\mu_u(\cdot), \nu_v(\cdot)$ prescribing probability one to the elements $u(\cdot) \in P, v(\cdot) \in E$. ■

As shown, the mixed strategies extend the players' strategy sets in the original game and preserve the value of the game and the equilibrium point, if any. In most cases, the equilibrium point in mixed strategies is found to be existent, while the equilibrium point may not exist.

We consider a special class of zero–sum two–person games Γ where there is the equilibrium point in mixed strategies.

Theorem 4 *Let $P \in CompR^k (E \in CompR^\ell)$ be a compact set in Euclidean spaces of corresponding dimension, and the function $K(u(\cdot),v(\cdot))$ be continuous on Cartesian product $P \times E$. Then in the game $\Gamma = < P, E, K(u(\cdot),v(\cdot)) >$ there is the equilibrium point in mixed strategies.*

For the proof this theorem you may refer to [8].

In some cases, using specific properties of the payoff function, we may provide a more precise definition for structure of mixed optimal strategies in the game Γ. Specially , this occurs in the games with convex and concave payoff function. In what follows we will be dealing with this type of games and, therefore, we will consider them in details in the following section.

1.4 Games with convex payoff function

Assume that the set E is compact in R^n, the set P is convex and compact in R^n, the function $K(u(\cdot),v(\cdot))$ is continuous in all its variables and convex in

$u(\cdot)$ for all $v(\cdot)$. Such game $\Gamma =< P, E, K(u(\cdot), v(\cdot)) >$ will be called the game with convex payoff function (or convex game).

Let us consider a mixed extension of the game Γ over spaces of various probability measures on P and E.

Investigation of such type of games is of interest to us because the solution of the family of pursuit–evasion type games with incomplete information reduces to them. The solution has simple geometrical structure. In this section we will define a general form of optimal strategies and indicate a method of computing the value of the game.

We hope that the reader has some knowledge of fundamentals of the theory of convex sets and convex functions, therefore the relevant results of this theory are given without proof.

Let I_z be a mixed strategy by which the point z is chosen with probability 1. The following theorem holds.

Theorem 5 *The value of the game* Γ *is determined by the formula*

$$v = \min_{u \in P} \max_{v \in E} K(u, v).$$

Player II has an optimal strategy of the form $v_0 = \sum_{i=1}^{n+1} \lambda_i L_{v_i}$. *(*$\sum_{i=1}^{n+1} \lambda_i = 1$, $\lambda_i \geq 0$*). At the same time, all pure strategies* I_{u_0}*, under which* $\min_u \max_v K(u, v)$ *is achieved, are optimal for Player I . In addition, if the function* $K(u, v)$ *with each fixed* $v(\cdot)$ *is strictly convex in* $u(\cdot)$*, then the optimal strategy for Player I is unique.*

In Theorem 5, ν_0 is a probability measure concentrated on not more than the $(n + 1)$ points in the set $E, v_i(\cdot)$.

Before turning immediately to the proof of Theorem 5, we shall prove some auxiliary lemma describing the properties of convex functions.

Let B be a compact convex region in the n–dimensional space whose elements are denoted by y. The function f is called linear if $f(\sum \lambda_i y_i) = \sum \lambda_i f(y_i)$ with $\sum \lambda_i = 1$ (the linear function is not necessarily homogenous). The function $f(y) \equiv 1$ denoted by w, is linear. The linear functions form the $(n + 1)$ dimentional linear space E. We denote by F an element of the conjugate space E^* .

Lemma 3 *If* $F(w) = 1$*, then there is such* y *that* $F(f) = f(y)$ *for all* $f \in E$.

Proof: It suffices to show that $n + 1$ equations $F(f_i) = f_i(y)$ have a solution with respect to y for $n + 1$ linearly independent elements $f_i \in E$. But w may be taken as f_1. The first equation then converts to identity. The other n equations are independent and have a solution, since y is an element of the n–dimensional space. ■

Lemma 4 *The set of all functions f, that are nonnegative on the set B, form a closed convex cone $P \subset E$ with its vertex at zero, and the function w is its interior point. Moreover, the region $P(y) = \{f(y) : f(y) \in P\}$, for which $P(y) \geq 0$, coincides with B. (We denote by $P(y)$ a collection of all $f(y)$ for $f \in P$, and by $f(B)$ a collection of all $f(y)$ for $y \in B$).*

Proof: First show that P is a closed convex cone. Let $f_1, f_2 \in P$, $\alpha \geq 0$, $\beta \geq 0$.Then $f = \alpha f_1 + \beta f_2$ also belongs to P, since $f(y) = \alpha f_1(y) + \beta f_2(y) \geq 0$ for all $y \in B$. That the P is closed is evident. If $f \equiv 0$, then $f \in P$.

We now show that the region, for which $P(y) \geq 0$, coincides with E. Let $y \notin B$. Then y may be separated from B by a hyperplane, i.e. it is possible to indicate such $f \in E$ that $f(y) < c < f(B)$. Consequently, the function $f - cw$ belongs to P and is negative in y.

Suppose w is not the interior point of P. Then there is a sequence $f_n \to w, f_n \in P$. For each $y_n \in B$ there is $y_n \in B$ such that $f_n(y_n) < 0$. Since B is compact region , then there exists a subsequens y_{n_k} converging to $y_0 \in B$ and such that $f_{n_k}(y_{n_k}) < 0$. Then, $f(y_0) < 0$, as well. In this case , however, $\lim_{n_k \to \infty} f_{n_k}(y_0) < 0$. At the same time, $\lim_{n \to \infty} f_n(y) = 1$ for all y's on B, i.e. w is the interior point of the set P. ■

Lemma 5 *Let Q be a compact convex set from E not intersecting P. Then there are such $y \in B$ and $\delta > 0$ that $Q(y) \leq -\delta$.*

Proof: Let $F \in E^*$ and shares P and Q, i.e.

$$F(f) \geq b \geq F(q) \tag{1.4.1}$$

for all $f \in P$ and $q \in Q$. From this it immediately follows that $F(f) \geq 0$ for all $f \in P$. Indeed, assume that the case is the contrary. Let f_1 be such that $F(f_1) < 0$. Then $\lambda f_1 \in P$ for any $\lambda > 0$ and we have $F(\lambda f_1) = \lambda F(f_1) < 0$. Since λ cannot be chosen arbitrarily large, this contradicts (1.4.1). Thus, $F(f) \geq 0$ and, additionally, there is $b \neq 0$ satisfying (1.4.1), since $0 \in P$, and $F(0) = 0$, and the sets P and Q are convex, closed and do n ot intersect. From (1.4.1) we obtain

$$F(f) \geq 0 \geq F(q) - b. \tag{1.4.2}$$

Setting $(-b) = \delta > 0$, we derive the inequality

$$F(Q) + \delta \leq F(P) \ \ (\delta \geq 0). \tag{1.4.3}$$

Since w is an interior point of P, then $F(w) > 0$ holds. We may choose F in such a manner that $F(w) = 1$. By Lemma 3, we find the point y satisfying the inequality $Q(y) + \delta \leq 0 \leq P(y)$. In accordance with the preceding lemma, $y \in B$. This completes the proof of Lemma 5. ■

Lemma 6 *If the function* $\sup_\alpha f_\alpha(B)$ *is positive for the family* $\{f_\alpha\}$ *of linear functions, then for the properly chosen* $\lambda_i \geq 0$, $\sum_{i=1}^{n+1} \lambda_i = 1$ *and* α_i, $i = 1, \ldots, n, n+1$, *the function* $f = \sum_{i=1}^{n+1} \lambda_i f_{\alpha_i}$ *is a member of* P, *i.e.* $f(B) > 0$.

Proof: If $f_\alpha(y) > 0$ for the fixed α and y, then there is an open set containing y where there is a strict inequality. Consequently, by Heine–Borel theorem on covers, we may find such finite subset $\{f_{\alpha_j}\}$ that $\max_j\{f_{\alpha_j}(y)\} > 0$ for each $y \in B$.

Denote by Q a convex hull of the subfamily $\{f_{\alpha_j}\}$. Because of Lemma 5 , the sets P and Q intersect. Since Q is bounded, and P is not bounded, then a boundary point of the set Q lies in P. This point belongs to a space of dimension not exceeding n, therefore it may be represented as a convex linear combination of not more than $n+1$ functions f_α. This proves the lemma. ■

Theorem 6 *Let* $\{\varphi_\alpha\}$ *be a family of continuous convex functions defined on the compact convex n–dimensional set* B. *The function* $\sup_\alpha \varphi_\alpha(y)$ *achieves a minimum value of* c *at a point* $y_0 \in B$. *Moreover, for any given* $\delta > 0$ *there are such indices* α_i *and such* λ:, *that for all* $y \in B$.

$$\sum_{i=1}^{n+1} \lambda_i \varphi_{\alpha_i} \geq c - \delta,$$

where $\lambda_i \geq 0$ *and* $\sum_{i=1}^{n+1} \lambda_i = 1$.

Proof: Suppose that for all $y \in B$, satisfying the condition of Theorem 5, the following inequality holds

$$b > \inf_{y \in B} \sup_\alpha \varphi_\alpha(y) = c.$$

The set $y \in B$, for which $\varphi_\alpha(y) \leq b$ holds for all α, is nonempty and closed. Moreover, this set decreases as b decreases. Therefore, if we choose a sequence $b_k \to c$, then the intersection of relevant sets must be non–empty because of the principle of embedded intervals. Each point y_0 of this intersection satisfies the first part of the theorem.

To prove the second part of the theorem, we will consider a family $\{f_\beta\}$ of all linear functions satisfying the condition

$$[\varphi_\alpha - f_\beta](B) \geq 0 \tag{1.4.4}$$

for a certain α. Any φ_α–tangent plane satisfies this condition, so that family f_β contains all planes tangent to φ_α. Consequently, for all y's

$$\sup_\beta f_\beta(y) = \sup_\alpha \varphi_\alpha(y) \geq c.$$

Consider for a fixed $\delta > 0$ the family

$$F_\beta = [f_\beta - (c-\delta)w].$$

Since $\sup_\beta f_\beta(B) > c$, we have

$$\sup_\beta [f_\beta - (c-\delta)w](B) \geq \delta > 0.$$

We shall apply Lemma 6 to the family F_β and choose $\beta_1, \ldots, \beta_{n+1}, \lambda_1, \ldots, \lambda_{n+1}$. Let φ_{α_i} be the functions corresponding to f_{β_i} from (1.4.4). Then

$$\sum_{i=1}^{n+1} \lambda_i \varphi_{\alpha_i}(y) \geq \sum_{i=1}^{n+1} \lambda_i f_{\beta_i}(y) \geq c - \delta$$

for all $y \in B$, which was stated in the theorem.

We now prove Theorem 5. Consequently, the function $\varphi(u) = \sup_v K(u,v)$ achieves a minimum value at some point $u_0 \in B$, and for every $\epsilon > 0$ we may choose $\{\lambda_i^\epsilon\}$ and $\{v_i^\epsilon\}$. Thus,

$$\sum_{i=1}^{n+1} \lambda_i^\epsilon K(u, v_i^\epsilon) \geq \varphi(u_0) - \epsilon$$

for all $u \in P$, where λ_i^ϵ, v_i^ϵ take values from compact sets, i.e. $\lambda_i^\epsilon \geq 0$ and $\sum_{i=1}^{n+1} \lambda_i^\epsilon = 1$. Therefore, as $\epsilon \to 0$, for every i we may choose the limit points λ_i^0, v_i^0, $i = 1, \ldots, n+1$, satisfying the conditions $\lambda_i^0 \geq 0$ and $\sum_{i=1}^{n+1} \lambda_i^0 = 1$, $v_i^0 \in B$,

$$\sum_{i=1}^{n+1} \lambda_i^0 K(u, v_i^0) \geq \varphi(u_0) \tag{1.4.5}$$

for all $u \in P$. The inequality (1.4.5) implies that if Player II adopts a mixed finite strategy $v^*(\cdot) = \sum_{i=1}^{n+1} \lambda_i^0 I_{v_i^0}$, then his payoff is at least $\varphi(u_0)$. At the same time, if Player I adopts a pure strategy u_0, then with any v Player II's payoff equals

$$K(u_0, v) \leq \sup_v K(u_0, v) = \varphi(u_0).$$

Thus, $\varphi(u_0)$ is the value of the game, and $\{I_{u_0}, v^*\}$ a pair of optimal strategies. Let us now assume that K is a strictly convex with respect to u for any v, and show that the optimal strategy for Player I here is unique. Also, for Player II we choose a fixed optimal strategy of the form $\sum_{i=1}^{n+1} \lambda_i^0 I_{v_0^i}$, define P as a set of such points u that

$$\sum_{i=1}^{n+1} \lambda_i^0 K(u, v_i^0) < \varphi(u_0) + \frac{1}{\nu}, \quad \nu = 1, 2, \ldots,$$

and denote by μ a probability measure related to any optimal strategy for Player I . It is readily seen that $\mu(P - P_\nu) = 0$ for any ν. But it follows from the strict convexity that the intersection $\bigcap P_\nu$ consists of a unique point. Consequently, the mixed optimal strategy I_{u_0} is unique. This completes the proof of the theorem. ■

The result of this section will be illustrated by referring to examples.

Example 8. Let us consider a special case given in Example 1 from 1. Let $S_1 = S_2 = S$ and the set S be a closed circle on the plane, with O as its center and R as radius.

The payoff function $K(u,v) = \rho(x,y), x \in S, y \in S$,is strictly convex with respect to x for fixed y. Indeed, let $y = (y_1, y_2) \in S$, $x^1 = (x_1^1, x_2^1) \in S$, $x^2 = (x_1^2, x_2^2) \in S$. If $x_1 \neq x_2$, then for all $0 \leq \lambda \leq 1$

$$\rho(y, \lambda x_1 + (1-\lambda)x^2) < \lambda\rho(y, x^1) + (1-\lambda)\rho(y, x^2). \tag{1.4.6}$$

In fact,

$$\begin{aligned}\{\lambda y_1 + (1-\lambda)y_1 - [\lambda x_1^1 + (1-\lambda)x_1^2]\}^2 = \\ \{\lambda(y_1 - x_1^1) + (1-\lambda)(y_1 - x_1^2)\}^2 = \\ = \lambda^2(y_1 - x_1^1)^2 + (1-\lambda)^2(y_1 - x_1^2)^2 + \\ +2\lambda(1-\lambda)(y_1 - x_1^1)(y_1 - x_1^2).\end{aligned} \tag{1.4.7}$$

Similarly

$$\begin{aligned}\{\lambda y_2 + (1-\lambda)y_2 - [\lambda x_2^1 + (1-\lambda)x_2^2]\}^2 = \\ = \lambda^2(y_2 - x_2^1)^2 + (1-\lambda)^2(y_2 - x_2^2)^2 + \\ +2\lambda(1-\lambda)(y_2 - x_2^1)(y_2 - x_2^2).\end{aligned} \tag{1.4.8}$$

At the same time,

$$\begin{aligned}\Big\{\lambda\sqrt{(y_1 - x_1^1)^2 + (y_2 - x_2^1)^2} + \\ + (1-\lambda)\sqrt{(y_1 - x_1^2)^2 + (y_2 - x_1^2)^2}\Big\}^2 = \\ = \lambda^2\left[(y_1 - x_1^1)^2 + (y_2 - x_2^1)^2\right] + \\ +(1-\lambda)^2\left[(y_1 - x_1^2)^2 + (y_2 - x_2^2)^2\right] + \\ +2\lambda(1-\lambda)\sqrt{[(y_1 - x_1^1)^2(y_1 - x_1^2)^2 + (y_2 - x_2^1)^2(y_2 - x_2^2)^2 +} \\ \overline{+(y_1 - x_1^1)^2(y_2 - x_2^2)^2 + (y_2 - x_2^1)^2(y_1 - x_1^2)^2]}.\end{aligned} \tag{1.4.9}$$

Adding up (1.4.7) and (1.4.8) and comparing with (1.4.7) we have that it suffices to prove the inequality

$$(y_1 - x_1^1)(y_1 - x_1^2) + (y_2 - x_2^1)(y_2 - x_2^2) \leq$$

$$\leq \sqrt{[(y_1 - x_1^1)^2(y_1 - x_1^2)^2 + (y_2 - x_2^1)^2(y_2 - x_2^2)^2 + +(y_1 - x_1^1)^2(y_2 - x_2^2)^2 + (y_2 - x_2^1)^2(y_1 - x_1^2)^2]}. \tag{1.4.10}$$

By squaring (1.4.10), we get

$$[2(y_1 - x_1^1)(y_1 - x_1^2)(y_2 - x_2^1)(y_2 - x_2^2)] \leq$$

$$\leq (y_1 - x_1^1)^2(y_2 - x_2^2)^2 + (y_2 - x_2^1)^2(y_1 - x_1^2)^2 \tag{1.4.11}$$

or, which is the same,

$$[(y_1 - x_1^1)(y_2 - x_2^2) - (y_2 - x_2^1)(y_1 - x_1^2)]^2 \geq 0 \tag{1.4.12}$$

However, (1.4.12) is always satisfied if $x^1 \neq x^2 \neq y$. Thus, we have completed the proof for the convexity of the function $\rho(x, y)$ in x, with each y being fixed. By the condition, the set S (circle) is also convex. Therefore, we can directly apply Theorem 1, from which it follows that the value of the game

$$v = \min_{x \in S} \max_{y \in S} \rho(x, y). \tag{1.4.13}$$

The example in 2 suggests that

$$v = \min_{x \in S} \max_{y \in S} \rho(x, y) = R. \tag{1.4.14}$$

In this case, the point $x \in S$, on which the minimum of expression $\max_{y \in S} \rho(x, y)$ is achieved, is unique and coincides with the center of the circle S (i.e., with the point O (see 2)). By Theorem 5, this point precisely is the optimal strategy for Player I (minimizer) in the given case. Theorem 1 states that Player II (maximizer) has a mixed optimal strategy prescribing a positive probability to not more three points in the set S. However, because of the symmetry of the set S, by using the optimal strategy, the player chooses with probability $(\frac{1}{2}, \frac{1}{2})$ any two diametrically opposite points on the boundary of S.

Denote such a strategy for Player II by $\nu^* = \frac{1}{2} I_{y_1} + \frac{1}{2} I_{y_2}$. To prove the optimality of ν^*, it suffices to establish that for all $x, y \in S$,

$$M(I_0, y) \leq R = M(I_0, \nu^*) \leq M(x, \nu^*). \tag{1.4.15}$$

Here M is the expectation of the payoff (see 3), I_0 is a pure strategy for Player I designed to choose a center of the circle S, and y_1, y_2 are arbitrary diametrically opposite points on the boundary of the circle $S, M(I_0, \nu^*) = \frac{1}{2}R + \frac{1}{2}R = R$. Since R is the value of the game, and I_0 an optimal strategy for Player I , then the left–hand side of (1.4.15) is apparent.

Now, let x be an arbitrary point from S (Fig.8). Then

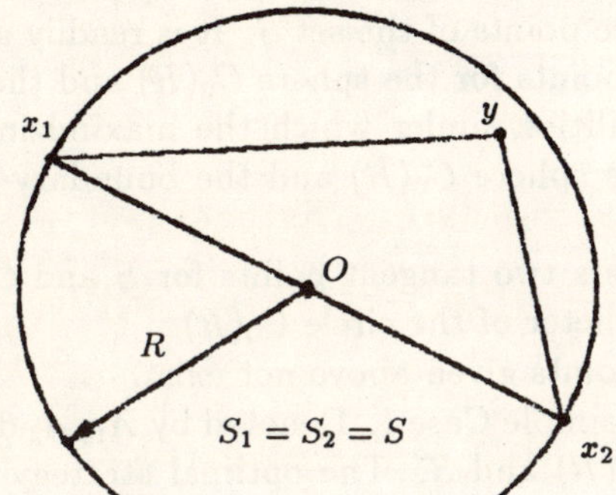

Fig. 8.

$$M(x,\nu^*) = \frac{1}{2}\rho(x,y_1) + \frac{1}{2}\rho(x,y_2) \geq \frac{1}{2}2R = R.$$

This completes the proof of the inequality (1.4.15).

Example 9. Consider the generalization of the preceding example to the case where the set S is a closed convex set on the plane. In other aspects this game coincides with the one from Example 1, Since all conditions of Theorem 1 are satisfied, the value of the game

$$v = \min_{x\in S}\max_{y\in S}\rho(x,y) \qquad (1.4.16)$$

and Player I has an optimal pure strategy.

We attempt now to define the quantity v and optimal strategies for the Players I and II. We construct a minimum sphere $C_0(R)$ containing the set S. Let O be a center of this sphere, and R its radius. We will show that $v = R$. Indeed, $\max_{y\in S}\rho(x,y) \geq R$ holds for each $x \in S$, since if x_1 would be such that $\max_{y\in S}\rho(x_1,y) = R^1 = R-\delta,(\delta > 0)$, then there would be the sphere $C_{x_1}(R^1)$ with its center at the point x_1 and R^1 as its radius, containing S. Since $R^1 > R$, then the above assumption contradicts the minimality of the sphere $C_0(R)$. At the same time, there are points $\bar{y}$ on which $\min_{x\in S}\max_{y\in S}\rho(x,y) = \rho(0,\bar{y})$ is achieved. These are the tangent points of the boundary of the sphere $C_0(R)$ and of the boundary of the set S. Thus,

$$v = \min_{x\in S}\max_{y\in S}\rho(x,y) = R.$$

Now let I_0 be Player I 's strategy prescribing probability 1 to the choice of the center O of the sphere $C_0(R)$ (evidently, $O \in S$). Employing Theorem 1, we have that the optimal strategy for Player II prescribes positive probability to not more than three points of the set S. It is readily seen that these points must be the tangent points for the sphere $C_0(R)$ and the boundary of the set S. To define probabilities, under which the maximizing player chooses the tangent points for the sphere $C_0(R)$ and the boundary of the set S, we will consider two cases.

Case 1. There exists two tangent points for S and $C_0(R)$ such that their joining chord is a diameter of the circle $C_0(R)$.

Case 2. The two points given above not exist.

Suppose we have a simple Case 1. Denoted by A_1, A_2 diametrically opposite tangent points for $C_0(R)$ and S. The optimal strategy for the maximizer is the choice of points A_1 and A_2 with equal probabilities $(\frac{1}{2}, \frac{1}{2})$. Denote this strategy for Player II by μ^*. It suffices to show that for all $A \in S$

$$R = E(I_0, \mu^*) \leq E(I_A, \mu^*).$$

The latter is proved as in (1.4.15).

In case 2, there are necessarily such three tangent points A_1, A_2, A_3 that the center O of the circle $C_0(R)$ belongs to a convex hull of these points. We choose a coordinate system, in such a way that the center O of the circle $C_0(R)$ is its origin. In this system, let (x_1, y_1), (x_2, y_2), (x_3, y_3) be coordinates for the points A_1, A_2, A_3. Then there exist such $\lambda_1 \geq 0, \sum_{i=1}^3 \lambda_i = 1, i = 1,2,3$ that $\sum_{i=1}^3 \lambda_i x_i = 0$, $\sum_{i=1}^3 \lambda_i y_i = 0$, since any point of the convex polyhedron is representable as a convex linear combination of its vertices.

We show that a mixed optimal strategy for the maximizer is the choice of one of the points A_1, A_2, A_3 with probabilities $\lambda_1, \lambda_2, \lambda_3$. Denote this strategy by ν^*. Consider the function $E(I_A, \nu*)$ for $A \in S$. It suffices to show that, based on (1.4.5), for all $A \in S$.

$$E(I_A, \nu^*) \geq E(I_0, \nu^*) = R.$$

However, the function

$$E(I_A, \nu^*) = \sum_{i=1}^{3} \lambda_i \sqrt{(x - x_i)^2 + (y - y_y i)^2},$$

where x, y are coordinates of the point $A \in S$, is strictly convex with respect to x, y and is continuously differentiable with respect to x, y, therefore it has a unique point of minimum that is determined from the condition

$$\frac{\partial E}{\partial x} = \frac{\partial}{\partial x} \sum_{i=1}^{3} \lambda_i \sqrt{(x - x_i)^2 + (y - y_i)^2} = 0,$$

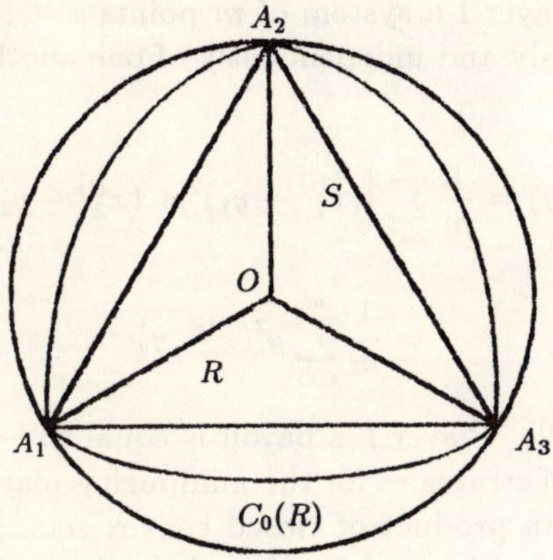

Fig. 9.

$$\frac{\partial E}{\partial y} = \frac{\partial}{\partial y}\sum_{i=1}^{3}\lambda_i\sqrt{(x-x_i)^2+(y-y_i)^2} = 0.$$

This yields

$$\sum_{i=1}^{3}\frac{\lambda_i(x-x_i)}{\sqrt{(x-x_i)^2+(y-y_i)^2}} = 0, \qquad (*)$$

$$\sum_{i=1}^{3}\frac{\lambda_i(y-y_i)}{\sqrt{(x-x_i)^2+(y-y_i)^2}} = 0.$$

It can be readily seen that the point $x = 0, y = 0$ is a solution to equations $(*)$, since, substituting $x = 0, y = 0$ in $(*)$, we obtain

$$\sum_{i=1}^{3}\frac{\lambda_i x_i}{\sqrt{(x_i^2+y_i^2)}} = \sum_{i=1}^{3}\frac{\lambda_i x_i}{R} = 0$$

$$\sum_{i=1}^{3}\frac{\lambda_i y_i}{\sqrt{(x_i^2+y_i^2)}} = \sum_{i=1}^{3}\frac{\lambda_i y_i}{R} = 0$$

Thus, I_0 is a point of minimum of the function $E(I_A, \nu^*)$, i.e. for all $A \in S$

$$R = E(I_0, \nu^*) \leq E(I_A, \nu^*).$$

which was to be shown (Fig.9).

Consider a generalization of the preceding game. Let Player I choose not one point $x \in S$, but m points $x^{(1)}, \ldots, x^{(m)} \in S$. Formally, the game proceeds as follows. The convex compact set S is given on the plane. Player II chooses a point $y \in S$, and Player I a system of m points $x^{(1)}, \ldots, x^{(m)} \in S$. Choices are made simultaneously and independently of one another. Player II's payoff is assumed to be

$$K(x,y) = \frac{1}{m}\sum_{i=1}^{m}\left[(x_1^{(i)} - y_1)^2 + (x_2^{(i)} - y_2)^2\right] =$$

$$= \frac{1}{m}\sum_{i=1}^{m}\rho^2(x^{(i)}, y)$$

here $x = (x^{(1)}, \ldots, x^{(m)})$. Player I 's payoff is equal to $-K$.

In this game a set of strategies for the minimizing player $P = [S]^m$ is convex and closed as Cartesian product of closed convex sets. The function $K(x,y)$ is strictly convex, in x with every fixed y. Indeed, the function $w(u) = u^2$ is strictly convex.

If the function $w(u)$ is strictly convex , then the function $w(u+c)$ is also strictly convex with any fixed c. If the function $w_1(u_1), w_2(u_2), \ldots, w_n(u_n)$ are convex, then the function $\Phi(u_1, \ldots, u_n) = \sum_i w_i(u_i)$ is also convex. Thus , we may apply Theorem 1 on the structure of optimal strategies for players, according to which the optimal strategy for Player I is pure, and the mixed strategy of Player II concentrates measure at no more than three points of the set S. The value of the game is equal to

$$v = \min_{x_{(1)},\ldots,x_{(m)} \in S} \max_{y \in S} \frac{1}{m}\sum_{i=1}^{m}\rho^2(x^{(i)}, y).$$

We compute the quantity ν and find an optimal pure strategy for Player I . Denote by $S_1(x^{(i)}, \ldots, x^{(m)})$ a geometric locus of points y on the plane for which $\frac{1}{m}\sum \rho^2(x^{(i)}, y) = \ell^2 = const.$ We show that this set is a sphere with its center at the point $\bar{x} = (\bar{x}_1, \bar{x}_2)$, where

$$\bar{x}_i = \frac{1}{m}\sum_{j=1}^{m} x_i^{(j)}, \quad i = 1, 2,$$

its radius being

$$R^2 = \ell^2 - \sum_{i=1}^{2}\left\{\frac{1}{m}\sum_{j=1}^{m}\left[\left(x_i^{(j)}\right)^2 - (\bar{x}_i)^2\right]\right\}. \tag{1.4.17}$$

Indeed

$$\frac{1}{m}\sum_{j=1}^{m}\left[(x_1^{(j)} - y_1)^2 + (x_2^{(j)} - y_2)^2\right] = \ell^2,$$

$$\frac{1}{m}\sum_{j=1}^{m}\left[\left(x_1^{(j)}\right)^2 - 2x_1^{(j)}y_1 + \left(x_2^{(j)}\right)^2 - 2x_2^{(j)}y_2 + y_1^2 + y_2^2\right] = \ell^2,$$

$$\frac{1}{m}\sum_{j=1}^{m}\left[-2x_1^{(j)}y_1 - 2x_2^{(j)}y_2 + y_1^2 + y_2^2\right] =$$

$$= \ell^2 - \frac{1}{m}\sum_{j=1}^{m}\left[\left(x_1^{(j)}\right)^2 + \left(x_2^{(j)}\right)^2\right],$$

$$\left\{\left[\frac{\sum_j x_2^{(j)}}{m}\right]^2 - 2\frac{\sum_j x_2^{(j)}}{m}y_2 + y_2^2\right\} +$$

$$+\left\{\left[\frac{\sum_j x_1^{(j)}}{m}\right]^2 - 2\frac{\sum_j x_1^{(j)}}{m}y_1 + y_1^2\right\} -$$

$$\left[\frac{\sum_j x_1^{(j)}}{m}\right]^2 - \left[\frac{\sum_j x_2^{(j)}}{m}\right]^2 =$$

$$= \ell^2 - \frac{1}{m}\sum_{j=1}^{m}\left[\left(x_1^{(j)}\right)^2 + \left(x_2^{(j)}\right)^2\right],$$

$$(\bar{x}_1 - y_1)^2 + (\bar{x}_2 - y_2)^2 =$$

$$= \ell^2 - \frac{1}{m}\sum_{j=1}^{m}\left[\left(x_1^{(j)}\right)^2 + \left(x_2^{(j)}\right)^2\right] + \frac{m}{m}(\bar{x}_1)^2 + \frac{m}{m}(\bar{x}_1)^2,$$

$$(\bar{x}_1 - y_1)^2 + (\bar{x}_2 - y_2)^2 =$$

$$\ell^2 - \frac{1}{m}\sum_{i=1}^{2}\sum_{j=1}^{m}\left[\left(x_i^{(j)}\right)^2 - (\bar{x}_i)^2\right].$$

Expression (1.4.17) and $\bar{x}_i$ correspond to sample variance and sample means of the quantities $x_i^{(j)}$, hence the expression

$$\frac{1}{m}\sum_{i=1}^{2}\sum_{j=1}^{m}\left[\left(x_i^{(j)}\right)^2 - (\bar{x}_i)^2\right]$$

may be represented as

$$\frac{1}{m}\sum_{i=1}^{2}\sum_{j=1}^{m}(x_i^{(j)})^2 - \bar{x}_i)^2 \geq 0.$$

Evidently, there must be

$$\ell^2 \geq \frac{1}{m}\sum_{i=1}^{2}\sum_{j=1}^{m}(x_i^{(j)} - \bar{x}_i)^2.$$

With the fixed center $\bar{x}$, a maximal radius is obtained from the condition of minimum for a sample variance, i.e.

$$\min_{x_i^{(j)}} \frac{1}{m}\sum_{j=1}^{m}(x_i^{(j)} - \bar{x}_i)^2, \qquad (1.4.18)$$

under condition $\frac{1}{m}\sum_{j=1}^{m} x_i^{(j)} = \bar{x}_i$, $i = 1, 2$.

The minimum in (1.4.18) is achieved with $x_i^{(j)} = \bar{x}_i$. i.e. to obtain a maximal radius, it is necessary to concentrate all points $x_i^{(j)}$ at the point $\bar{x}_i$ (at the center of the sphere).

Consider a minimal sphere containing the set S. Let $x_0 = (x_1^0, x_2^0)$, R^0 be a center and radius of this sphere. Then, with $\bar{x}_i = x_i^0$, $x_i^{(j)} = x_i^0$, $i = 1, 2$, $j = 1, \ldots, m$, $\ell^2 = (R^0)^2$, the sphere $S_\ell(x^{(1)}, \ldots, x^{(m)})$ also coincides with the minimal sphere containing the set S. Therefore, for any point $y \in S$

$$\frac{1}{m}\sum_{j=1}^{m}\rho^2(x^{(j)}, y) \leq (R^0)^2 = \ell^2 \qquad (1.4.19)$$

(here all $x^{(j)} = x_0$, $j = 1, \ldots, m$). Evidently, there are no points $x^{(j)}$ such that the inequality (1.4.19) is satisfied for all $y \in S$ with $\ell^1 < \ell$ (since S_ℓ is the minimal sphere containing S).

Thus, the optimal strategy for Player I is the choice of m identical points $(x^0)^{(i)} = x^0$, $i = 1, \ldots, m$, that belong to S and are centers of the minimal spheres containing the set S (the point x^0 being unique), and the optimal strategies for Player II prescribe positive probabilities to not more than three tangent points y_1, y_2, y_3 for the minimal sphere, containing the set S, and this set. The value of the game is equal to the radius of minimal sphere containing the set S.

1.5 One class of games with complete information

Let us consider a class of multistage games with complete which are according a generalizations of Example 2 from 1. The game will take place in the Euclidean n-dimensional space R^n. We denote Player I 's position by $x \in R^n$, and

Player II's position by $y \in R^n$. For each $x \in R^n, y \in R^n$ we determine, respectively, the sets U_x, V_y which are assumed to be compact sets of Euclidean spaces R^n.

The game starts from positions x_0, y_0. At the initial time instant the Player I and II make a choice of points $x_1 \in U_{x_0}$ and $x_1 \in V_{y_0}$. In this case, Player II's choice is communicated to Player I before he chooses points $x_1 \in U_{x_0}$ (i.e. Player II leads). At the points x_1, y_1 , the Player I and II choose $x_2 \in U_{x_1}$ and $y_2 \in V_{y_1}$, and Player II's choice is communicated to Player I before he chooses the point $x_2 \in U_{x_1}$, etc. In position x_{k-1}, y_{k-1} the players choose $x_k \in U_{x_{k-1}}, y_k \in V_{y_{k-1}}$, and Player II's choice is communicated to Player I before he chooses the point $x_k \in U_{x_{k-1}}$. The process terminates at the N-th step by choosing $x_N \in U_{x_{N-1}}, y_N \in V_{y_{N-1}}$ and passing to state x_N, y_N.

The families of sets $U_x, V_y, x \in R^n, y \in R^n$ are assumed to be continuous in Hausdorff metric with respect to x, y. This means that for any $\epsilon > 0$ there will be such $\delta > 0$ that for $\|x - x_0\| < \delta$ $\|y - y_0\| < \delta$.

$$(U_{x_0})_\epsilon \supset (U_x),\ (U_x)_\epsilon \supset U_{x_0}$$

$$((V_{y_0})_\epsilon \supset V_y),\ (V_y)_\epsilon \supset V_{y_0}).$$

Here $U_\epsilon, (V_\epsilon)$ is the ϵ–neighbourhood of the set $U(V)$.

We shall not provide the proof for the following Lemma, since it is well known in analysis.

Lemma 7 *Let $f(x', y')$ be a continuous function on the Cartesian product $U_x \times V_y$. If the families $\{U_x\}, \{V_y\}$ are Hausdorff continuous in x, y, then the functionals*

$$F_1(x, y) = \max_{y' \in V_y} \min_{x' \in U_x} f(x', y'),$$

$$F_2(x, y) = \min_{x' \in U_x} \max_{y' \in V_y} f(x', y')$$

are continuous in x, y.

Let $\bar{x} = (x_0, \dots, x_N)$ and $\bar{y} = (y_0, \dots, y_N)$ be trajectories for the Player I and II realized during the game, Player II's payoff is the quantity

$$\max_{0 \geq k \geq N} f(x_k, y_k, k) = F(\bar{x}, \bar{y}), \tag{1.5.1}$$

where $f(x, y, k)$ is a continuous function of x, y. Player I's payoff is equal to $-F$ (zero–sum two–person game).

Assume that the game is a perfect information game, i.e. at each time instant (at each step) the players know positions x_k, y_k and time instant k. Additionally, Player II choice of y_{k+1} is known to Player I.

Player I's strategies are all possible functions $u(x,y,t)$ such that $u(x_k,y_{k+1},k) \in U_{x_k}$. Player II's strategies are all possible functions $v(x,y,t)$ such that $v(x_k,y_k,k) \in V_{y_k}$. These strategies will be called pure strategies (as distinguished from mixed strategies).

Suppose the Players I and II adopt pure strategies $u(x,y,t), v(x,y,t)$. In the situation $(u(\cdot), v(\cdot))$ the play proceeds as follows. At the first step, Player II passes from the state y_0 to the state $y_1 = v(x_0, y_0, 0)$, and Player I from x_0 to $x_1 = U(x_0, y_1, 0) = u(x_0, v(x_1, y_1, 0), 0)$, (since Player I knows Player II's choice). At the second step, the players pass to the states $y_2 = v(x_1, y_1, 1)$, $x_2 = u(x_1, y_2, 1) = u(x_1, v(x_1, y_1, 1), 1)$, etc. At the k-th step, the Players I and II pass from the states x_{k-1}, y_{k-1} to the states $y_k = v(x_{k-1}, y_{k-1}, k-1)$, $x_k = u(x_{k-1}, y_k, k-1) = u(x_{k-1}, v(x_{k-1}, y_{k-1}, k-1), k-1)$, etc.

Thus, each situation $(u(\cdot), v(\cdot))$ uniquely defines the pair of trajectories of the Players I and II $\bar{x} = (x_0, \ldots, x_N)$ and $\bar{y} = (y_0, \ldots, y_N)$, and hence with the payoff $K(u(\cdot), v(\cdot)) = F(\bar{x}, \bar{y})$ determined from the formula (1.5.1).

The games depend on two parameters, namely, initial positions x_0, y_0 and duration N, hence it is denoted by $\Gamma(x_0, y_0, N)$. To continue our studies, each game $\Gamma(x_0, y_0, N)$ may be conveniently placed into the family of games $\Gamma(x, y, T)$ that are dependent on parameters x, y, T.

The following theorem, which is a generalization of Zermello theorem for the finite games with complete information, is valid.

Theorem 7 *In the game $\Gamma(x_0, y_0, N)$ there is an equilibrium point in pure strategies.*

Proof: The proof will be carried out by the induction method with respect to the number of steps in the game N. With $N = 0$ the theorem is evident. Assume that the theorem holds for all $N < k$ and prove its validity for $N = k$.

Let $N < k$. Consider the family of the games $\Gamma(x, y, N)(N < k)$. In this game, let $\bar{u}^N_{x,y}(\cdot, t), \bar{v}^N_{x,y}(\cdot, t)$ be optimal strategies for the Players I and II, and $v(x, y, N)$ the value of the game $\Gamma(x, y, N)$. The function $F(x, y, k)$ is defined as follows:

$$F(x,y,k) = \max_{y' \in V_y} \min_{x' \in U_x} \{\max[f(x',y',0), V(x',y',k-1)]\} =$$

$$= \max[f(x,y,0), \max_{y' \in V_y} \min_{x' \in U_x} V(x',y',k-1)] = \qquad (1.5.2)$$

$$= \max[f(x,y,0), V(\tilde{x}, \tilde{y}, k-1)]$$

Strategies $(u^*)^k_{x,y}, (v^*)^k_{x,y}$ are defined by the rule

$$(u^*)^k_{x,y}(\cdot,t) = \begin{cases} \bar{u}^{k-1}_{x',y'}(\cdot,t), \text{ at } t > 0,\ x' \in U_x,\ y' \in V_y, \\ \bar{x} \in \{x' : V(x',y',k-1) = \min_{x'} V(x',y',k-1)\}, t = 0, \end{cases}$$

$$(v^*)^k_{x,y}(\cdot,t) = \begin{cases} \bar{v}^{k-1}_{x',y'}(\cdot,t) & \text{at } t>0,\ x' \in U_x,\ y' \in V_y \\ \tilde{y}, & \text{at t=0.} \end{cases}$$

We will show that $F(x,y,k)$ is the value of the game $\Gamma(x,y,k)$, and $(u^*)^k_{x,y}(\cdot,t)$ and $(v^*)^k_{x,y}(\cdot,t)$ are optimal strategies for the players I and II in this game. By definition, the payoff K in the situation $(u^*)^k_{x,y}(\cdot,t),(v^*)^k_{x,y}(\cdot,t)$ is equal to $F(x,y,k)$.

Now let one of the players, say Player II, depart from the strategy $(v^*)^k_{x,y}(\cdot,t)$, choosing a strategy $(v)^k_{x,y}(\cdot,t)$ instead of it. Then, from (1.5.2) we obtain[1]

$$F(x,y,k) \ge \max\left[f(x,y,0), \min_{x' \in U_x} V(x',\bar{y},k-1)^*\right]. \qquad (1.5.3)$$

Generally, $\bar{y}$ does not coincide here with $\tilde{y}$ from (1.5.2), since the strategy $(v)^k_{x,y}(\cdot,t)$ differs from the strategy $(v^*)^k_{x,y}(\cdot,t)$.

Let $\min_{x \in U_x} V(x,\bar{y},k-1)$ is achieved at the point $\bar{x} = \tilde{x}(\tilde{y})$. Consider the game $\Gamma(\bar{x},\bar{y},k-1)$. Let $K(u,v)$ be the payoff function, and $\bar{x}^{k-1}_{\bar{x},\bar{y}}$. $\bar{v}^{k-1}_{\bar{x},\bar{y}}$ the optimal strategies for the Player I and II in this game. Then

$$V(\bar{x},\bar{y},k-1) = K(\bar{u}^{k-1}_{\bar{x},\bar{y}}, \bar{v}^{k-1}_{\bar{x},\bar{y}}) \ge K(\bar{u}^{k-1}_{\bar{x},\bar{y}}, v^{k-1}_{\bar{x},\bar{y}}) \qquad (1.5.4)$$

for any strategy $v^{k-1}_{\bar{x},\bar{y}}$ of Player II in the $(k-1)$–stage game.

In (1.5.4), let $v^{k-1}_{\bar{x},\bar{y}}$ be a restriction of the strategy $v^k_{x,y}(\cdot,t)$ on a position set in the game $\Gamma(\bar{x},\bar{y},k-1)$. Then, using (1.5.3), (1.5.4), we may write the following chain of inequalities:

$$\begin{aligned} F(x,y,k) &\ge K((u^*)^k_{x,y},(v^*)^k_{x,y}) \ge \\ &\ge \max[f(x,y,0), V(\bar{x},\bar{y},k-1)] \ge \qquad (1.5.5)\\ &\ge \max[f(x,y,0), K(\bar{u}^{k-1}_{\bar{x},\bar{y}}, v^{k-1}_{\bar{x},\bar{y}})] = \\ &= K((u^*)^k_{x,y},(v)^k_{x,y}). \end{aligned}$$

The inequality (1.5.5) is satisfied for all strategies $(v)^k_{x,y}$. Similarly, it may be shown that for all $(u)^k_{x,y}$

$$\begin{aligned} F(x,y,k) &= K((u^*)^k_{x,y},(v^*)^k_{x,y}) \le \\ &\le K((u)^k_{x,y},(v^*)^k_{x,y}). \qquad (1.5.6) \end{aligned}$$

is satisfied. Thus, $F(x,y,k)$ is the value of the game $\Gamma(x,y,k)$. This proves Theorem 1. ■

[1] Using Lemma 7 and induction method, the continuity of the function $V(x,y,k)$ in x,y may be readily proved.

Corollary 1. The value of the game $\Gamma(x, y, k)$ is a solution of the functional equation

$$V(x, y, k) = \max[f(x, y, 0), \max_{y' \in V_y} \min_{x' \in U_x} V(x', y', k - 1)] \tag{1.5.7}$$

with the initial condition $V(x, y, 0) = f(x, y, 0)$.

Now consider the game $\bar{\Gamma}(x_0, y_0, N)$ differing from the game $\Gamma(x_0, y_0, N)$ in that here Player I is acting as the leader. Thus, in the game $\Gamma(x_0, y_0, N)$ at each step k Player II knows, apart from states x_k, y_k and time k, the state $x_{k+1} \in U_{x_k}$ choosen by Player I at the instant k. At each step k Player I knows x_k, y_k, k only. As in Theorem 1, we may show that the value of the game $\bar{\Gamma}(x_0, y_0, N)$ exists and the value function of the game $\bar{V}(x_0, y_0, N)$ satisfies the functional equation.

$$\bar{V}(x, y, k) = \max[f(x, y, 0), \max_{y' \in V_y} \min_{x' \in U_x} \bar{V}(x', y', k - 1)] \tag{1.5.8}$$

under the initial condition $\bar{V}(x, y, 0) = f(x, y, 0)$.

Corollary 2. Consider the games $\Gamma'(x_0, y_0, N)$ and $\bar{\Gamma}(x_0, y_0, N)$ differing from the games $\Gamma(x_0, y_0, N)$ and $\bar{\Gamma}(x_0, y_0, N)$ only in the form of the payoff function. Assume that in these games the payoff of Player II equals the distance between him and Player I at the last step of game, i.e. $\rho(x_N, y_N)$. The proof of Theorem 1 and Corollary 1 then holds if instead of functional equations (1.5.7), (1.5.8) we consider equations

$$V'(x, y, k) = \max_{y' \in V_y} \min_{x' \in U_x} V'(x', y', k - 1) \tag{1.5.7'}$$

under the initial condition $V'(x, y, 0) = \rho(x, y)$ and

$$\bar{V}'(x, y, k) = \min_{x' \in U_x} \max_{y' \in V_y} \bar{V}'(x, y, k - 1) \tag{1.5.8'}$$

under the initial condition $\bar{V}'(x, y, 0) = \rho(x, y)$. Indeed, the above equations are derived directly from (1.5.7), (1.5.8), if we set $f(x, y, k) = 0$ for $k < N$ and $f(x, y, N) = \rho(x, y)$.

Example 10. Consider a discrete game of pursuit in which the sets U_x are the circles with α as a radius and their center at the point x, and the sets V_y are the circles with β as a radius and their center at the point $y (\alpha > \beta)$. This corresponds to the game of pursuit in which Player II (the evader) moves on the plane at a speed of not more than β, and Player I (the pursuer) at a speed of not more than α. The pursuer's speed exceeds that of the evader , and Player I is second in turn to play. Such type game is the discrete game of "simple pursuit" with discrimination for the evader. The game duration is N

steps, and Player I I's payoff equals a distance between the players at the final step.

This payoff is a special case of the payoff defined as the maximum value of some function along a realized trajectory of the game. Indeed, if $f(x_k, y_k, k) = 0$ is set for all $k < N, f(x_N, y_N, N) = \rho(x_N, y_N)$, then the payoff $F(\bar{x}, \bar{y})$ in the game is defined as

$$F(\bar{x}, \bar{y}) = \max_{0 \leq k \leq N} f(x_k, y_k, k).$$

We shall find the value of game and optimal strategies for the players by solving the equation (1.5.7). Since $f(x_k, y_k, k) = 0$ with $k < N$, we get the equation

$$V(x, y, k) = \max_{y' \in V_y} \min_{x' \in U_x} V(x', y', k-1) \tag{1.5.9}$$

with the initial condition $V(x, y, 0) = \rho(x, y)$. Hence we have

$$V(x, y, 1) = \max_{y' \in V_y} \min_{x' \in U_x} \rho(x', y')$$

Since U_x and V_y are the circles with centers at x and y, and α and β as radii, then if $U_x \supset V_y$, the function $V(x, y, 1) = 0$,if, however, $U_x \subset V_y$ then $\max_{y' \in V_y}$ $\min_{x' \in U_x} \rho(x', y') = \rho(x, y) + \beta - \alpha = \rho(x, y) - (\alpha - \beta)$ (see analysis of Example 1 in 2). We have

$$V(x, y, 1) = \begin{cases} 0, & \text{if } U_x \supset V_y, \text{ or } (\rho(x, y) - (\alpha - \beta) \leq 0), \\ \rho(x, y) - (\alpha - \beta), & \text{if } U_x \not\supset V_y, \end{cases} \tag{1.5.10}$$

or, which is the same,

$$V(x, y, 1) = \max[0, \rho(x, y) - (\alpha - \beta)].$$

Let $\rho(x', y')$ be a continuous function defined on $U_x \times V_y$ and c a real number. Then

$$\max_{y' \in V_y} \min_{x' \in U_x} [f(x', y') + c] = c + \max_{y' \in V_y} \min_{x' \in U_x} f(x', y').$$

Using the induction by a number of steps k, we prove that the following formula holds: or, which is the same,

$$V(x, y, k) = \begin{cases} 0, & \text{if } \rho(x, y) - k(\alpha - \beta) \leq 0, \\ \rho(x, y) - k(\alpha - \beta), & \text{in the opposite case,} \end{cases} \tag{1.5.11}$$

otherwise,

$$V(x, y, k) = \max[0, \rho(x, y) - k(\alpha - \beta)].$$

Let (1.5.11) be satisfied for $k < m$. We prove that it holds for $k = m$. Let us employ the (1.5.9). This yields

$$V(x, y, m) = \max_{y' \in V_y} \min_{x' \in U_x} V(x', y', m-1)$$

or

$$V(x,y,m) = \max_{y' \in V_y} \min_{x' \in U_x}$$

$$\begin{cases} 0, & \text{if } \rho(x',y') - (m-1)(\alpha-\beta) \le 0, \\ \rho(x',y') - (m-1)(\alpha-\beta), & \text{in the opposite case.} \end{cases}$$

Consider the function $\Phi(x',y') = \max[0, \rho(x',y') - (m-1)(\alpha-\beta)]$ for $x' \in U_x, y' \in V_y$:

$$\max_{y' \in V_y} \min_{x' \in U_x} \{\max[0, \rho(x',y') - (m-1)(\alpha-\beta)]\} =$$

$$= \max[0, \max_{y' \in V_y} \min_{x' \in U_x} [\rho(x',y') - (m-1)(\alpha-\beta)]] =$$

$$= \max[0, \rho(x,y) - (\alpha-\beta) - (m-1)(\alpha-\beta)] =$$

$$= \max[0, \rho(x,y) - m(\alpha-\beta)].$$

However,

$$V(x,y,m) = \max_{y' \in V_y} \min_{x' \in U_x} \Phi(x',y'),$$

i.e.

$$V(x,y,m) = \rho(x,y) - m(\alpha-\beta),$$

which was to be proved.

If $V(x,y,m) = \rho(x,y) - m(\alpha-\beta)$, i.e $\rho x, y - m(\alpha-\beta) > 0$, then by using the optimal strategy, Player II chooses, at each time instance, the intersection point m for the line of centers x, y and the boundary of V_y which is farthest from x. Similarly, by the optimal strategy, at each time instant Player I chooses from the set U_x the point which is nearest to the point chosen by Player II. Moreover, if Player II's action is optimal (i.e., he chooses M), then Player I chooses the point m' on the interval xM which belongs to U_x and is nearest to M.

As a result, if both players act optimally, the sequences of the chosen points $x_0, x_1, \ldots, x_N, y_0, y_1, \ldots, y_N$ lie on the same line passing through x_0, y_0 (Fig.10). If $V(x,y,m) = 0$, then the optimal strategy for Player II is arbitrary. However, Player I 's strategy is the same as in the previous case until the players first pass to the states for which $\max_{y' \in V_y} \min_{x' \in U_x} \rho(x',y') = 0$. Beginning with this state, Player I repeats Player II's choice.

1.6 Simultaneous games of pursuit with non–convex payoff functions

In this section a solution is given to some simultaneous games of pursuit in which the payoff function and the strategy sets for players are non–convex. The

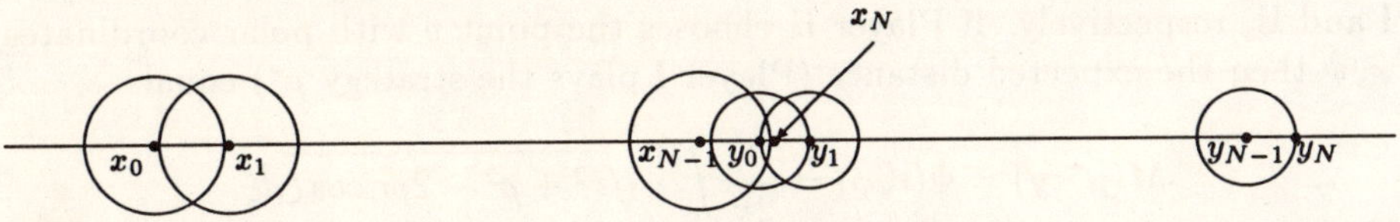

Fig. 10.

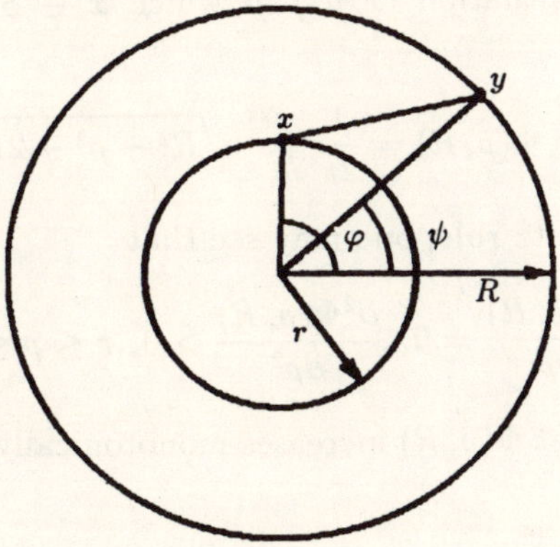

Fig. 11.

results obtained in 4 are not applicable to such games, therefore a solution for both players belongs to the class of mixed strategies. The existence of solution in this class follows from Theorem 3.

Example 11 (simultaneous game of pursuit in a ring). This game is a special case of the game from Example 1 where the sets $S_1 = S_2 = S$ and S a ring. The radii of outer and inner circles of the ring S are denoted respectively by R and r (Fig.11).

We will show that the optimal strategies for the Players I and II are a choice of the points with uniform distrbution on the inner (for Player I) and the outer (for Player II) circles of the ring S. Denote these strategies by μ^* (for Player I) and ν^* (for Player II).

For the above strategies the mean value of the payoff (distance) is equal to

$$M(\mu^*, \nu^*) = \frac{1}{4\pi^2} \int_0^{2\pi} \int_0^{2\pi} \sqrt{R^2 + r^2 - 2Rr\cos(\varphi - \psi)} d\varphi d\psi =$$

$$= \frac{1}{2\pi} \int_0^{2\pi} \sqrt{R^2 + r^2 - 2Rr\cos\xi d\xi} = \Phi(r, R),$$

where φ and ψ are vectorial (polar) angles of pure strategies for the Players I and II, respectively. If Player II chooses the point y with polar coordinates φ, ψ then the expected distance (Player I plays the strategy μ^*) equals

$$M(\mu^*, y) = \Phi(r, \phi) = \frac{1}{2\pi}\int_0^{2\pi} \sqrt{r^2 + \rho^2 - 2\rho r \cos\xi d\xi}.$$

It is readily seen that for $r \leq \rho \leq R$ the function $\varphi(\rho) = \pi^2 + r^2 - 2r\rho\cos\xi$ increases monotonically. In particular, $\varphi(\rho) \leq \varphi(R)$ for $r \leq \rho \leq R$. Therefore, for any strategy of Player II the expected distance does not exceed $\Phi(r, R)$.

Now consider the situation (x, ν^*) in which $x \in S$. Then the expected distance equals

$$M(x, \nu^*) = \Phi(\rho, R) = \frac{1}{2\pi}\int_0^{2\pi} \sqrt{R^2 + \rho^2 - 2R\rho\cos\xi d\xi}.$$

By applying the Leibnitz rule, one may see that

$$\frac{\partial\Phi(0, R)}{\partial\rho} = 0, \ \frac{\partial^2\Phi(\rho, R)}{\partial\rho^2} > 0, \ r \leq \rho \leq R.$$

From this it follows that $\Phi(\rho, R)$ increases monotonically in ρ, hence $\Phi(r, R) \leq \Phi(\rho, R)$ and we have

$$M(x, \nu^*) \geq M(\mu^*, \nu^*) \geq M(\mu^*, y)$$

for all $x, y \in S$. We have thus proved the optimality of strategies (see Theorem 3).

Thus, the value of the game is equal to the integral:

$$M(\mu^*, \nu^*) = \frac{1}{2\pi}\int_0^{2\pi} \sqrt{R^2 + r^2 - 2Rr\cos\xi d\xi} = \Phi(r, R).$$

If S is the circle with he radius R ($r = R$), then the value of the game equals $\frac{4}{\pi}R$.

Example 12. Consider a simultaneous zero–sum game of pursuit between the detail of pursuers $P = \{P_1, P_2\}$ (Player I) and the evader E (Player II). Player $P = \{P_1, P_2\}$ chooses a point $x = (x_1, x_2)$, $x_1 \in S$, $x_2 \in S$, and Player E, who does not know Player P's choice (simultaneously with him), chooses the point $y \in S$. Player E's payoff is set to be equal $\min_{i=1,2} \rho(x_i, y)$.

Thus, Player E aims to maximize the distance between himself and Detail P (maximization of a minimum distance to one of the detail partners). Player P pursues the opposite object. We shall provide a solution for both players.

Case A. The set S is the interval $c = [-1, +1]$. We show that mixed optimal strategies are: for Player P the choice of points $(-1, 1/3)$, $(-1/3, 1)$

with equal probabilities, and for Player E the choice of points $-1, 0, 1$ also with equal probabilities. The value of the game is equal to $1/3$.

The mixed strategies are denote by μ^* and ν^*, respectively. Let Player E play the strategy ν^*, and Player P chooses an arbitrary pure strategy $x = (x_1, x_2)$. We may always assume that $x_1 \leq x_2$. The situation (x, ν^*) corresponds to the payoff

$$M(x, \nu^*) = \frac{1}{3}(1 + x_1) + \frac{1}{3}(1 - x_2) + \frac{1}{3}\min(|x_1|, |x_2|). \tag{1.6.1}$$

Evidently, if $x_1 \leq x_2 \leq 0$, then

$$\min_x M(x, \nu^*) = \min_{x_1, x_2} \frac{2 + x_1 - 2x_2}{3} = \frac{1}{3} \tag{1.6.2}$$

is achieved with $x_1 = -1$, $x_2 = 0$. If $0 \leq x_1 \leq x_2$, then

$$\min_x M(x, \nu^*) = \min_{x_1, x_2} \frac{2 + 2x_1 - x_2}{3} = \frac{1}{3} \tag{1.6.3}$$

is achieved with $x_1 = 0$, $x_2 = 1$.

Consider the case $x_1 \leq 0 \leq x_2$. Clearly,

$$\min_M(x, \nu^*) =$$

$$\begin{cases} \min_{x_1} \frac{2+x_1}{3} = \frac{1}{3} & \text{with } -x_1 \geq x_2 \text{ and is achived with } x_1 = -1, \\ \min_{x_2} \frac{2-x_2}{3} = \frac{1}{3} & \text{with } -x_1 \leq x_2 \text{ and is achived with } x_2 = 1. \end{cases} \tag{1.6.4}$$

In particular, with $x_1 = -1$, $x_2 = 1/3$ and with $x_1 = -1/3$, $x_2 = 1$ the payoff of Player E equals $1/3$.

Thus, Eq.(1.6.1)–Eq.(1.6.4) are satisfied for any pure strategy $x = (x_1, x_2)$. Player E receives the payoff $M(x, \nu^*) \geq 1/3$.

Now let Player P play the strategy μ^*, and Player E an arbitrary pure strategy $y \in c$.

Assume that $y \geq 0$. Consider two possiblities: $0 \leq y \leq 1/3$ and $1/3 \leq y \leq 1$. In the first case Player E receives the payoff

$$M(\mu^*, y) = \frac{1}{2}\left(\frac{1}{3} - y\right) + \frac{1}{2}\left(\frac{1}{3} + y\right) = \frac{1}{3}, \tag{1.6.5}$$

in the second

$$M(\mu^*, y) = \frac{1}{2}(1 - y) + \frac{1}{2}\left(y - \frac{1}{3}\right) = \frac{1}{3}. \tag{1.6.6}$$

In view of the symmetry for $y \leq 0$, the payoff of Player E turns out to be $1/3$. Consequently, playing any pure strategy, Player E cannot a payoff below $1/3$.

From (1.6.5) and (1.6.6) we may conclude that $M(\mu^*, \nu^*) = 1/3$.

Case B. S is a circle with radius r, its center being at the origin of coordinates. In what follows, we denote by $D_r(S_r)$ a circle circumference of the circle with radius r, its center being at the origin of coordinates.

Consider the function

$$\Phi(r, \rho) = r^2 + \rho^2 - \frac{4}{pi} r\rho, \; 0 \le r, \; \rho \le R.$$

Lemma 8 *The function $\Phi(r, \rho)$. (as a function of the variable r) is strictly convex and achieves the absolute minimum at a unique point $r_0 = 2R/\pi$.*

Proof: We have

$$\frac{\partial^2 \Phi(r, \rho)}{\partial r^2} = \frac{\partial^2 \Phi(r, \rho)}{\partial \rho^2} = 2 > 0. \tag{1.6.7}$$

Consequently, the function $\Phi(r, \rho)$, $r \in [OR]$, is strictly convex, and the derivative

$$\frac{\partial \Phi}{\partial r} = 2r - \frac{4R}{\pi} \tag{1.6.8}$$

is strictly monotone. Evidently, the function (1.6.8) vanishes at a unique point $r_0 = 2R/\pi$. Because of the strict convexity of the function $\Phi(r_0, R)$ (1.6.7), the point r_0 is a unique point of absolute minimum. This proves the lemma. ■

Lemma 9 *The function $\Phi(r, \rho)$ is strictly convex in ρ and achieved absolute maximum at the point $\rho_0 = R$.*

Proof: Because of the symmetric entry of variables r and ρ into the function $\Phi(r, \rho)$, the strict convexity of $\Phi(r, \rho)$ follows from Lemma 8.

We have

$$\Phi(r_0, R) - \Phi(r_0, 0) = r_0^2 + R^2 - \frac{4}{\pi} r_0 R - r_0^2 =$$

$$= R^2 - \frac{4}{\pi} \cdot \frac{2R}{\pi} R = \frac{R^2}{\pi} (\pi^2 - \rho) > 0.$$

We have thus proved the lemma completely. ■

Remark. Lemmas 8 and 9 suggest that the point (r_0, R) is the saddle point of the function $\Phi(r, \rho)$. Indeed, $\Phi(r_0, \rho) < \Phi(r_0, R) < \Phi(r, R)$.

Theorem 8 *Mixed optimal strategies are: for Player P the choice of x_1 with uniform destribution on S_{r_0}, $x_2 = -x_1$, for Player E the choice of y with uniform destribution on S_2. The value of the game is equal to $\Phi(r_0, R)$.*

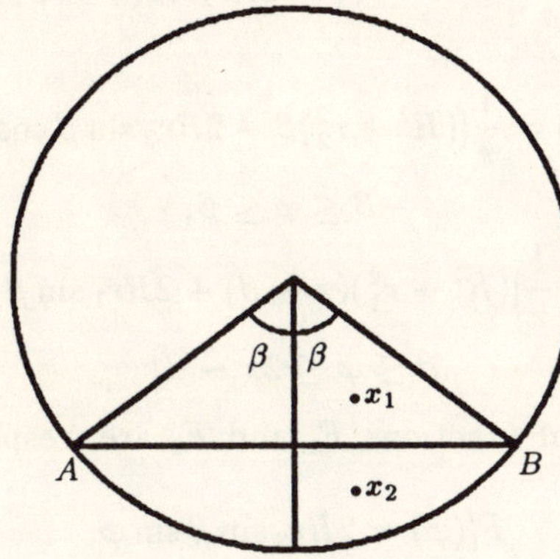

Fig. 12.

Proof: Denote the stated strategies by μ^*, ν^*.

Let Player E adopt the strategy ν^*, and Player P an arbitrary pure strategy $x = (x_1, x_2)$; $x_i = (r_1 \cos\varphi_i, r_1 \sin\varphi_i)$, $i = 1, 2$.

Consider the case $x_1 = x_2$. Denote by r a number $r_1 = r_2$, and by φ an angle $\varphi_1 = \varphi_2$. Player E's payoff is equal to

$$M(x, \nu^*) = \frac{1}{2\pi}\int_0^{2\pi} [R^2 + r^2 - 2Rr\cos(\psi - \varphi)]d\psi =$$

$$= R^2 + r^2 \geq R^2 + r^2 - \frac{4}{\pi}Rr = \Phi(r, R). \tag{1.6.9}$$

Consequently, when adopting a pure strategy $x = (x_1, x_2)$, $x_1 = x_2$, Player E receives the payoff $M(x, \nu^*) \geq \Phi(r, R) \geq \Phi(r_0, R)$ (see Lemma 8). In what follows, we assume that $x_1 \neq x_2$.

On the plane R^2, we introduce a polar system of coordinates. As the origin of coordinates, we take the point O, and as the polar axis the ray starting from the point O normal to the chord AB - the set of the points of the circle D_r that are equidistant from x_1 and x_2 (Fig.12).

For simplicity of notations, we propose that with respect to a new system of coordinates, the point x_i has the same coordinates $(r_i \cos\varphi_i, r_i \sin\varphi_i)$, $i = 1, 2$. Player E's payoff is equal to

$$M(x, \nu^*) = \frac{1}{2\pi}\int_0^{2\pi} \min_i[R^2 + r_i^2 - 2Rr_i\cos(\psi - \varphi_i)]d\psi =$$

$$= \frac{1}{2\pi}\int_{-\beta}^{\beta}[R^2 + r_2^2 - 2Rr_2\cos(\psi - \varphi_2)]d\psi +$$

$$+\frac{1}{2\pi}\int_{\beta}^{2\pi-\beta}[R^2 - r_1^2 - 2Rr_1 cos(\psi-\varphi_1)]d\psi.$$

Let

$$F_1(\varphi) = \frac{1}{\pi}[(R^2 + r_2^2)\beta - 2Rr_2\sin\beta\cos\varphi],$$
$$-\beta \le \varphi \le \beta,$$
$$F_2(\varphi) = \frac{1}{\pi}[(R^2 + r_1^2)(\pi-\beta) + 2Rr_1\sin\beta\cos\varphi],$$
$$\beta \le \varphi \le 2\pi - \beta.$$

The stationary points of functions F_1 and F_2 are, respectively, 0 and π, for $0 < \beta < \pi/2$ and

$$F_1'(\varphi) = 2Rr_2\sin\beta\sin\varphi,$$
$$F_2'(\varphi) = -2Rr_1\sin\beta\sin\varphi,$$

0 and π are the points of absolute minimum of the functions $F_1(\varphi)$ and $F_2(\varphi)$ ($F_1'(\varphi) < 0$ for $\varphi \in (-\beta, 0)$, $F_1'(\varphi) > 0$ for $\varphi \in (0, \beta)$. Similarly, $F_2'(\varphi) < 0$ for $\varphi \in (\beta, \pi)$ and $F_2'(\varphi) > 0$ for $\varphi \in (\pi, 2\pi - \beta)$).

Consequently,

$$M(x, \nu^*) = F_1(\varphi_1) - F_2(\varphi_2) \ge F_1(0) + F_2(\pi) =$$
$$\frac{1}{2\pi}\int_{-\beta}^{\beta}(R^2 + r_2^2 - 2Rr_2\cos\varphi)d\psi +$$
$$+\frac{1}{2\pi}\int_{-\beta}^{2\pi-\beta}(R^2 + r_1^2 + 2Rr_1\cos\varphi)d\psi \qquad (1.6.10)$$

i.e. using the strategy $x_1 = (-r_1, 0)$, $x_2 = (r_2, 0)$ Player P receives a larger payoff than using the strategy $x_i = (r_i\cos\varphi_i, r_i\sin\varphi)$.

Now let x_1 and x_2 lie on the diameter of the circle D_r, the distance between them being $2r$, and the chord AB (for this case) forms the central angle 2α. Assume that $x_1 = (R\cos\alpha - r, 0)$, $x_2 = (R\cos\alpha + r, 0)$. Player E's payoff is equal to

$$\Psi(\alpha, r) = \frac{1}{2\pi}\int_{-\alpha}^{\alpha}[(R\cos\psi - R\cos\alpha - r)^2 + R^2\sin^2\psi]d\psi +$$
$$+\frac{1}{2\pi}\int_{-\alpha}^{2\pi-\alpha}[(R\cos\psi - R\cos\alpha + r)^2 + R^2\sin^2\psi]d\psi =$$
$$= \frac{1}{2\pi}\int_{-\alpha}^{\alpha}[R^2 - 2R\cos\psi(R\cos\alpha + r) + (R\cos\alpha + r)^2]d\psi +$$
$$+\frac{1}{2\pi}\int_{\alpha}^{2\pi-\alpha}[R^2 - 2R\cos\psi(R\cos\alpha - r) + (R\cos\alpha - r)^2]d\psi =$$

$$= \frac{1}{\pi}\{[R^2 + (R\cos\alpha + r)^2]\alpha - 2R\sin\alpha(R\cos\alpha + r) +$$
$$+ [R^2 + (R\cos\alpha - r)^2](\pi - \alpha) + 2R\sin\alpha(R\cos\alpha - r)\}.$$

We show that the function $\Psi(\alpha, r)$ for a fixed r with respect to α a achieves minimum when $\alpha = \pi/2$. Performing elementary computations, we obtain

$$\frac{\partial\Psi}{\partial\alpha} = \frac{2}{\pi}R\sin\alpha[(\pi - 2\alpha)r - \pi R\cos\alpha],$$

therefore for sufficiently small α $\frac{\partial\Psi(\alpha,r)}{\partial\alpha} < 0$, because $\sin\alpha > 0$, $r(\pi - 2\alpha) - \pi R\cos\alpha < 0$ (in the limiting case $r\pi - \pi R = \pi(r - R) < 0$, $r < R$). At the same time, $\frac{\partial\Psi(\pi/2,r)}{\partial\alpha} = \frac{2}{\pi}R \cdot 0 = 0$.

With each fixed r, the function $\frac{\partial\Psi(\alpha,r)}{\partial\alpha}$ has no zeros in α exept for $\alpha = \pi/2$. Suppose the opposite is true. Let $\alpha_1 \in (0, \pi/2)$ be a zero of the function $\frac{\partial\Psi(\alpha,r)}{\partial\alpha}$. Then, for $\alpha = \alpha_1$, the function $G(\alpha) = (\pi - 2\alpha)r - \pi R\cos\alpha$ also equals to zero. Thus, $G(\alpha_1) = G(\pi/2) = 0$. Evidently, $G(\alpha)$, as the difference of a linear function $(\pi - 2\alpha)r$ and the function $\pi R\cos\alpha$, has no zeros: $G(\alpha) > 0$ for $\alpha \in (\alpha_1, \pi/2)$. We compute its dervatives:

$$G'(\alpha) = -2r + \pi R\sin\alpha, \quad G''(\alpha) = \pi R\cos\alpha > 0$$

i.e. the function $G(\alpha)$ is convex. Since $G(\alpha) > 0$, $\alpha \in (\alpha_1, \pi/2)$, this contradicts the above inequality. Thus, the function $\frac{\partial\Psi(\alpha,r)}{\partial\alpha} < 0$ for $\alpha \in (\alpha_1, \pi/2)$ and $\frac{\partial\Psi(\pi/2,r)}{\partial\alpha} = 0$. Consequently, the function $\Psi(\alpha, r)$ with respect to α achieves absolute minimum for $\alpha = \pi/2$:

$$\Psi(\alpha, r) \geq \Psi(\pi/2, r).$$

This means that in this case the payoff of Player E

$$M(x, \nu^*) = \Psi(\alpha, r) \geq \Psi(\pi/2, r) \geq \Phi(r, R) \geq \Phi(r_0, R). \tag{1.6.11}$$

Based on (1.6.9)–(1.6.11), it turns out that for any pure strategy $x = (x_1, x_2)$ Player E has the payoff

$$M(x, \nu^*) \geq \Phi(r_0, R). \tag{1.6.12}$$

Suppose Player P adopts the strategy μ^*, and Player E an arbitrary pure strategy $y = (\rho\cos\psi, \rho\sin\psi)$. Then Player E obtains the payoff

$$M(\mu^*, y) = \frac{1}{2\pi}\int_0^{2\pi} \min[\rho^2 + r_0^2 - 2\rho r_0\cos(\psi - \varphi), \rho^2 + r_0^2 + 2\rho r_0\cos(\psi - \varphi)]d\varphi =$$

$$= \frac{1}{2\pi}\int_0^{2\pi} \min(\rho^2 + r_0^2 - 2\rho r_0\cos\xi, \rho^2 + r_0^2 + 2\rho r_0\cos\xi)d\xi = \Phi(r_0, \rho),$$

and, because of Lemma 9,

$$M(x,\nu^*) = \Phi(r_0,\rho) \le \Phi(r_0,R). \tag{1.6.13}$$

Inequalities (1.6.12) and (1.6.13) imply that μ and v are mixed optimal strategies respectively for the Player P and E, and $\Phi(r_0,R)$ is the value of the game. This proves theorem completely. ■

Example 13. Consider a generalization of the previous game in this case where pursuit is carried out by a pursuer detail involving m participants. Now, suppose that a simultaneous zero–sum game between the pursuer detail $P = \{P_1,\dots,P_m\}$ (Player I) and the evader E (Player II) is given. Player $P = \{P_1,\dots,P_m\}$ chooses the point $x = (x_1,\dots,x_m)$, $x_i \in S$, $i = 1,\dots,m$ and Player E, without any knowledge of Player P's choice (simultaneously with him), chooses the point $y \in S$. The payoff of Player E is assumed to be equal $\min_{i\in\{1,\dots,m\}} \rho(x_i,y)$.

Thus, as in the previous case, Player E aims to maximize the distance between himself and the detail (maximization of a minimum distance from one of the partners of the detail P). Player P pursues the opposite object.

We shall give a solution for the case where the set S coincides with the interval $[-1,+1]$. At present, the solution of the game in the general case is unknown.

Theorem 9 *The mixed optimal strategy for Player E is the choice of the points*

$$-1,\ -\frac{2m-3}{2m-1},\dots,-\frac{2m-2i-1}{2m-1},\dots,$$

$$-\frac{1}{2m-1},\dots,-\frac{1}{2m-1},\frac{1}{2m-1},\dots,1$$

with probabilities $1/2m$. *The optimal strategy for Player P is the the equiprobable choice of two sets of m points*

$$-1,\ -\frac{2m-5}{2m-1},\dots,-\frac{1}{2m-1},\ \frac{3}{2m-1},\dots,\frac{2m-3}{2m-1},$$

$$-\frac{2m-3}{2m-1},\dots,-\frac{3}{2m-1},\ \frac{1}{2m-1},\dots,\frac{2m-5}{2m-1},\ 1.$$

The value of the game is equal to $1/(2m-1)$.

Proof: Introduce the following notation:

$$\ell_{i-1} = \left[\frac{2m-4i+3}{2m-1},\frac{2m-4(i-1)+1}{2m-1}\right],$$

$$\ell_i = \left[\frac{2m-4i+1}{2m-1}, \frac{2m-4i+3}{2m-1}\right].$$

We show that $M(F^*, y) \leq 1/(2m-1)$ for all $y \in [-1,1]$, where F is a strategy of Player P. Indeed, for $y \in \ell_i$

$$M(F^*, y) = \frac{1}{2}\min_i \left|\frac{2m-4i+3}{2m-1} - y\right| +$$

$$+\frac{1}{2}\min_i \left|\frac{2m-4i+1}{2m-1} - y\right| = \frac{1}{2m-1}.$$

A similar equality is derived for $y \in \ell_{i-1}$.

We now prove the inequality

$$M(x_1, \ldots, x_m; G^*) \geq \frac{1}{2m-1}$$

where G^* is a strategy of Player E. We restrict ourselves to the cases when the points x_i are lying on the interval (in the remaining cases the proof is similar):

1. $x_i \in \ell_i$, $i = 1, 2, \ldots, m$;
2. $x_i \in \ell_i$, $i = 1, 2, \ldots, k-1, k+1, \ldots, m$, $x_k \in \ell_p$, $k \neq p$.

In the first case,

$$\frac{2m-4i+1}{2m-1} \leq x_i \leq \frac{2m-4i+3}{2m-1},$$

$$M(x_1, x_2, \ldots, x_m; G^*) =$$

$$= \frac{1}{2m}\left[\sum_{j=1}^{m}\left(\frac{2m-4j+3}{2m-1} - x_j\right) + \sum_{j=1}^{m}\left(x_j - \frac{2m-4j+1}{2m-1}\right)\right] = \frac{1}{2m-1}.$$

The second case means that two points x_i fall on one of the intervals and we have

$$M(x_1, \ldots, x_m; G^*) =$$

$$= \frac{1}{2m}\Bigg[\sum_{j\neq k, j\neq p}\left(\frac{2m-4j+3}{2m-1} - x_j\right) + \sum_{j\neq k, j\neq p}\left(x_j - \frac{2m-4j+3}{2m-1}\right) +$$

$$+\min\left(\frac{2m-4p+3}{2m-1} - x_p, \frac{2m-4p+3}{2m-1} - x_k\right) +$$

$$+\min\left(x_p - \frac{2m-4p+1}{2m-1}, x_k - \frac{2m-4p+1}{2m-1}\right) +$$

$$+\left(x_{k+1} - \frac{2m-4k+3}{2m-1}\right) + \left(\frac{2m-4k+1}{2m-1} - x_{k-1}\right)\Bigg] \geq$$

$$\geq \frac{1}{2m}\left(\frac{2m}{2m-1} - \frac{4}{2m-1} + \frac{4}{2m-1}\right) = \frac{1}{2m-1}.$$

The latter inequalities show the optimality of strategies F^*, G^* and prove the theorem. ■

Example 14. Two players – the pursuing detail $P = \{P_1, \ldots, P_m\}$ and the evading detail $E = \{E_1, \ldots, E_n\}$ – choose the systems of points $x = (x_1, \ldots, x_m)$, $x_i \in [-1, 1]$, $i = 1, \ldots, m$, and $y = (y_1, \ldots, y_m)$, $y_j \in [-1, 1]$, $j = 1, \ldots, n$, respectively. The choice is made simultaneously and independently, i.e., in making his choice, P has no information about the choice of E and vice versa.

Player E's payoff is assumed to equal

$$\frac{1}{2}(\max_i \min_j |x_i - y_j| + \max_j \min_i |x_i - y_j|).$$

The game is zero–sum. We will prove the following simple theorem.

Theorem 10 *The optimal strategy for each partner of the detail $P_i \in P$, $i = 1, 2, \ldots, m$, is the choice of the point $\{0\}$ with probability 1 (strategy F^*), and the optimal strategy for the detail $E = \{E_i, \ldots, E_n\}$ – is the choice of the points $-1, +1$ with equal probabilities and subsequent concentration of the entire detail E there in (strategy G^*). The value of the game equals 1.*

Proof: Suppose Player E chooses a strategy G^*, and Player P an arbitrary pure strategy $x = (x_1, \ldots, x_m)$. For the purposes of definiteness, assume that $x_1 \leq x_2 \leq \ldots \leq x_m$. Then

$$M(x, G^*) = \frac{1}{4}\left(\min_i |-1 - x_i| + \max_i |-1 - x_i|\right) +$$

$$+\frac{1}{4}\left(\min_i |1 - x_i| + \max_i |1 - x_i|\right) +$$

$$\frac{1}{4}\left(|1 + x_1| + |1 + x_m| + |1 - x_1| + |1 - x_m|\right) = 1.$$

Suppose now Player P choses the strategy F^*, and Player E the arbitrary pure strategy $y = (y_1, \ldots, y_n)$, $y_1 \leq y_2 \leq \ldots \leq y_n$. Then

$$M(F^*, y) = \frac{1}{2}\left(\max_j |y_j| + \min_j |y_j|\right) \leq 1.$$

Thus, the strategies F^*, G^* are actually optimal, and the value of the game is equal to 1. ■

Chapter 2

Definition of differential game of pursuit and existence theorem of equilibrium points

2.1 Nonformal description

Differential games of pursuit are an extension of the multistage games of pursuit discussed in Ch. 1, to the case where the number of steps in the game becomes infinite (continuum) and the players have an opportunity to make decisions continuously. Therefore, such formulation naturally suggests that the players' trajectories represent solution of the system of differential equations, the right–hand sides of which depend on the parameters under the players' control.

Let $x \in R^n$, $y \in R^n$, $u \in U \subset R^\ell$, $v \in V \subset R^\ell$, $f(x,u), g(y,v)$ be the n–dimensional vector functions given on $R^n \times U$, $R^n \times V$ respectively. Consider two systems of ordinary differential equations

$$\dot{x} = f(x,y), \tag{2.1.1}$$

$$\dot{y} = g(y,v) \tag{2.1.2}$$

with the initial conditions x_0, y_0.

Player $P(E)$ starts the motion from the point (position) x_0, y_0 and moves in the phase space R^n, choosing at each instant a value of the parameter $u \in U$, $(v \in V)$ in accordance with his objectives and information accessible in each current state.

The most readily describable is the complete information case. In the differential game this means that the phase states $x(t), y(t)$ are known to the players at each time instant t when the parameters $u \in U, v \in V$. are chosen. Sometimes, at each current instant t, one of the players, say, Player P

requires additional information about the value of the parameter $v \in V$ chosen by Player E at the same instant. In such a situation, Player E is said to be discriminated, and the game itself is called the game with discrimination for Player E. This definition of complete information suggests an analogy with the multistage games of pursuit with complete information from Ch. 1.

The parameters $u \in U, v \in V$ are called the controls for the Player P and E, respectively. Let the motions $x(t), y(t)$, $(x(0) = x_0, y(0) = y_0)$ be realized. The trajectory $x(t)$ $(y(t))$ is called the motion trajectory of the Player P (E). The Players' objectives in the game are determined with the help of the payoff, that is dependent on the realized trajectories $x(t)$ $(y(t))$. In some cases the game is assumed to have a prescribed duration T.

Let $x(T), y(T)$ be phase states for the Players P and E at moment of the termination of the game T. The payoff of Player E is then assumed to be equal $M(x(T), y(T))$, where $M(x, y)$ is a continuous function given on $R^n \times R^n$. In a special case, when

$$M(x(T), y(T)) = \rho(x(T), y(T)) \tag{2.1.3}$$

where $\rho(x(T), y(T)) = \sqrt{\sum_{i=1}^{n}(x_i(T) - y_i(T))^2}$ is the Euclidean distance between the points $x(T), y(T)$, the game describes the process of pursuit, in which Player E is aiming to recede from Player P by the game termination instant for a maximum distance. In all cases, the game is assumed to be zero–sum, i.e. the payoff of Player P is equal to $-M(x(T), y(T))$. If condition (2.1.3) is satisfied, then Player P is aiming to ensure a maximum approach to Player E by the game termination instant T.

The payoff thus defined depends only on the finite position of the players and the player, say Player P, scores the results he has obtained by the game termination instant T. Therefore, quite reasonable is the problem formulation where Player E's payoff is defined as the minimum distance between the players existing in the process of the game

$$\min_{0 \leq t \leq T} \rho(x(t), y(t)) \tag{2.1.4}$$

Sometimes a limitation on the game duration is not essential, and the game proceeds until the Players P and E arrive at a particular result.

Let the m–dimensional surface S be given in R^{2n} and called terminal. Set

$$t_\Pi = \min\{(t : (x(t), y(t)) \in S\}$$

i.e. t_Π is the first instant for the point $x(t), y(t)$ to fall on S. If for all the $t \geq 0$ the point $x(t), y(t) \notin S$ then t_Π is assumed to be equal $+\infty$. With the trajectories of $x(t), y(t)$ realized, the payoff of Player E is assumed to be equal t_Π (Player P's payoff is equal to $-t_n$). If S is a sphere with radius $\ell \geq 0$.

$$\sqrt{\sum_{i=1}^{n}(x_i - y_i)^2} = \ell.$$

then we are dealing with the problem of pursuit, in which Player P is aiming to ensure the fastest approach to Player E for a distance $\ell \geq 0$. If $\ell = 0$, then by the capture is meant a coincidence of phase coordinates for the Players P and E. Player E seeks to delay the capture instant. Such type of games of pursuit are called the response games of pursuit (or the time optimal games of pursuit).

At times, it is essential to define the set of initial states for the Players P and E, from which Player P can ensure the capture of Player E for a distance ℓ (ℓ -capture) and the set of initial states, from which Player E can ensure that such a capture does not take place for the distance ℓ (ℓ–capture) in a finite time. The first set is called the capture zone (C, Z), and the second the escape zone (E, Z).

Evidently, these zones do not meet. However, of importance is the question whether the closure of the union for the capture and escape zones covers the entire phase space. Although this issue will be addressed in what follows, we now note that for adequate description of the process it suffices to define the payoff as follows.

Suppose there is $t_\Pi \leq \infty$ (see (2.1.4)). Then the Payoff of Player E is assumed to be -1. If, however, $t_\Pi = \infty$ then his payoff is assumed to be $+1$ (Player P's payoff is assumed to be equal Player E's payoff with opposite sign). The games of pursuit with such a payoff are called the pursuit–type games of degree.

Phase constraints. If we require additionally that during the game the point (x, y) would not leave a set $S \in R^{2n}$, then we have the phase–constrained differential game of pursuit. A special case of such a game is the "Life–line" game.

The "life–line" game is the zero–sum two–person game of kind, in which the payoff of Player E is assumed to be $+1$, if he succeeds in reaching the boundary of the set S ("life– line") before he has been captured by Player P. Thus, the aim of Player E is to reach the boundary of the set S prior to his rendezvous with Player P (i.e. before he has been approached by Player P for a distance $\ell, \ell \geq 0$). The aim of Player P, however, is to approach Player E for the distance ℓ while the latter is in the set S. It is assumed that Player P cannot abandon the set S.

Example 1 (simple motion). The game takes place on the plane. The motion of the Players P and E is described by a system of differential equations

$$\dot{x}_2 = u_1,$$
$$\dot{x}_2 = u_2, u_1^2 + u_2^2 \leq \alpha^2$$
$$\dot{y}_1 = v_1,$$
$$\dot{y}_2 = v_2, v_1^2 + v_1^2 \leq \beta^2, \alpha \geq \beta \tag{2.1.5}$$

Physically, the implication of equations (2.1.5) is that the Player P and E are moving in the plane at limited velocities, the maximum velocities α and β being constant in value, and the velocity of Player E does not exceed that of Player P.

By choosing at each time instant a control $u = (u_1, u_2)$ constrained by $u_1^2 + u_1^2 \leq \alpha^2$ (the set U), the player can change the velocity direction (the direction of the velocity vector). Similarly, by choosing at each time instant a control $v = (v_1, v_2)$ constrained by $v_1^2 + v_2^2 \leq \beta^2$. (the set V), Player E can also change the velocity direction at each time instant. Evidently, if $\alpha > \beta$, then the capture zone (CZ) coincides with the entire space, i.e. Player P can always ensure the capture of Player E for any distance ℓ. To this end, it suffices to choose the motion at the maximum velocity α and to direct the velocity vector at each time instant t towards the pursued point $y(t)$, i.e. to carry out pursuit along the pursuit line.

If $\alpha \leq \beta$, then the set (EZ) coincides with the entire space of the game less the points (x, y), for which $\rho(x, y) \leq \ell$ Indeed, if at the initial instant $\rho_0(x_0, y_0) > \ell$, then Player E can always ensure capture avoidance by receding from Player P along the straight line joining the starting position x_0, y_0 and the maximum velocity β.

A remarkable fact established here, will be encountered on what follows. To form the control ensuring capture avoidance for Player E, it suffices to know only the initial states, while in the case $\alpha > \beta$, to form the control ensuring capture of a player E, Player P needs information on his own and the opponent's state at each current time instant.

Example 2. The Players P and E are the material points with unit masses moving on the plane under the action of the modulus–constrained and the frictional forces. The motion equations are of the form

$$\dot{x}_1 = x_3,$$
$$\dot{x}_2 = x_4,$$
$$\dot{x}_3 = u_1 - k_P x_3, \dot{x}_4 = u_2 - k_P x_4, u_1^2 + u_2^2 \leq \alpha^2,$$
$$\dot{y}_1 = y_3, \tag{2.1.6}$$
$$\dot{y}_2 = y_4,$$

$$\dot{y}_3 = v_1 - k_E y_3, v_1^2 + v_2^2 \le \beta^2,$$
$$\dot{y}_4 = v_2 - k_E y_4$$

where $(x_1, x_2), (y_1, y_2)$ are geometric coordinates, and $(x_3, x_4), (y_3, y_4)$ .... are, respectively, the moments of the points P and E, k_p and k_E the friction coefficients, α and β the maximum forces applicable to the material points P and E. The motion starts from the states $x_i(0) = x_i^0, y_i(0) = y_i^0, i = 1, 2, 3, 4$ Here, by the state is meant the momentum and coordinate space of the Players P and E rather than their geometric position. The sets U, V are the spheres

$$U = \left\{u = (u_1, u_2) : u_1^2 + u_2^2 \le \alpha^2\right\},$$

$$V = \left\{v = (v_1, v_2) : v_1^2 + v_2^2 \le \beta^2\right\}$$

The implication here is that the Players P and E may choose the direction of applied forces at each time instant. However, the maximum values of these forces are restricted by the constants α and β. In this formulation, as shown below, the condition $\alpha > \beta$ (power superiority) is inadequate for Player P to complete the game.

2.2 Game of pursuit in normal form

In 1 we did not provide the method of choosing the controls $u \in U, v \in V$ by the Players P and E during the game according to the incoming information. In other words, we gave no definition for the concept of strategy in the differential game.

This concept may be defined by employing different approaches. We focus only on the intuitively evident game – theoretical properties to be possessed by the concept.

As noted in Ch. 1, the strategy must describe the player's behavior in all information states, in which he may be during the game. In the game with the prescribed duration T, the information state of each player is determined by the phase vectors of the states $x(t), y(t)$ at the current instant t and by the time t from the start of the game. Therefore, it would be natural to consider the strategy of Player $P(E)$ as a function $u(x, y, t)(v(x, y, t))$ with values in the control set U (V). That is how the strategy is defined in [1]. Such type strategies are called synthesizing. However, this method of defining a strategy suffers from some grave disadvantages.

Indeed, let the Players P and E choose the strategies $u(x, y, t), v(x, y, t)$. Then to determine the players' motion strategy and, consequently, the payoff (which is trajectory–dependent), we substitute functions $u(x, y, t), v(x, y, t)$, instead of control parameters u, v, into (2.1.1), (2.1.2), and integrate them on the time interval $[0, T]$ under the initial conditions x_0, y_0. Thus we get the following system of ordinary differential equations:

$$\begin{aligned} \dot{x} &= f(x, u(x, y, t)), \\ \dot{y} &= g(y, v(x, y, t)). \end{aligned} \tag{2.2.1}$$

For the existence and uniqueness of a solution to the system (2.2.1) it is essential that some conditions be imposed on the functions $f(x, u), g(y, v)$ and the strategies $u(x, y, t), v(x, y, t)$. The first group of conditions (for $f(x, u)$, $g(y, v)$) places no limitations on the strategical capabilities of players, refers to the problem statement and are justified by the physical nature of the process involved. The case is different with the constraints on the class of functions (strategies) $u(x, y, t), v(x, y, t)$. Such constraints for the players' capabilities are inconsistent with the notions (adopted in the game theory) above the players' freedom to choose a behavior, and, in some cases, lead to substantial impoverishment of the strategy sets. For example, if we restrict ourselves only to the continuous functions. $u(x, y, t), v(x, y, t)$, then the problems are encountered where there are no solutions in the class of continuous functions. The assumption of a wider strategic spaces makes it impossible to ensure the existence of a unique solution to the system (2.2.1) on the interval $[0, T]$. At times, to overcome this difficulty, one considers the sets of strategies $u(x, y, t), v(x, y, t)$, for which the system (2.2.1) has a unique solution that is capable of continuation on the interval $[0, T]$. Aside from the nonconstructiveness of definition for the strategy sets, such on approach is not adequately substantiated, since the set of all strategy points $u(x, y, t), v(x, y, t)$, for which the system (2.2.1) has a unique solution, proves to be nonrectangular. The latter circumstance, as shown below, leads to transformation of the equilibrium point concept making it game–theoretically meaningless.

Another approach, proposed by N.N.Krasovsky [10,12], enables the players to make on ambiguous choice of controls in each information state.

In this monograph we choose piecewise open–loop strategies as strategies in the differential game.

The piecewise open–loop strategy $u(\cdot)$ for Player P is composed of a pair $\{\sigma, \alpha\}$, where σ is some partitioning $0 = t'_0 < t'_1 < \ldots < t'_n < \ldots$ of the time interval $[0, \infty]$ by the points t'_k without finite points of accumulation; α is the mapping which places each point y'_k and phase states $x(t'_k), y(t'_k)$ in correspondence with a measurable open–loop control $u(t)$ for $t \in [t'_k, t'_{k+1})$ (a measurable function $u(t)$). Similarly, the piecewise open–loop strategy $v(\cdot)$ for Player E is composed of a pair $\{\tau, \beta\}$, where τ is some partitioning $0 = t''_0 < t''_1 < \ldots < t''_n < \ldots$ of the time interval $[0, \infty)$ by the points t''_k without finite points of accumulation; β is the mapping which places each point t''_k and positions $x(t''_k), y(t''_k)$ in correspondence with a measurable open–loop control

$v(t)$ on the interval $[t''_k, t''_{k+1})$ (a measurable function $v(t)$).

Denote the set of all piecewise open–loop strategies for Player P by $\mathbf{P}$, and set of all possible piecewise open–loop strategies for Player E by $\mathbf{E}$. Let $u(t), v(t)$ be a pair of measurable open–loop controls for the Players P and E (measurable function) with values in the set of controls U, V

Consider a system of ordinary differential equations

$$\begin{aligned} \dot{x} &= f(x, u(t)), \\ \dot{y} &= g(y, v(t)) \end{aligned} \tag{2.2.2}$$

on the time interval $[0, \infty]$ with the initial conditions $x(0) = x_0, y(0) = y_0$.

Condition A. With any pair of measurable open–loop controls $v(t)$, $u(t)$ and any initial conditions x_0, y_0, the system of differential equations (2.2.2) has a unique solution $x(t)$, $y(t)$, $(x(0) = x_0, y(0) = y_0)$, which can be continued on the time interval $t \in [0, \infty)$.

Let x_0, y_0 be a pair of initial conditions for equations (2.2.2). The system $S = \{x_0, y_0; u(\cdot), v(\cdot)\}$, where $u(\cdot) \in \mathbf{P}, v(\cdot) \in \mathbf{E}$ is called the situation in the differential game.

Using Condition A, it can readily be shown that to each situation $S = \{x_0, y_0; u(\cdot),\ v(\cdot)\}$ uniquely corresponds a pair of trajectories $x(t)$, $y(t)$ such that $x(0) = x_0, y(0) = y_0$ and

$$\begin{aligned} \dot{x} &= f(x(t), u(\cdot)), \\ \dot{y} &= g(y(t), v(\cdot)). \end{aligned} \tag{2.2.3}$$

Indeed, let $u(\cdot) = \{\sigma, \alpha\}$, $v(\cdot) = \{\tau, \beta\}$. Moreover, let $0 = t_0 \leq t_1 \leq \ldots \leq t_k \leq \ldots$ be a partitioning of the interval $[0, \infty)$ which is a union of partitionings σ, τ. The solution of system (2.2.3) is constructed as follows. On each interval $[t_k, t_{k+1})$, $k = 0, 1, \ldots$, the images of the mappings α, β are the measurable open–loop controls $u(t)$, $v(t)$, hence, by Condition A, the system of equations

$$\dot{x} = f(x(t), u(t)),$$

$$\dot{y} = g(y(t), v(t))$$

has a unique solution $x(t)$, $y(t)$, such that $x(0) = x_0, y(0) = y_0$, on the interval $[t_0, t_1)$. Taking $x(t_1), y(t_1)$ as initial conditions, on the interval $[t_1, t_2)$ we construct a solution to (2.2.4) by reusing the measurability of controls $u(t), v(t)$ as the mapping images α, β on the interval $[t_k, t_{k+1})$, $k = 0, 1, \ldots$. Obtaining $x(t_2), y(t_2)$ we continue the process with the result that we find a unique solution $x(t), y(t)$, so that $x(0) = x_0, y(0) = y_0$. Any trajectory $x(t)$ $(y(t))$ corresponding to $\{x_0, y_0; u(\cdot), v(\cdot)\}$ is called the trajectory of the pursuer P (the evader E).

In what follows, we consider the games with the four types of payoff functions.

Terminal payoff. Given are: a number $T > 0$ and the function $M(x, y)$ continuous on $\{x, y\}$. The payoff in the situation $S = \{x_0, y_0; u(\cdot), v(\cdot)\}$ is defined as follows:

$$K(x_0, y_0; u(\cdot), v(\cdot)) = M(x(T), y(T)),$$

where $x(T) = x(t)|_{t=T}, y(T) = y(t)|_{t=T}$ (here $x(t), y(t)$ are the trajectories of Players P and E corresponding to the situation S). We are dealing with the problem of pursuit when the function $M(x, y)$ is the Euclidean distance between the points x and y.

Maximum result. Let $M(x, y)$ be a continuous real function. In the situation $S = \{x_0, y_0; u(\cdot), v(\cdot)\}$, the payoff of Player E is assumed to be $\min_{0 \le t \le T} M(x(t), y(t))$, where $T > 0$ is a given number. If $M(x, y) = \rho(x, y)$, then the game describes the process of pursuit.

Integral payoff. Some m–dimensional manifold S and a continuous function $M(x, y) > 0$ are given in $R^n \times R^n$. Assume that in the situation $S = \{x_0, y_0; u(\cdot), v(\cdot)\}$, t_Π is the first instant at which the trajectory $x(t), y(t)$ falls on S. Then $K(x_0, y_0; u(\cdot), v(\cdot)) = \int_0^{t_\Pi} M(x(t), y(t))dt$ (if $t_\Pi = \infty$, then $K \stackrel{df}{=} \infty$), where $x(t), y(t)$ are the trajectories, respectively, for the Players P and E corresponding to the situation S. In the case $M \equiv 1, K = t_\Pi$, we face the problem of pursuit.

Qualitative payoff. The payoff function K may take one of the three values $+1, 0, -1$ depending on a position of $x(t_\Pi)$ in $R^n \times R^n$.

After defining the strategy set for the Players P and E and the payoff function, we can define the differential game of pursuit as the game in normal form. In Ch. 1, by the normal form is meant the triple (X, Y, K), where (X, Y) are the spaces of pairs of all possible strategies in the game, K is the payoff function defined on $X \times Y$.

In the case involved, the payoff function is defined not only on the set of pairs of all possible strategies in the game but also on the set of all pairs of the initial position x_0, y_0. Therefore, to each pair $(x_0, y_0) \in R^n$ corresponds its own game in normal form i.e. we actually define a family of games in normal form which depend on the parameters $x_0, y_0 \in R^n \times R^n$.

By the normal form of the differential game $\Gamma(x_0, y_0)$ given on the space of strategy pairs $P \times E$ is meant the expression

$$\Gamma(x_0, y_0) = (x_0, y_0; P, E, K(x_0, y_0; u(\cdot), v(\cdot)))$$

where $K(x_0, y_0; u(\cdot), v(\cdot))$ is the payoff function defined by any one of the four methods given above.

If the payoff function K in the game Γ is terminal, then the relevant game Γ is called the game with terminal payoff. If the function K is defined in the second way, then the game is called the game for achievement of the maximum result. If the function K in the game Γ has the integral form, then the relevant game Γ is called the integral payoff game. When the payoff function in the game Γ is qualitative, the relevant game Γ is called the qualitative game (the game of kind)[1].

Naturally, optimal strategies cannot exist in the class of piecewise open–loop strategies (in view of the open structure of the class). However, we can show that, in a sufficiently large number of cases, for any $\epsilon > 0$ there is the ϵ–equilibrium point (see 2, Ch. 1).

Recall the definition for the ϵ–equilibrium point.

Let an $\epsilon > 0$ be given. The situation $S_\epsilon^* = \{x_0, y_0; u_\epsilon^*(\cdot), v_\epsilon^*(\cdot)\}$ is called the ϵ–equilibrium point in the game $\Gamma(x_0, y_0)$ if for all $u(\cdot) \in P, v(\cdot) \in E$ the inequality holds.

$$K(x_0, y_0; u(\cdot), v_\epsilon^*(\cdot)) + \epsilon \geq K(x_0, y_0; u_\epsilon^*(\cdot), v_\epsilon^*(\cdot)) \geq$$

$$K(x_0, y_0; u_\epsilon^*(\cdot), v(\cdot)) - \epsilon. \tag{2.2.5}$$

The strategies $u_\epsilon^*(\cdot)$, $v_\epsilon^*(\cdot)$ defined in (2.2.5), are called the ϵ-optimal strategies for the Players P and E.

The following lemma is a rephrased version of the lemma in Ch. 1 for differential games.

Lemma 1 *Suppose that in the game* $\Gamma(x_0, y_0)$ *for every* $\epsilon > 0$ *there exist a limit*

$$\lim_{\epsilon \to \infty} K(x_0, y_0; u_\epsilon^*(\cdot), v_\epsilon^*(\cdot)).$$

Assume that for any initial conditions x_0, y_0 from the region $X \subset R^n \times R^n$, any $\epsilon > 0$ there is the ϵ–equilibrium point. We offer the following definition.

Defenition 1. The function $V(x, y)$ defined at each point $x_0, y_0 \in R^n \times R^n$ as

$$\lim_{\epsilon \to \infty} K(x_0, y_0; u_\epsilon^*(\cdot), v_\epsilon^*(\cdot)) = V(x_0, y_0),$$

is called the value of the game $\Gamma(x_0, y_0)$ from the initial conditions $x_0, y_0 \in X$.

The existence of the equilibrium point in the game $\Gamma(x_0, y_0)$ for any $\epsilon > 0$ is equivalent to the fulfillment of the equality

$$\sup_{v(\cdot) \in E} \inf_{u(\cdot) \in P} K(x_0, y_0; u(\cdot), v(\cdot)) = \inf_{u(\cdot) \in P} \sup_{v(\cdot) \in E} K(x_0, y_0; u(\cdot), v(\cdot)).$$

[1]Such type games are treated in some detail in Ch. 5 which deals with the "Life line" games where the function K is specified.

If in the game $\Gamma(x_0, y_0)$ for any $\epsilon > 0$ there are ϵ–optimal strategies for the Players P and E, then the game $\Gamma(x_0, y_0)$ is said to have a solution.

Definition 2. Let $u^*(\cdot), v^*(\cdot)$ be a pair of strategies such that

$$K(x_0, y_0; u(\cdot), v^*(\cdot)) \geq K(x_0, y_0; u^*(\cdot), v^*(\cdot)) \geq K(x_0, y_0; u^*(\cdot), v(\cdot)) \quad (2.2.7)$$

for all $u(\cdot) \in P$ and $v(\cdot) \in E$. The situation $(x_0, y_0; u^*(\cdot), v^*(\cdot))$ is then called the equilibrium point in the game $\Gamma(x_0, y_0)$.

The strategies $u^*(\cdot) \in P$ and $v^*(\cdot) \in E$ from (2.2.7) are called optimal strategies for the Players P and E.

The existence of the equilibrium point in the game $\Gamma(x_0, y_0)$ is equivalent to the fulfillment of the equality

$$\max_{v(\cdot)\in E} \min_{u(\cdot)\in P} K(x_0, y_0; u(\cdot), v(\cdot)) = \min_{u(\cdot)\in P} \max_{v(\cdot)\in E} K(x_0, y_0; u(\cdot), v(\cdot)).$$

Evidently, if there is an equilibrium point, then for any $\epsilon > 0$ it is also the ϵ–equilibrium point, i.e. the function $V(x_0, y_0)$ here just coincides with $K(x_0, y_0; u^*(\cdot), v^*(\cdot))$ (see 2, Ch. 1).

Consider the synthesizing strategies.

Definition 3. The situation $u^*(x, y, t), v^*(x, y, t)$ is called the equilibrium point in the game in synthesizing strategies if the inequality holds.

$$\begin{aligned} &K(x_0, y_0; u(x, y, t), v^*(x, y, t)) \geq \\ &\geq K(x_0, y_0; u^*(x, y, t), v^*(x, y, t)) \geq \quad (2.2.8) \\ &\geq K(x_0, y_0; u^*(x, y, t), v(x, y, t)). \end{aligned}$$

Note that the equilibrium point in the ordinary sense in the class of functions $u(x, y, t)$, $v(x, y, t)$ is not possible because of the nonrectangularity of the situation space (the spaces of admissible pairs $u(x, y, t)$, $v(x, y, t)$, for which the system (2.2.1) has a unique solution). In fact, the situation $u^*(x, y, t)$, $v^*(x, y, t)$ is the equilibrium point in the game Γ in the class of functions $u(x, y, t)$, $v(x, y, t)$, if the inequality (2.2.8) holds for all situations $(u(x, y, t), v^*(x, y, t))$, and $(u^*(x, y, t), v(x, y, t))$, for which there is a unique solution of the system (2.2.1) from the initial states x_0, y_0 on $[0, \infty)$.

The strategies $u^*(x, y, t), v^*(x, y, t)$ are called the optimal strategies for the Players P and E.

The function $V(x_0, y_0) = K(x_0, y_0; u^*(x, y, t), v^*(x, y, t))$ is called the value function of the game.

A distinction is made between the concepts of equilibrium point in the piecewise open–loop and the synthesizing strategies. In the synthesizing strategies, we cannot require the inequality (2.2.7) to be satisfied for all strategies

$(u^*(x,y,t), v(x,y,t))$, since for some u,v the pairs $(u^*,v),(u,v^*)$ cannot be admissible (in an appropriate situation the equation system (2.2.1) may have, generally, no solution, or may have no unique solution). Such type of games were studied in the theory of games over the finite strategy sets and were called games with prohibited Situations (see [3]).

In what follows, we consider the classes of piecewise open–loop strategies unless otherwise specified.

2.3 Existence of the ϵ–equilibrium point in differential games with prescribed duration

In this section we will prove the existence of the ϵ–equilibrium point in the differential games of pursuit with prescribed duration in the class of piecewise open–loop strategies described in 2.

Consider in what follows the case where Player E's payoff is a distance $\rho(x(T),\ y(T))$ at the last instant T of the game. Start with the game with terminal payoff. Let the dynamics of the game be given by the following differential equations:

$$\text{for } P \quad \dot{x} = \rho(x,u), \tag{2.3.1}$$

$$\text{for } E \quad \dot{y} = g(y,v). \tag{2.3.2}$$

Here $x,y \in R^n$, $u \in U$, $v \in V$, where U,V are respectively the compact sets of Euclidean spaces R^k and $R^\ell, t \in [0,\infty)$ (see [16, 17, 23, 48, 52–56, 63–65]).

Definition 4. Denote by $P(x_0,t_0,t)$ the set of all points $x \in R^n$, for which there is a measurable open–loop control $u(t)$ transferring the point x_0 to the point x on the time interval $[t_0,t]$ (here the phase point is assumed to be in the position x at the instant t). The set $P(x_0,t_0,t)$ is the reachability set for Player P. Likewise, the reachability set $E(y_0,t_0,t)$ is defined for Player E.

Assume that the functions f,g are such that the reachability sets $P(x_0,t_0,t)$, $E(y_0,t_0,t)$ for the Players P and E, respectively, satisfy the following conditions:

1. $P(x_0,t_0,t)$ is determined for any $x_0 \in R^n$, $t_0,t \in [0,\infty]$ $(t_0 \le t)$ and is a compact set of the space R^n;

2. the function $P(x_0,t_0,t)$ is continuous in all of its arguments in Hausdorf metric, i.e. for any $\epsilon > 0$, $x_0 \in R^n$, $t \in [0,\infty)$ $(t_0 \le t)$ there is such $\delta > 0$ that $|t_0 - t_0| < \delta$, $|t - t'| < \delta$, $\rho(x_0,x_0') < \delta$, then $\rho^*(P(x_0,t_0,t), P(x',t_0',t')) < \epsilon$. Recall that the Hausdorf metric ρ^* in the

space of compact subsets R^n is given as follows : $\rho^*(A,B) = \max(\rho'(A,B), \rho'(B,A))$. The same is for $E(y_0,t_0,t)$. Here $\rho'(A,B) = \max_{\alpha\in A}\rho(\alpha,B)$, where ρ is standard metric in R^n.

The existence theorem will be proved for the game of pursuit with prescribed duration $\Gamma(x_0,y_0,T)$, where x_0, $y_0 \in R^n$ are respectively the initial positions of the Players P and E, and T is the duration of the game which proceeds as follows. The Players P and E at the time instant $t_0 = 0$ start moving from the positions x_0, y_0 in compliance with the chosen piecewise open–loop strategy. The games terminates at the time instant $t = T$, and Player E receives from Player P the payoff equal to $\rho(x(T), y(T))$. (see 1, Ch. 2).

At each time instant $t \in [0,T]$ of the game $\Gamma(x_0,y_0,T)$, each player knows the time instant t, his own position, and the opponent's position. Denote by $\hat{P}(x_0,t_0,t)$ $(\hat{E}(y_0,t_0,t))$. a set of the trajectories of system (2.3.1), (2.3.2) emanating from the point $x_0(y_0)$ and defined on the interval $[t_0,t]$.

Let $\delta = \delta_n = \frac{T}{2^n}$ and introduce the games $\Gamma_i^\delta(x_0,y_0,T), i = 1,2,3$, that are auxiliary with respect to the game $\Gamma(x_0,y_0,T)$.

The game $\Gamma_1^\delta(x_0,y_0,T)$ proceeds as follows. At the first step, Player E, being in the position y_0, chooses a trajectory v_1^δ from the set $\hat{E}(y_0,0,\delta)$, and Player P, being in the position x_0 and having knowledge of Player E's choice at this step, in turn chooses a trajectory u_1^δ from the set $\hat{P}(y_0,0,\delta)$. At the k–th step ($k = 2,\ldots,2^n$), Player E knows the position of Player P–the point x_{k-1}^δ and own position – the point y_{k-1}^δ and chooses trajectory v_k^δ from the set $\hat{E}(t_{k-1}^\delta,(k-1)\delta,k\delta)$, Player P knows additionally Player E's choice at this step and chooses a trajectory u_k^δ from the set $\hat{P}(t_{k-1}^\delta,(k-1)\delta,k\delta)$. The game terminates at the 2^n–th step and Player E receives from Player P the payoff equal to $\rho(x(T), y(T))$.

The game $\Gamma_2^\delta(x_0,y_0,T)$ differs from the game $\Gamma_1^\delta(x_0,y_0,T)$. in that, as the k-th step ($k = 1,\ldots,2^n$), Player P chooses a trajectory u_k^δ from the set $\hat{P}(t_{k-1}^\delta,(k-1)\delta,k\delta)$ and knows the position x_{k-1}^δ and y_{k-1}^δ at this instant; here Player E makes his choice with the additional knowledge of the choice made by Player P at this step.

The game $\Gamma_3^\delta(x_0,y_0,T)$ differs from the game $\Gamma_2^\delta(x_0,y_0,T)$ in that, at the 2^n–th step, Player P chooses a trajectory $U_{2^n}^\delta$ from the set $\hat{P}(x_{2^n-1}^\delta,(2^n-1)\delta,2^n\delta)$, whereupon Player E receives the payoff that is equal to $\rho(x(T), y(T-\delta))$.

To be noted is that the games $\Gamma_i^\delta(x_0,y_0,T)$ $i = 1,2$ are a special cases of the games from Example 2, Ch. 1.

In fact, as a family of sets $U_x, x \in R^n$, it suffices to take $U(x) = P(x,0,\delta)$, and as a family of sets $V_y - V(y) = E(y,0,\delta)$ i.e. the family of reachability sets from the initial states x and y in the time $\delta = \frac{T}{2^n}$.

Denote the values of the games $\Gamma_i^\delta(x_0,y_0,T)$ by $Val\Gamma_i^\delta(x_0,y_0,T)$

Lemma 2 *Let $\{A(r)\}$, $\{B(s)\}$ be the family of compact sets in R^n continuously dependent on parameters $r, s \in R^n$ in the Hausdorf metric. Then the functions*

$$F(r,s) = \max_{x\in A(r)} \min_{y\in B(s)} f(x,y),$$

$$G(r,s) = \min_{y\in B(s)} \max_{x\in A(r)} f(x,y),$$

are continuous on the product $R^n \times R^n$ if $f(x,y)$ is continuous on $R^n \times R^n$ (this lemma is similar to Lemma 7, Ch. 1).

Proof: Let us prove the lemma for the function $F(r,s)$. A similar proof is provided for the function $G(r,s)$.

The function $g(x) = \min_{y\in B(s)} f(x,y)$ is the envelope of the continuous function family and is continuous on the set $A(r)$. Let a number $\epsilon > 0$ be given. By the continuity of the function $g(x)$ on the space R^n there is such $\epsilon > 0$ that if the point x' belonging to the set $A(r)\Delta A(r')$[2], with a particular r' is such that there is such point $x'' \in A(r) \bigcap A(r') < \delta$ that $\rho(x'', x') < \delta$, then the inequality $|g(x'') - g(x')| < \epsilon$ is valid. The definition of Haysdorf metric suggests that, to satisfy this condition, it is sufficient that the inequality $\rho^*(A(r)A(r')) < \delta$ be satisfied, because of the continuity of the family of sets $A(r)$, by the parameter r in the Haysdorff metric for any $\delta > 0$ it is possible to select such η that for $\rho(r', r) < \eta$ the inequality $\rho(A(r), A(r'))$[3]$< \delta$ is satisfied. Consequently, by any $\epsilon > 0$ it is possible to find such $\eta > 0$ that for r', r such that $\rho(r', r) < \eta$ the following inequality holds

$$| \max_{x\in A(r)} \min_{y\in B(s)} f(x,y), - \max_{x\in A(r')} \min_{y\in B(s)} f(x,y), |< \epsilon$$

The continuity of the function F in $s \in S$ is proved in much the same way. ■

Lemma 3 *In the games $\Gamma_i^\delta(x_0, y_0, T), i = 1,2,3,$ there exists the equilibrium points for every $x_0, y_0, T < \infty$, the functions $Val\Gamma_i^\delta(x_0, y_0, T))$ are continuous in $x_0, y_0 \in R^n$, and T, and for every $n \geq 0$ the following inequality holds*

$$Val\Gamma_1^\delta(x_0, y_0, T) \leq Val\Gamma_2^\delta(x_0, y_0, T), T = 2^n\delta. \tag{2.3.3}$$

Proof: The existence of the equilibrium point in the games $\Gamma_i^\delta(x_0, y_0, T))$, $i = 1,2,3$ follows from Theorem 7, Ch. 1, and its corollaries. To prove

---

[2] $A\Delta B = (A \setminus B) \bigcup (B \setminus A)$

[3] To be compared with the equations (1.5.7'),(1.5.8').

the inequality (2.3.3), we write the functional equations for the functions $Val\Gamma_i^\delta(x_0, y_0, T))$, $i = 1, 2$:

$$Val\Gamma_1^\delta(x_0, y_0, T) =$$

$$\max_{v_1^\delta \in \hat{E}(y_0,0,\delta)} \min_{u_1^\delta \in \hat{P}(x_0,0,\delta)} Val\Gamma_i^\delta(u_1^\delta(\delta), v_1^\delta(\delta), T - \delta)) \tag{2.3.4}$$

$$\dots\dots\dots$$

$$Val\Gamma_1^\delta(x^\delta_{2^n-1}, y^\delta_{2^n-1}, \delta) =$$

$$= \max_{v^\delta_{2^n} \in \hat{E}(y^\delta_{2^n-1}, T-\delta, T)} \min_{u^\delta_{2^n} \in \hat{P}(x^\delta_{2^n-1}, T-\delta, T)} \rho(u^\delta_{2^n}(T), v^\delta_{2^n}(T)),$$

$$Val\Gamma_2^\delta(x_0, y_0, T) =$$

$$= \min_{u_1^\delta \in \hat{P}(x_0,0,\delta)} \max_{v_1^\delta \in \hat{E}(y_0,0,\delta)} Val\Gamma_2^\delta(u_1^\delta(\delta), v_1^\delta(\delta), T - \delta), \tag{2.3.5}$$

$$\dots\dots\dots$$

$$Val\Gamma_2^\delta(x_0, y_0, T) =$$

$$= \min_{u^\delta_{2^n} \in \hat{P}(x^\delta_{2^n-1}, T-\delta, T)} \max_{v^\delta_{2^n} \in \hat{E}(y^\delta_{2^n-1}, T-\delta, T)} \rho(u^\delta_{2^n}(T), v^\delta_{2^n}(T)).$$

The inequality (2.3.3) is obtained by applying sequentially the inequalities (1.2.4) to the functional equations (2.3.4) and (2.3.5).

The continuity of the functions $Val\Gamma_i^\delta(x_0, y_0, T)$, $i = 1, 2, 3$ is proved by applying sequentially Lemma 2 to the functional equations (2.3.4), (2.3.5). (The sets $\hat{E}$, $\hat{P}$ in (2.3.4), (2.3.5) may be replaced by suitable reachability sets which enables utilization of Lemma 2.) ■

Lemma 4 *For any integer $n > 0$, the following inequalities hold:*

$$Val\Gamma_1^{\delta_n}(x_0, y_0, T) \le Val\Gamma_1^{\delta_{n+1}}(x_0, y_0, T),$$

$$Val\Gamma_2^{\delta}(x_0, y_0, T) \ge Val\Gamma_2^{\delta_{n+1}}(x_0, y_0, T)$$

Proof: We show the validity of the first inequality. The second inequality is proved in the same manner.

Suppose that Player E receives a guaranteed payoff $Val\Gamma_1^{\delta_n}(x_0, y_0, T)$ by choosing on optimal strategy in the game $\Gamma_1^{\delta_n}(x_0, y_0, T)$. Let us restrict the class of admissible strategies for Player E in the game $\Gamma_1^{\delta_{n+1}}(x_0, y_0, T)$ to the class of strategies in the game $\Gamma_1^{\delta_n}(x_0, y_0, T)$. It can be readily seen that here he ensures for himself the payoff equal to $Val\Gamma_1^{\delta_n}(x_0, y_0, T)$,this proves the validity of the required inequality. ■

Theorem 1 *For any $x_0, y_0 \in R^n, T < \infty$ the next limiting equality holds:*

$$\lim_{n\to\infty} Val\Gamma_1^{\delta_n}(x_0, y_0, T) = \lim_{n\to\infty} Val\Gamma_2^{\delta_n}(x_0, y_0, T)$$

Proof: Consider the games $\Gamma_2^{\delta_n}(x_0, y_0, T)$, $\Gamma_3^{\delta_n}(x_0, y_0, T)$. To each pair of strategies $u(\cdot), v(\cdot)$ in the game $\Gamma_2^{\delta_n}(x_0, y_0, T)$ corresponds a pair of strategies $u'(\cdot), v'(\cdot)$ in the game $\Gamma_3^{\delta_n}(x_0, y_0, T)$ which are truncations of the pair $u(\cdot), v(\cdot)$ at the last step $(u(\cdot) = u'(\cdot))$. Denote the payoff function in the games $\Gamma_2^{\delta_n}(x_0, y_0, T)$ $\Gamma_3^{\delta_n}(x_0, y_0, T)$ by $K_2(u(\cdot), v(\cdot))$ and $K_3(u'(\cdot), v'(\cdot))$ respectively. The inequality

$$K_2(u(\cdot), v(\cdot)) \le K_3(u'(\cdot), v'(\cdot)) + \rho(x_{2^n}^{\delta_n}, y_{2^n-1}^{\delta_n}),$$

is satisfied. This means that the inequality

$$Val\Gamma_2^{\delta_n}(x_0, y_0, T) \le Val\Gamma_3^{\delta_n}(x_0, y_0, T) + \\ + \max_{y\in E(y_0,0,T-\delta)} \max_{y'\in E(y,T-\delta_n,T)} \rho(y, y'). \tag{2.3.6}$$

is also satisfied.

Write (2.3.6) for the games with other initial data:

$$Val\Gamma_2^{\delta_n}(x_0, y_1^{\delta_n}, T) \le Val\Gamma_3^{\delta_n}(x_0, y_1^{\delta_n}, T) + \max_{y\in E(y_1^{\delta_n},0,T-\delta)} \max_{y'\in E(y,T-\delta_n,T)} \rho(y, y').$$

Be the construction of the set $E(y_1^{\delta_n}, 0, T-\delta_n)$, the inclusion

$$E(y_1^{\delta_n}, 0, T-\delta_n) \subset E(y_0, 0, Y) \;\; (y_1^{\delta_n} \in E(y_0, 0, \delta_n)),$$

is valid and, because of the continuity of the function $\rho(x, y)$ and the compactness of the sets involved, the inequality

$$Val\Gamma_2^{\delta_n}(x_0, y_1^{\delta_n}, T) \le Val\Gamma_3^{\delta_n}(x_0, y_1^{\delta_n}, T) + \\ + \max_{y\in E(y_0,0,T)} \max_{y'\in E(y,T-\delta_n,T)} \rho(y, y'). \tag{2.3.7}$$

is satisfied. The definition of the games $\Gamma_1^{\delta_n}(x_0, y_0, T)$, $\Gamma_3^{\delta_n}(x_0, y_0, T)$ implies the equality

$$Val\Gamma_3^{\delta_n}(x_0, y_0, T) = \max_{y_1^{\delta_n}\in E(y_0,0,\delta_n)} Val\Gamma_3^{\delta_n}(x_0, y_1^{\delta_n}, T). \tag{2.3.8}$$

in view of the continuity of the function $E(y, 0, T)$ and initial condition $E(y, 0, 0) = 0$, the addend in (2.3.7) tends to zero as $n \to \infty$. Denote it by $\epsilon_1(n)$. From (2.3.7), (2.3.8) we get

$$Val\Gamma_1^{\delta_n}(x_0, y_0, T) \ge Val\Gamma_2^{\delta_n}(x_0, y_1^{\delta_n}, T) - \epsilon_1(n). \tag{2.3.9}$$

Because of the continuity of $Val\Gamma_2^{\delta_n}(x_0, y_0, T)$, (2.3.9) implies the inequality

$$Val\Gamma_1^{\delta_n}(x_0, y_0, T) \geq Val\Gamma_2^{\delta_n}(x_0, y_0, T) - \epsilon_1(n) + \epsilon_2(n), \tag{2.3.10}$$

where $\epsilon_2(n) \to 0$ as $n \to \infty$. When making a passage in (2.3.10) to the limit as $n \to \infty$, which is possible by Lemma 4 and the theorem on the bounded monotone sequence, we have

$$\lim_{n\to\infty} Val\Gamma_1^{\delta_n}(x_0, y_0, T) \geq \lim_{n\to\infty} Val\Gamma_2^{\delta_n}(x_0, y_0, T). \tag{2.3.11}$$

From Lemma 3 follows the opposite inequality. Consequently, both limits in (2.3.11) coincide.

The statement of Theorem 1 has been proved on the assumption that the sequence partitioning the interval $[0, T]$

$$\sigma_n = \{t_0 = 0 < t_1 < \ldots < t_N = T\}, n = 1, \ldots,$$

satisfies the condition $t_{j+1} - t_j = \frac{T}{2^n}$, $j = 0, 1, \ldots, 2^n - 1$. It can readily be seen that the statement of Theorem 1 holds for any sequence σ_n of partitioning the interval $[0, T]$ such that

$$\gamma(\sigma_n) = \max_i (t_{j+1} - t_j) \overset{n\to\infty}{\Rightarrow} 0. \tag{2.3.12}$$

Consider any such two sequences of partitioning the interval $[0, T]$, $\{\sigma_n\}$ and $\{\sigma_n'\}$. The following lemma holds. ■

Lemma 5

$$\lim_{n\to\infty} Val\Gamma_1^{\sigma_n}(x_0, y_0, T) = \lim_{n\to\infty} Val\Gamma_1^{\sigma_n'}(x_0, y_0, T),$$

where $x_0, y_0, \in R^n$, $T < \infty$.

Proof: We give the proof by contradiction. Assume that the statement of the lemma is not true, and assume that the following inequality will be a satisfied

$$\lim_{n\to\infty} Val\Gamma_1^{\sigma_n}(x_0, y_0, T) > \lim_{n\to\infty} Val\Gamma_1^{\sigma_n'}(x_0, y_0, T).$$

Then the natural numbers n_1, m_1 may be found such that the following inequality will be satisfied

$$Val\Gamma_2^{\sigma_{m_1}}(x_0, y_0, T) \geq Val\Gamma_1^{\sigma_{m_1}}(x_0, y_0, T) >$$

$$> Val\Gamma_2^{\sigma_{n_1}'}(x_0, y_0, T) \geq Val\Gamma_1^{\sigma'}(x_0, y_0, T).$$

Denote by $\bar{\sigma}$ partitioning of the interval $[0, T]$ by the points belonging to both the partitioning σ_{m_1} and σ_{n_1}. The following inequality is satisfied for $\bar{\sigma}$

$$Val\Gamma_2^{\bar{\sigma}}(x_0, y_0, T) \leq Val\Gamma_1^{\sigma'_{n_1}}(x_0, y_0, T) <$$

$$< Val\Gamma_1^{\sigma_{m_1}}(x_0, y_0, T) \leq Val\Gamma_1^{\bar{\sigma}}(x_0, y_0, T),$$

whence

$$Val\Gamma_2^{\bar{\sigma}}(x_0, y_0, T)) < Val\Gamma_1^{\bar{\sigma}}(x_0, y_0, T).$$

This contradicts (2.3.3) from Lemma 3. Consequently, our assumption is false and the statement of Lemma 5 is proved. ∎

Theorem 2 *For all $x_0, y_0, T < \infty$ in the game $\Gamma(x_0, y_0, T)$ there is the ϵ – equilibrium point for any $\epsilon > 0$. Moreover,*

$$Val\Gamma(x_0, y_0, T) = \lim_{n\to\infty} Val\Gamma_1^{\sigma_n}(x_0, y_0, T),$$

where $\{\sigma_n\}$ is any decreasing sequence of partitions of the interval $[0, T]$ (to be compared with Theorem 7, Ch. 1).

Proof: Let $\epsilon > 0$ be an arbitrary given number. We show that such strategies $u_\epsilon(\cdot)$ and $v_\epsilon(\cdot)$ will be found for the players P and E respectively that for all strategies $u(\cdot) \in P, v(\cdot) \in E$ the following inequalities are satisfied:

$$K(x_0, y_0, u_\epsilon(\cdot), v(\cdot)) - \epsilon \leq K(x_0, y_0, u_\epsilon(\cdot), v_\epsilon(\cdot)) \leq$$

$$\leq K(x_0, y_0, u(\cdot), v_\epsilon(\cdot)) + \epsilon.$$

Because of Theorem 1, there exist such partition σ_u (σ_v) of the interval $[0, T]$ that

$$Val\Gamma_2^{\sigma_u}(x_0, y_0, T) - \lim_{n\to\infty} Val\Gamma_2^{\sigma_n}(x_0, y_0, T) < \epsilon,$$

$$\lim_{n\to\infty} Val\Gamma_1^{\sigma_n}(x_0, y_0, T) - Val\Gamma_1^{\sigma_n}(x_0, y_0, T) < \epsilon.$$

Let

$$u^\epsilon(\cdot) = (\sigma_v^\epsilon, \alpha_u^\epsilon), \ \ v^\epsilon(\cdot) = (\sigma_v^\epsilon, \alpha_v^\epsilon),$$

where $\alpha_u^\epsilon, \beta_v^\epsilon$ are optimal strategies for the players P and E in the game $\Gamma_2^{\sigma_{u^\epsilon}}(x_0, y_0, T)$ and $\Gamma_1^{\sigma_{v^\epsilon}}(x_0, y_0, T)$, respectively. In this case, based on the choice of maps α_u^ϵ and β_v^ϵ, Player P ensures the payoff at least $(-\lim_{n\to\infty} Val\Gamma_1^{\sigma_n}(x_0, y_0, T) + \epsilon)$, and Player E the payoff $\lim_{n\to\infty} Val\Gamma_1^{\sigma_n}(x_0, y_0, T) - \epsilon$. Consequently $u^\epsilon(\cdot)$, $v^\epsilon(\cdot)$ are the ϵ–optimal strategies for the players P and E, respectively. The function

$$Val\Gamma(x_0, y_0, T) = \lim_{n\to\infty} Val\Gamma_1^{\sigma_n}(x_0, y_0, T)$$

is the value function of the game $\Gamma(x_0, y_0, T)$ ■

Corollary. In the proof of the existence theorem, we did not use the specific form of the payoff $\rho(x(T), y(T))$. It was important only the continuous dependence of the payoff from the realized trajectories. Therefore, Theorem 2 holds also if instead of $\rho(x(T), y(T))$ any continuous functional of trajectories $x(t)$, $y(t)$ is considered. In particular, such a functional may be equal too $\min_{0 \le t \le T} \rho(x(t), y(t))$, i.e. the minimum distance between the players during the game. Therefore, the result of this section also holds for the differential games of pursuit for achievement of the maximal result and prescribed duration.

2.4 Existence of ϵ–equilibrium points in optimal time differential pursuit games

Optimal time differential pursuit games constitute a special case of differential games with integral payoff, as defined in 2, Ch. 1. The classes of strategies P and E are the same as in the game of prescribed duration. The set

$$S = \{(x, y) : \rho(x, y) \le l, l > 0\}.$$

is given in $R^n \times R^n$.

Let $x(t), y(t)$ be trajectories for the players P and E in the situation $(u(\cdot), v(\cdot))$ from the initial states x_0, y_0. Denote

$$t_\Pi(x_0, y_0; u(\cdot), v(\cdot)) = \min\{t : (x(t), y(t)) \in S\}$$

(if such t that $(x(t), y(t)) \in S$ does not exist, then $t_\Pi(x_0, y_0; u(\cdot), v(\cdot))$ is assumed to be $+\infty$).

In the differential optimal time pursuit game, Players E's payoff is assumed to be

$$K(x_0, y_0; u(\cdot), v(\cdot)) = t_\Pi(x_0, y_0; u(\cdot), v(\cdot)).$$

Player P's payoff is equal too $-K$.

The game depends on the initial states x_0,y_0, hence it is denoted by $\Gamma(x_0,y_0)$.

From the definition of the payoff function it follows that in the game $\Gamma(x_0, y_0)$ Player E aims to maximize the time of approaching Player P on a given distance $\ell \ge 0$. On the other hand, Player P seeks to minimize this time.

The optimal time pursuit game $\Gamma(x_0, y_0)$ is related directly to the prescribed duration pursuit game for achievement of the maximum result. Let $\Gamma(x_0, y_0, T)$ be the game of pursuit with prescribed duration T for achievement of the maximum result (Player E's payoff is equal to $\min_{0 \le t \le T} \rho(x(t), y(t))$). As shown in 3 of this chapter, for such type of games for every $\epsilon > 0$ in the class of piecewise open–loop strategies there is an ϵ–equilibrium point. Let $V(x_0, y_0, T)$ be the value of such a game, and $V(x_0, y_0)$ the value of the game $\Gamma(x_0, y_0)$.

Lemma 6 *With the fixed x_0, y_0 the function $V(x_0, y_0, T)$ decreases monotonically in T on the time interval* $[0, \infty)$.

Proof: Let $T_1 > T_2 > 0$. Denote by $v_\epsilon^{T_1}(\cdot)$ an ϵ–optimal strategy for Player E in the game $\Gamma(x_0, y_0, T_1)$. For Player E, this strategy ensures that the distance between him and Player P on the interval $[0, T_1]$ is at least $\max[0, V(x_0, y_0, T_1) - \epsilon]$ Consequently, it also ensures the distance between them $\max[0, V(x_0, y_0, T_1) - \epsilon]$ on the interval $[0, T_2]$, where $T_2 < T_1$. Hence

$$V(x_0, y_0, T_2) \ge \max[0, V(x_0, y_0, T_1) - \epsilon] \tag{2.4.1}$$

(the strategy $v_\epsilon^{T_1}(\cdot)$, which is ϵ–optimal in the game $\Gamma(x_0, y_0, T_1)$, is not necessarily ϵ–optimal in the game $\Gamma(x_0, y_0, T_2)$). Since ϵ may be chosen arbitrarily, then the statement of the lemma follows from (2.4.1).

Consider the equation

$$V(x_0, y_0, T) = \ell \tag{2.4.2}$$

with respect to T. Three cases are possible here:

1. the (2.4.2) has no root;

2. it has only one root;

3. it has more than one root.

In case 3, it follows from the monotonicity and continuity of the function $V(x_0, y_0, T)$ in T that the (2.4.2) has the whole segment of roots, i.e. the function $V(x_0, y_0, T_1)$, as the function of T, has a constancy interval (Fig.13).

Let us consider each case separately.

In case 1 it is possible that:

a) $V(x_0, y_0, T_1) < \ell$ for all $T \ge 0$;

b) $\inf_{T>0} V(x_0, y_0, T) > \ell$;

c) $\inf_{T>0} V(x_0, y_0, T) = \ell$.

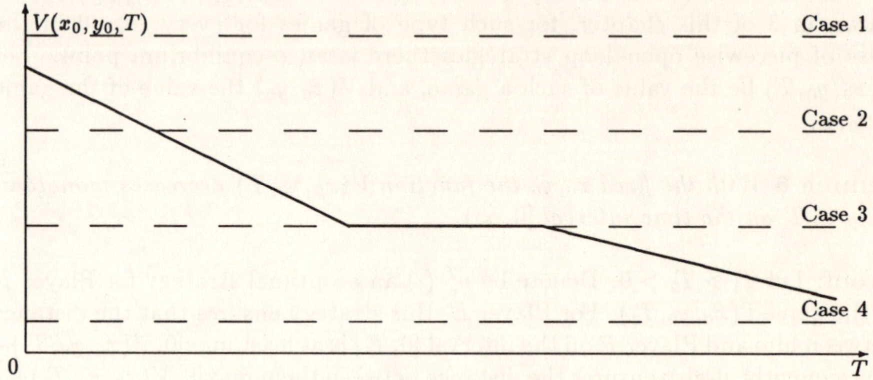

Fig. 13

In case 1a, we have

$$V(x_0, y_0, T) = \rho(x_0, y_0) < \ell$$

i.e. $t_\Pi(x_0, y_0; u(\cdot), v(\cdot)) = 0$ for all $u(\cdot), v(\cdot)$. Then the value of the game $\Gamma(x_0, y_0)$ is equal to $V(x_0, y_0) = 0$. In case 1b

$$\inf_{T \geq 0} V(x_0, y_0, T) = \lim_{T \to 0} V(x_0, y_0, T) > \ell.$$

For every (sufficiently large) $T > 0$, Player E has the strategy $v(\cdot, T)$, which ensures the avoidance of ℓ–capture on the interval $[0, T]$. Player P has no strategy which ensures the ℓ–capture of Player E in finite time. Indeed, if such a strategy existed and ensured a particular capture time $V(x_0, y_0)$, then it would be better than the optimal strategy for Player P in the game $\Gamma(x_0, y_0; V(x_0, y_0))$, since, by assumption the value of this game $V(x_0, y_0; V(x_0, y_0)) > \ell$. At the same time, it cannot be stated that Player E has a strategy which ensures the avoidance of ℓ–capture for any time. The problem of finding the initial states, in which such a strategy exists, reduces to a solution of the game of kind for Player E. Thus, for $\ell < \lim_{T\to\infty} V(x_0, y_0, T)$ it may be stated that the value of the game, if any, is larger than any fixed T, i.e. it is equal to $+\infty$.

Consider case 1c together with case 3.

Consider case 2. Let T_0 be a unique root of (2.4.2). The monotonicity and continuity of the function $V(x_0, y_0, T)$ in T imply the existence of such $\delta > 0$ that for all $T_0 > T > T_0 - \delta$

$$V(x_0, y_0, T) > V(x_0, y_0, T_0) \tag{2.4.3}$$

and for all $T_0 + \delta > T > T_0$.

$$V(x_0, y_0, T) < V(x_0, y_0, T_0) \tag{2.4.4}$$

Consider the game of pursuit $\Gamma(x_0, y_0, T)$ $(T_0 + \delta > T > T_0)$. It has the ϵ–equilibrium point in the class of piecewise open–loop strategies for any $\epsilon > 0$. This means that there exists the strategy $u_{\epsilon_1}(\cdot)$ for Player P which guarantees him approaching Player E on a distance $V(x_0, y_0, T) + \epsilon_1$. From (2.4.4) it follows that $\epsilon_1 > 0$ may be chosen in such a way that $\epsilon_1 < V(x_0, y_0, T_0) - V(x_0, y_0, T) = \ell - V(x_0, y_0, T)$. Choosing ϵ_1 in this way, we see that the strategy $u_{\epsilon_1}(\cdot)$ for Player P guarantees him approaching Player E on the distance

$$V(x_0, y_0, T) + \epsilon_1 < \ell - V(x_0, y_0, T) + V(x_0, y_0, T) = \ell. \tag{2.4.5}$$

From (2.4.5) it follows that the strategy $u_{\epsilon_1}(\cdot)$ ensures the ℓ–capture of Player E in the time not exceeding $T_0 + \delta$. Here ϵ_1 is found by δ and can be defined for any $\delta > 0$. Evidently, by adopting the ϵ_2–optimal strategy $v_{\epsilon_2}(\cdot)$ in the game $\Gamma(x_0, y_0, T)$ $(T_0 > T > T_0 - \delta)$ where ϵ_2 is chosen from the condition $\epsilon_2 < V(x_0, y_0, T) - V(x_0, y_0, T_0) = V(x_0, y_0, T) - \ell$, Player E can ensure that in the time T Player P will not approach him on a distance

$$V(x_0, y_0, T) - \epsilon_2 \geq V(x_0, y_0, T) - V(x_0, y_0, T) + \ell = \ell,$$

i.e. in the time T $(T_0 - \epsilon < T < T_0)$ Player E can make the ℓ–capture impossible. This means that in the optimal time pursuit game $\Gamma(x_0, y_0)$there is the ϵ–equilibrium point in piecewise open–loop strategies for any $\epsilon > 0$, with the value of the game being equal to $T_0(x_0, y_0)$. Here T_0 is a unique root of the Equation (2.4.2).

In case 3, denote the minimum root of (2.4.2) by T_0. Generally, we cannot claim that the value of the game $Val\Gamma(x_0, y_0) = T_0$. Indeed, from $V(x_0, y_0, T_0) = \ell$ it only follows that in the game $\Gamma(x_0, y_0, T_0)$ for any $\epsilon > 0$, there is the strategy $u_\epsilon(\cdot)$ for Player P which guarantees him the $\epsilon + \ell$–capture of Player E in the time T. The existence of more than one roots of (2.4.2) implies (the monotonicity of $V(x_0, y_0, T)$) the existence of the interval of constancy in $T[T_0, T_1]$ of the function $V(x_0, y_0, T)$ Therefore, an increase in the duration of the game $\Gamma(x_0, y_0, T)$ on δ, where $\delta < T_1 - T_0$, does not involve a decrease in the guaranteed approach to Player E, i.e. for all $T \in [T_0, T_1]$. player P can only ensure an approach to Player E on a distance $\ell + \epsilon$ (for any $\epsilon > 0$), and we have no reason to say that for a particular $T \in [T_0, T_1]$ the value of ϵ turns out to be zero. Evidently, if the equilibrium point (instead of the ϵ–equilibrium) existed in the game $\Gamma(x_0, y_0, T_0)$, then the value of the game $\Gamma(x_0, y_0)$ would correspond exactly to T_0.

We modify the concept of the equilibrium point in the game $\Gamma(x_0, y_0)$. Henceforth in this section it is more convenient to employ the notation $\Gamma(x_0, y_0, \ell)$ instead of $\Gamma(x_0, y_0)$ with the implication that in the game $\Gamma(x_0, y_0, \ell)$ the play terminates when the players have approached one another on a distance ℓ. Let $t^{\ell}_{\Pi}(x_0, y_0; u(\cdot), v(\cdot))$ be the necessary time until the players approach one another on the distance ℓ in the situation $(u(\cdot), v(\cdot))$. ■

Definition 5. For $\epsilon > 0$ $\delta > 0$ the pair of strategies $\bar{u}_\epsilon^\delta(\cdot), \bar{v}_\epsilon^\delta(\cdot)$ is said to constitute the ϵ, δ–equilibrium situation in the game $\Gamma(x_0, y_0, \ell)$ if it constitutes the ϵ–equilibrium point in the game $\Gamma(x_0, y_0, \ell + \delta)$, i.e. if

$$t_\Pi^{\ell+\delta}(x_0, y_0; u(\cdot), \bar{v}_\epsilon^\delta(\cdot)) + \epsilon \geq t_\Pi^{\ell+\delta}(x_0, y_0; \bar{u}_\epsilon^\delta(\cdot), \bar{v}_\epsilon^\delta(\cdot)) \geq$$

$$\geq t_\Pi^{\ell+\delta}(x_0, y_0; \bar{u}_\epsilon^\delta(\cdot), v_\epsilon^\delta(\cdot)) - \epsilon \tag{2.4.6}$$

for all $u(\cdot) \in P$, $v(\cdot) \in E$.

Definition 6. Suppose there is a sequence $\{\delta_k\}$, $\delta_k \to 0$ such that in all games $\Gamma(x_0, y_0, \ell + \delta_k)$ for any $\epsilon > 0$ there is the ϵ–equilibrium point. Then the limit

$$\lim_{k \to 0} V(x_0, y_0, \ell + \delta_k) = V'(x_0, y_0, \ell) \tag{2.4.7}$$

is called value of the game $\Gamma(x_0, y_0, \ell)$.

Note that the definition does not depend on the choice of a sequence δ_k since the function $V(x_0, y_0, \ell)$ decreases monotonically in ℓ.

Definition 7. We say that the game $\Gamma(x_0, y_0, \ell)$ has the value in the generalized sense, if there is a sequence $\{\delta_k\}$, $\delta_k \to 0$, such that for any $\epsilon > 0$ and $\delta_k \in \{\delta_k\}$ in the game $\Gamma(x_0, y_0, \ell)$ there exist equilibrium situation.

From (2.4.6) and (2.4.7) it follows that if the game has the value in the ordinary sense, then its value $V'(x_0, y_0, \ell)$ (in a generalized sense) exists and equals

$$\lim_{\substack{\epsilon \to 0 \\ \{\delta_k\} \to 0}} t_\Pi^{\ell+\delta_k}(x_0, y_0, \bar{u}_\epsilon^\delta, \bar{v}_\epsilon^\delta(\cdot)) = V'(x_0, y_0, \ell) \tag{2.4.8}$$

Next, from the definition of the value of the game $\Gamma(x_0, y_0, \ell)$ (in a generalized sense) it follows that if in the game $\Gamma(x_0, y_0, \ell)$ for any $\epsilon > 0$ there is the ϵ–equilibrium point in the ordinary sense (i.e. a solution in the ordinary sense), then it is exactly the solution, and $V(x_0, y_0, \ell) = V'(x_0, y_0, \ell)$ (it suffices to take a sequence $\{\delta_k\}$, $\delta_k \equiv 0$,for all k).

Theorem 3 *Suppose there exist the least finite root of the (2.4.2), $T_0 < \infty$. Then there is the value (in the generalized sense) of the optimal time game of pursuit $\Gamma(x_0, y_0, \ell)$, and $V'(X_0, y_0, \ell) = T_0$.*

Proof: The monotonicity and continuity of the function $V(x_0, y_0, T)$ in T implies the existence of such sequence $T_k \to T_0$ from the left, that $V(x_0, y_0, T_k) \to V(x_0, y_0, T_0) = \ell$ and the function $V(x_0, y_0, T_k)$ is strictly monotone at the points T_k. Let

$$\delta_k = V(x_0, y_0, T_k) - \ell \geq 0 \tag{2.4.9}$$

The strict monotonicity of the function $V(x_0, y_0, T)$ at the points T_k suggests that the equation $V(x_0, y_0, T) = \ell + \delta_k$ has a unique root T_k. This means that for any $\delta \in \{\delta_k\}$ in the game $\Gamma(x_0, y_0, \ell + \epsilon)$ there is the ϵ–equilibrium point for any $\epsilon > 0$ (see case 2). This, however, means that in the game $\Gamma(x_0, y_0, \ell)$ there is a solution in a generalized sense and

$$\lim_{k \to 0} V(x_0, y_0, \ell + \delta_k) = \lim_{k \to 0} T_k = T_0 = V'(x_0, y_0, \ell)$$

This proves the theorem. ■

Now consider case 1c. We have $\inf_T V(x_0, y_0, T) = \ell$. Let $T_k \to \infty$. Then $\lim_{k\to\infty} V(x_0, y_0, T_k) = \ell$.The monotonicity and continuity of $V(x_0, y_0, T)$ in T suggests that a sequence T_k may be chosen in such a way that the function $V(x_0, y_0, T)$ is strictly monotone at the points T_k. Then, as in the proof of Theorem 3, we may show that there exists such a sequence $\{\delta_k\}$ that

$$\lim_{k \to \infty} V(x_0, y_0, \ell + \delta_k) = \lim_{k \to \infty} T_k = T_0 = \infty$$

Thus, in this case a generalized solution also exists, and a generalized value of the game $\Gamma(x_0, y_0, \ell)$ equals infinity.

2.5 Alternative

In many cases it is preferable to find out weather player P can guarantee the ℓ–capture from the given initial positions x, y in a fixed time T. And if not we have to determine whether Player E can guarantee the avoidance of ℓ–capture in the prescribed time T.

Let $V(x, y, T)$ be the value of the game with prescribed duration T from the initial states $x, y \in R^n$, the payoff being $\min_{0 \le t \le T} \rho(x(t), y(t))$. Then two cases (two alternatives) are possible:

1. $V(x, y, T) > \ell$;
2. $V(x, y, T) \le \ell$.

Let us consider case 1. It follows from the definition of the function $V(x, y, T)$ that for any $\epsilon > 0$ there is such a strategy $v_\epsilon^*(\cdot)$ for Player E that

$$K(x, y; u(\cdot), v_\epsilon^*(\cdot)) \ge V(x, y, T) - \epsilon.$$

holds for all strategies $u(\cdot)$. By choosing a sufficiently small ϵ, we may ensure that

$$K(x, y; u(\cdot), v^*\epsilon(\cdot)) \ge V(x, y, T) - \epsilon > \ell$$

for all strategies $u(\cdot)$ of player P. The form of the payoff function K suggests that by adopting the strategy $v_\epsilon^*(\cdot)$, Player E can ensure that the inequality

$\min_{0\leq t\leq T}\rho(x(t), y(t)) > \ell$ be satisfied independent of Player P's actions, i.e. in case 1 Player E ensures the avoidance of ℓ–capture on the time interval $[0, T]$ independent of Player P's actions.

Consider case 2. Let T_0 be a minimal root of the equation ($T_0 < T$) $V(x, y, t) = \ell$ with the fixed x, y (if $\rho(x, y) < \ell$, then T_0 is assumed to be 0). The definition of $V(x, y, T_0)$ then implies that in the game $\Gamma(x, y, T_0)$ for any $\epsilon > 0$ Player P has the strategy $u_\epsilon^*(\cdot)$ which ensures that the inequality

$$K(x, y; u_\epsilon^*(\cdot), v(\cdot)) \leq V(x, y, T_0) + \epsilon = \ell + \epsilon$$

is satisfied for all strategies $v(\cdot)$ of Player E. The form of the payoff function K suggests that, by adopting the strategy $u_\epsilon^*(\cdot)$, Player P can ensure that the inequality $\min_{0\leq t\leq T_0}\rho(x(t), y(t)) \leq \ell + \epsilon$ is satisfied independent of Player E's actions. Making the prolongation of the strategy $u_\epsilon^*(\cdot)$, arbitrarily on the interval $[T_0, T]$, we get in case 2 that for any $\epsilon > 0$ Player P can ensure the $\epsilon + \ell$–capture of Player E in the time T, whatever the action of the latter may be.

We have actually proved the following theorem (about alternative).

Theorem 4 *For any $x, y, \in R^n$ and $T > 0$ at least one of the following two alternatives holds.*

1. *From the initial states x, y Player E can ensure the avoidance of ℓ–capture during the time T independent of Player P's actions.*

2. *For every $\epsilon > 0$ Player P can ensure the $\ell + \epsilon$–capture of Player E from the initial states x, y in the time T independent of Player E's actions. For each fixed $T > 0$ the entire space $R^n \times R^n$ is divided into three nonintersecting regions: the region $A = \{x, y : V(x, y, T) < \ell\}$, called the capture zone, the region $B = \{x, y : V(x, y, T) > \ell\}$ naturally called the escape zone, and the region $C = \{x, y : V(x, y, T) = \ell\}$, the neutral outcome zone.*

Let $x, y \in A$. By definition of A, for any $\epsilon > 0$ Player P has such strategy $u_\epsilon^*(\cdot)$ that

$$K(x, y; u_\epsilon^*(\cdot), v_\epsilon^*(\cdot)) \leq V(x, y, T) + \epsilon$$

for all strategies $v(\cdot)$ of Player E. By properly choosing of $\epsilon > 0$, it is possible to ensure that

$$K(x, y; u_\epsilon^*(\cdot), v_\epsilon^*(\cdot)) \leq V(x, y, T) + \epsilon < \ell$$

is satisfied. The latter means that the strategy $u_\epsilon^*(\cdot)$ of Player P guarantees him the ℓ–capture of Player E from the initial states x, y in the time T. As a result, we get the following improvement of Theorem 4.

Theorem 5 *For any fixed $T > 0$ the entire space is divided into three nonintersecting regions A, B, C possessing the following properties :*

1. *for any $x, y \in A$, Player P has the strategy $u_\epsilon^*(\cdot)$ which ensures the ℓ–capture of Player E on the interval $[0, T]$ independent of the latter's actions;*

2. *for any $x, y \in B$ Player P has the strategy $v_\epsilon^*(\cdot)$ which ensures the avoidance of ℓ–capture independent of Player P 's actions;*

3. *for any $x, y \in C$ and $\epsilon > 0$, Player P has the strategy $u_\epsilon^*(\cdot)$ which ensures the $\ell + \epsilon$–capture of Player E independent of the latter's actions.*

The explicit implication of Theorem 5 is that if $x, y \in A$, then Player P can always ensure that the trajectory $x(t), y(t)$ emanating from the initial states x, y does not leave the set $A \bigcup C$ during the time T. Similarly, if $x, y \in B$, then Player E can always ensure that the trajectory $x(t), y(t)$ emanating from the initial states x, y does not leave the set $B \bigcup C$ during the time T.

2.6 Differential games with dependent motions

So far we have discussed the differential games where the players' motions were described by (2.1.1), (2.1.2). A special feature of such games is that the control and phase states of the opponent did not affect explicitly players motion, since the variables y, v did not appear in the right–hand side of (2.1.2), and the variables x, u did not appear in the right–hand side of (2.1.2). In this, and in the following sections of the present chapter, we study the games, where the motion equations are of a general form, which does not enable us to interpret them as differential games of pursuit. In general, the theorems on the existence of a solution in the class of piecewise open–loop strategies, that are similar to theorems from 3–5 of the present chapter, do not hold for this case.

Let $z \in R^n$, $u \in U \subset CompR^k$, $v \in V \subset CompR^\ell$, $f(z, u, v)$ be the n–dimensional vector function given on $R^n \times U \times V$. Consider the system

$$z' = f(z, u, v) \tag{2.6.1}$$

with the initial condition $z(0) = z_0$. Then z moves in the phase space R^n under the action of the controls chosen at each instant respectively by the players P and E in accordance with their objectives and the information accessible in each state of the game. The players' objectives in the game are determined with the help of the payoff that is dependent on the realized trajectory $z(t)$. The game is assumed to be with prescribed duration T.

Let $M[z_0, z(t)]$ be a continuous functional of the trajectory $z(t)$ realized during the game. Assume that Player E's payoff is $M[z_0, z(t)]$, and Player P's payoff $-M[z_0, z(t)]$. As before, such type of differential games are called the games with prescribed duration. Sometimes the restriction on the duration of the game is of no importance, and the game proceeds until the players P and E achieve some particular result.

Let the m–dimensional surface S called the terminal surface, be given in R^n. Set $t_\Pi = \min\{t : z(t) \in S\}$, i.e. t_Π is the first instant at which the point $z(t)$ penetrates S. If for all $t \geq 0$ the point $z(t) \notin S$, then t_Π is assumed to be equal $+\infty$. Player E's payoff is assumed to be equal t_Π (Player P's payoff being $-t_\Pi$). As before, such type of differential games are called the optimal time games.

In many cases it is essential to define a set of initial states from which Player P can ensure that the point z falls on S independent of Player E's action, and, conversely, it is essential to define a set of initial states, from which Player E can ensure the avoidance of a rendezvous with the set S. Such type of differential games will be called the differential games of kind.

As before (see 2), we choose the classes of piecewise open–loop strategies. For the existence of the unique solution of system (2.6.1), which may be continued on the interval $[0, \infty)$ under any initial conditions $z_0 \in R^n$and any situation $(u(\cdot), v(\cdot))$ in piecewise open–loop strategies we impose on the right–hand sides of equations (2.6.1) the condition that is similar to Condition A from 2 of this chapter.

Condition B. The system of ordinary differential equations $\dot{z} = f(z, u(t), v(t))$ with any pair of measurable open–loop controls $u(t), v(t)$ and any initial condition z_0 has a unique solution $z(t)$ which can be extended to the time interval $[0, \infty)$. As in 2, Ch. 2, by employing Condition B, we can construct for any initial conditions the solution of system $\dot{z} = f(z, u(t), v(t))$, which can be extended to the interval $[0, \infty)$. Moreover, we may define the payoff function $K(z_0, u(\cdot), v(\cdot))$, whose form depends of the form of the game.

In contrast to the case of independent motions, in the differential games with dependent motions the ϵ–equilibrium points for any $\epsilon > 0$ in the class of piecewise open–loop strategies may not exist. This may be illustrated by referring to the following simple example.

Example 3. The game has a prescribed duration $T > 0$, $\dot{z} \in R^1$, $u \in [0, 1]$, $v \in [0, 1]$, $\dot{z} = (u - v)^2$, $z(0) = z_0$ $(z_0 > 0)$, $M(z_0, z(t)) = z(T)$.

Assume that the ϵ–equilibrium point in the class of piecewise open–loop strategies exists for all $\epsilon > 0$, and let $\{(u^*_\epsilon(\cdot), v^*_\epsilon(\cdot))\}$ be a family of the ϵ–equilibrium points for various $\epsilon > 0$, and V be the value of the game. Then

$$K(z_0; u(\cdot), v_\epsilon^*(\cdot)) + \epsilon \geq V \geq K(z_0; u_\epsilon^*(\cdot), v(\cdot)) - \epsilon \tag{2.6.2}$$

for all $u(\cdot), v(\cdot)$.

Since there always exists the strategy $\bar{u}_\epsilon(\cdot)$, coincident with $v_\epsilon^*(\cdot)$ $(\bar{u}_\epsilon(\cdot) = v_\epsilon^*(\cdot))$, by choosing the strategy $\bar{u}_\epsilon(\cdot)$ against the strategy $v_\epsilon^*(\cdot)$ Player P can ensure that the velocity of motion of the point z will be zero, which leads to the payoff $K(z_0; \bar{u}_\epsilon(\cdot), v_\epsilon^*(\cdot)) = z_0$ in the situation $(\bar{u}_\epsilon(\cdot), v_\epsilon^*(\cdot))$ Since the payoff $K(z_0; u(\cdot), v(\cdot))$ for any $u(\cdot), v(\cdot)$ cannot be less than z_0, and the inequality $z_0 + \epsilon \geq V$must hold for all $\epsilon > 0$, then $V = z_0$ and for all $\epsilon > 0$ and $v(\cdot)$ the following inequality is to be satisfied

$$z_0 \geq K(z_0; u_\epsilon^*(\cdot), v(\cdot)) - \epsilon. \tag{2.6.3}$$

Now let $u_\epsilon^*(\cdot) = \{\sigma^\epsilon, \alpha^\epsilon\}$ be the ϵ–optimal strategy for Player P. We construct the strategy $\bar{v}(\cdot) = \{\tau, \beta\}$, $\tau = \sigma^\epsilon$ for Player E. For any t_k, $z(t_k)$, the mapping β^ϵ associates the state $t_k, z(t_k)$, with the following open–loop control on the time interval $[t_k, t_{k+1})$:

$$v(t) = \begin{cases} 1 & \text{for all } t \text{ for which } u^\epsilon(t) \leq \frac{1}{2}, \\ 0 & \text{for all } t \text{ for which } u^\epsilon(t) > \frac{1}{2}, \end{cases}$$

where $u^\epsilon(t)$ is the open–loop control dictated by the strategy $u_\epsilon^*(t)$ on the interval $[t_k, t_{k+1})$ from the state t_k, $z(t_k)$. Then, in the situation $(u_\epsilon^*(\cdot), \bar{v}(\cdot))$ the velocity of the point z is $\frac{1}{4}$ or more, and the payoff $K(z_0; u_\epsilon^*(\cdot), \bar{v}(\cdot))$ in this situation is equal to $K(z_0; u_\epsilon^*(\cdot), \bar{v}(\cdot)) \geq z_0 + \frac{T}{4}$ for any ϵ–optimal strategy adopted by Player P.

If $\epsilon = \frac{T}{8}$, then (see (2.6.3))

$$z_0 \geq K(z_0; u_{\frac{T}{8}}^*(\cdot), \bar{v}(\cdot)) - \frac{T}{8} = \frac{T}{4} - \frac{T}{8} + z_0 = \frac{T}{8} + z_0.$$

The obtained contradiction means that, in the case involved, the ϵ–equilibrium point exists not for all $\epsilon > 0$, and we cannot talk about a solution of the game in the class of piecewise open–loop strategies.

The fundamental results for the games with dependent motions were obtained by N.N. Krasovsky [12] and his followers, who extended the classes of strategies to include mixed strategies and provide the proof of the existence of a solution in this extended classes.

Since the above mentioned are discussed in detail in [12], we just dwell here on the differential games with dependent motions and discrimination of one of the players where we can prove the existence of the ϵ–equilibrium points for any $\epsilon > 0$ in the class of piecewise open–loop strategies.

Denote by $\Gamma(z_0, T)$ the terminal payoff game of prescribed duration T from the initial state z_0. Introduce the upper and lower games with discrimination.

Let $\Gamma^-(z, T)$ be the lower game with discrimination for Player E. Let σ, τ be the partitions of the interval $[0, T]$ dictated by the piecewise open–loop strategies $u(\cdot), v(\cdot)$ for the players P and E in the game $\Gamma(z_0, T)$, and $w = \sigma \bigcup \tau$ be a union of these partitions. In the game $\Gamma(z, T)$, at each point $t_k \in w$ Player P knows t_k and $z(t_k)$, and Player E knows $t_k, z(t_k)$. Assume that in the game $\Gamma^-(z_0, T)$ Player P knows the partition w and, at the instants $t_k \in w$, knows t_k, $z(t_k)$ and a control $v(t)$ of Player E on the time interval $[t_k, t_{k+1}]$. Changes in the information state naturally involve changes in the strategy definition of Player P.

The strategy of Player P is defined in the game $\Gamma^-(z_0, T)$ as follows. Let $v(\cdot) = \{\tau, \beta\}$ be a strategy of Player E in the game $\Gamma(z_0, T)$ (the strategy classes of Player E in the games $\Gamma(z_0, T)$ and $\Gamma^-(z_0, T)$ coincide). Then, by the strategy $u(\cdot)$ for Player P in the game $\Gamma^-(z_0, T)$ is meant a pair $\{\sigma, \alpha\}$, where α is a map, which places each point $z(t_k)$ and control for Player E on the corresponding interval, in correspondence with a measurable open–loop control $u(t), t \in [t_k, t_{k+1})$.

Denote the strategy classes of the players in the game $\Gamma^-(z_0, T)$ by $P^- = \{u(\cdot)\}$ and $E = \{v(\cdot)\}$.

Let $\Gamma^+(z, T)$ be the upper game with discrimination of Player P. Let σ, τ be partitions of the time interval $[0, T]$ dictated by the piecewise open–loop strategies $u(\cdot), v(\cdot)$ for the players P and E in the game $\Gamma^+(z, T)$ and $w = \sigma \bigcup \tau$ be a union of these partitions. Assume that in the game $\Gamma^+(z_0, T)$ Player E knows the partition w and at the instants $t_k \in w$ he knows $t_k, z(t_k)$ and the control $u(t)$ of Player P on the time interval $[t_k, t_{k+1})$. The information state for Player P is the same as in the game $\Gamma(z_0, T)$. Changes in the information state of Player E involve changes in the definition of his strategy.

The strategy of Player E in the game $\Gamma^+(z, T)$ is defined as follows. Let $u(\cdot) = \{\sigma, \alpha\}$ be a strategy of Player P in the game $\Gamma^+(z, T)$ (the strategy classes of Player P in the games $\Gamma(z, T)$ and $\Gamma^+(z, T)$ coincide). Then by the strategy $v(\cdot)$ of Player E in the game $\Gamma^+(z, T)$ is meant a pair $\{\tau, \beta\}$. Here τ is a partition of the interval $[0, T]$ and β is a map which places a measurable open–loop control $v(t)$, $t \in [t_k, t_{k+1})$ in correspondence with each point t_k, $z(t_k)$ and control $u(t)$ of Player P on the time interval $[t_k, t_{k+1})$, where $t_k \in w$ and $u(t)$ is the control dictated by the open–loop strategy $u(\cdot)$ of Player P on the corresponding interval.

Denote the strategy classes for the players in the game $\Gamma^+(z_0, T)$ by $P = \{u(\cdot)\}$ and $E^+ = \{v(\cdot)\}$. We show that in the games $\Gamma^+(z_0, T)$ and $\Gamma^-(z_0, T)$ for the strategy classes under considering there on the ϵ–equilibrium point $(\epsilon > 0)$.

In order to prove the existence of ϵ–equilibrium points in the games $\Gamma^+(z_0,T)$ and $\Gamma^-(z_0,T)$, we will construct the approximating sequence of games $\Gamma^+_\delta(z_0,T)$ and $\Gamma^-_\delta(z_0,T)$ with discrete time as shown in this chapter for the games with independent motions.

The game $\Gamma^-_\delta(z_0,T)$ differs from the game $\Gamma^-(z_0,T)$ only by the strategy sets for the players. Let $\delta = \{t_k\}$ be a fixed partition of the time interval $[0,T]$. The set of strategies E_δ in the game $\Gamma^-_\delta(z_0,T)$ is a subset of strategies E in the game $\Gamma^-(z_0,T)$ and is defined as follows:

$$E_\delta = \{(\tau,\beta) : \tau = \delta,\ (\tau,\beta) \in E\},$$

i.e. this is the set of all such pairs (τ,β) for which the partition τ is the same. Likewise, the set of strategies P^-_δ in the game $\Gamma^-_\delta(z_0,T)$ is a subset of strategies P^- in the game $\Gamma^-(z_0,T)$ and is defined as follows:

$$P^-_\delta = \{(\delta',\beta) : \delta' = \delta, (\delta',\alpha) \in P^-\},$$

i.e. this is the set of all such pairs (δ',α) for which the partition $\delta' = \delta$ is the same.

Similarly, by fixing the partitions of the time interval $[0,T]$ for all strategies of the players P and E, it is possible to construct a discrete approximation $\Gamma^+_\delta(z_0,T)$, of the game $\Gamma^+(z_0,T)$

Generalization of Lemmas 2 and 3 to the case of games $\Gamma^+_\delta(z_0,T)$, $\Gamma^-_\delta(z_0,T)$ is elementary and we get the lemma that follows.

Lemma 7 *In the games $\Gamma^+_\delta(z_0,T)$ and $\Gamma^-_\delta(z_0,T)$ there are equilibrium points in the classes of strategies P^-_δ, E_δ and P_δ, E^+_δ respectively. The values of those games $V^+_\delta(z_0,T)$ and $V^-_\delta(z_0,T)$ are continuous functions of z,T. (The reachability sets of system (2.6.1) are assumed to be compact for any $z_0 \in R^n$, $T \geq 0$ and continuously dependent on z_0,T).*

Suppose the partition δ' contains one break point more than the partition δ, and the other break points t_k coincide with the break points of the partition δ.

Consider the games $\Gamma^-_{\delta'}(z_0,T)$ and $\Gamma^-_\delta(z_0,T)$. Let $t_k < t' < t_{k+1}$ by definition, at the point $z(t_k)$ in the game $\Gamma^-_{\delta'}(z_0,T)$ Player P receives from Player E a smaller amount of information that in the game $\Gamma^-_{\delta'}(z_0,T)$ (in the first case he knows $v(t)$ for $t \in [t_k,t')$, and in the second $v(t)$ for $t \in [t_k,t_{k+1})$). Player E, however, receives at the instant t' the information about the state $z(t')$ that is additional as against the one in the game $\Gamma^-_{\delta'}(z_0,T)$. This means that Player P's position in the game $\Gamma^-_\delta(z_0,T)$ becomes worse as against the one in the game $\Gamma^-_\delta(z_0,T)$, which leads to the inequality $V^-_{\delta'}(z_0,Y) \geq V^-_\delta(z_0,T)$. The same inequality follows from such explicit reasoning as $E_{\delta'} \supset E_\delta$. Restrict

Player E in the game $\Gamma^-_{\delta'}(z,T)$ to the class of strategies E_δ in the game $\Gamma^-_\delta(z,T)$ Then, in the game $\Gamma^-_{\delta'}(z,T)$, he can always ensure the payoff equal to $V^-_\delta(z,T)$, from which we get the inequality

$$V^-_{\delta'}(z_0,T) \geq V^-_\delta(z_0,T). \tag{2.6.4}$$

Similarly, we may get the inequality

$$V^+_{\delta'}(z_0,T) \leq V^+_\delta(z_0,T). \tag{2.6.5}$$

Let $\delta_1 \subset \delta_2 \subset \ldots \subset \delta_n \subset \ldots$, be a decreasing sequence of partitions of the interval $[0.T]$. Since the terminal payoff $M[z_0, z(T)]$ is bounded on the reachability set of the system (2.6.1) from the initial state z_0 in the time T, then the sequences $V^-_{\delta_k}(z_0,T)$ and $V^+_{\delta_k}(z_0,T)$ are bounded monotone sequences of continuous functions. Let $|\delta| = \max_{t_k\in\delta} |t_{k+1} - t_k|$ be a break rank.

Lemma 8 *There exist the limits*

$$V^+(z_0,T) = \lim_{|\delta_k|\to 0} V^+_{\delta_k}(z_0,T),$$

$$V^-(z_0,T) = \lim_{|\delta_k|\to 0} V^-_{\delta_k}(z_0,T),$$

that are independent of the choice of a partition sequence $\{\delta_k\}$.

The existence of the limits V^+, V^- for each fixed decreasing sequence $\{\delta_k\}$, $\delta_{k+1} \supset \delta_k$, $|\delta_k| \to 0$ follows from the monotonicity and boundlessness of sequences $\{V^+_{\delta_k}\}$, $\{V^-_{\delta_k}\}$.

We show that for all uniformly decreasing sequences $\{\delta_k\}^\infty$ and $\{\delta'_k\}^\infty$ of the interval partitions the following relationship holds

$$\lim_{k\to\infty} V^-_{\delta_k}(z_0,T) = \lim_{k\to\infty} V^-_{\delta'_k}(z_0,T)$$

Denote $\lim_{k\to\infty} V^-_{\delta_k} = V^-(\{\delta_k\})$ and $\lim_{k\to\infty} V^-_{\delta'} = V^-(\{\delta'_k\})$ Set $L = \sup_{\delta_k} V^-(\{\delta_k\})$ (here supremum is taken with respect to a set of all uniformly decreasing breaks). It is fairly easy to see that $L < \infty$, since all $V^-(\{\delta_k\})$ are bounded by the number $\sup M(z_0, z)$, $z \in C^T(z_0)$. Here $M(z)$ is the terminal payoff that is a continuous function, and $C^T(z_0)$ is the reachability set of system (2.6.1). Let an arbitrary $\eta > 0$ be given. By the definition of L, there is such a partition that

$$V^-_\delta(z_0,T) > L - \frac{\eta}{2}.$$

Let m be a number of interior points of the partition δ. Because of the continuity of the reachability set $C^T(z_0)$ (in the Hausdorff metric) and the continuity of the payoff M for any natural number m and any $\epsilon > 0$, there is such $\delta(\epsilon) > 0$ that, if in the partition $\delta' |t'_i - t'_{i-1}| < \delta(\epsilon)$ for all i, then for any partition δ'' containing m interior points the following inequality is satisfied

$$V^-_{\delta' \bigcup \delta''}(z_0, T) - V^-_{\delta'}(z_0, T) < \epsilon.$$

We choose δ' in such a way that $|t_i - t_{i-1}| < \frac{\delta}{2}(\frac{\eta}{2})$. Then

$$V^-_{\delta' \bigcup \delta''}(z_0, T) - V^-_{\delta'}(z_0, T) < \epsilon = \frac{\eta}{2}.$$

Moreover, $V^-_{\delta \bigcup \delta''}(z_0, T) > L - \frac{\eta}{2}$, whence we have that

$$V^-_{\delta'}(z_0, T) > L - \eta.$$

The latter inequality shows that for any decreasing sequence of partitions of the interval $[0, T]$

$$\lim_{k \to \infty} V^-_{\delta_k}(z_0, T) = L.$$

This exactly proves the first part of the statement in Lemma 8. Its second part is proved in much the same way.

Theorem 6 *In the games $\Gamma^+(z_0, T)$ and $\Gamma^-(z_0, T)$ for any $\epsilon > 0$ there are ϵ–equilibrium points and $V^+(z_0, T), V^-(z_0, T)$ are the values of the games $\Gamma^+(z_0, T)$ and $\Gamma^-(z_0, T)$, respectively.*

Proof: Let us prove the statement of the theorem for the game $\Gamma^-(z_0, T)$. The proof for the game $\Gamma^+(z_0, T)$ is made in much the same way. Fix an arbitrary $\epsilon > 0$ using Lemma 8, we may find such a partition δ that

$$V^-(z_0, T) - V^-_\delta(z_0, T) \leq \epsilon. \tag{2.6.6}$$

Let $\bar{u}_\delta$,$\bar{v}_\delta(\cdot)$ be optimal strategies for the players P and E in the game $\Gamma^-_\delta(z_0,T)$. We construct the following strategy for the Player P in the game $\Gamma^-(z_0, T)$. Let $v(\cdot) = (\tau, \beta)$ be a strategy for Player E, then by $u^*(\cdot) = (\delta, \alpha)$ is meant the strategy for Player P which places his every information state at the time instants $t_k \in \delta \bigcup \tau$ in correspondence with the same open–loop control $u(t)$ for $t \in [t_k, t_{k+1})$ as the strategy $\bar{u}_{\delta \bigcup \tau}$, that is optimal in the game $\Gamma^-_{\delta \bigcup \tau}(z(t_k), T - t_k)$, from the initial state $z(t_k)$, duration $T - t_k$ with the partition $\delta \bigcup \tau$.

We show that the strategy $u^*(\cdot)$, is optimal for Player P in the game $\Gamma^-(z_0, T)$. Evidently, as follows from (2.6.6), for any $\epsilon > 0$ by adopting the strategy $\bar{v}_\delta$, that is optimal in the game $\Gamma^-_\delta(z_0, T)$, Player E can ensure the payoff at least $V^-(z_0, T) - \epsilon$. We show that, by playing the strategy $u^*(\cdot)$, Player P guarantees the payoff for Player E not exceeding $V^-(z_0, T) = L$.

Suppose the opposite is true. Let $v(\cdot) = \{\tau', \beta\}$ be a strategy for Player E, under which the payoff in the situation $(u^*(\cdot), v(\cdot))$ is $L+c > L$. Consider a discrete game $\Gamma^-_{\delta \bigcup \tau'}(z_0, T)$ with the partition $\delta \bigcup \tau'$. In the situation $(u^*(\cdot), v(\cdot))$, the game $\Gamma^-(z_0, T)$ is realized as a discrete game with the partition $\delta \bigcup \tau'$. In addition, as follows from the definition of the strategy $u^*(\cdot)$, Player P chooses the same controls $u(t)$ for $t \in [t_k, t_{k+1})$ as if he plays optimally in the game $\Gamma^-_{\delta \bigcup \tau'}(z_0, T)$,i.e. he ensures that Player E's payoff will not exceed $V_{\delta \bigcup \tau'}(z_0, T)$ From (2.6.4), however, we have

$$V_{\delta}^-(z_0, T) \leq V^-_{\delta \bigcup \tau'}(z_0, T) \leq L = V^-(z_0, T).$$

The latter inequality means that Player P can ensure that Player E's payoff will not exceed L. This proves the theorem. ■

2.7 Alternative for games with dependent motions and discrimination

Consider the games with dependent motions from 6.

Let a compact set S be given in R^n. In some cases we have to find out whether a player can ensure that a trajectory $z(t)$ penetrates the terminal set S in a fixed time T from the initial position $z_0 \in R^n$. If this is not possible, then we have to find out whether Player E can ensure the avoidance of the point $z(t)$ falling on the set S during time T.

The reasoning given in the previous section point out the way of solving this problem for the upper $\Gamma^+(z_0, T)$ and the lower $\Gamma^-(z_0, T)$ games with discrimination.

Theorem 6 can be readily generalized to the case when the payoff of Player E is a continuous function of the realized trajectory: in particular, it can be generalized to the case when it is of the form

$$\min_{0 \leq t \leq T} \rho(z(t), S), \tag{2.7.1}$$

where $z(t)$ is a realized trajectory, ρ a distance to the set S. This section does not provide appropriate proofs since they are exactly as in 6.

Now, let $\Gamma^-(z_0, T)$ and $\Gamma^+(z_0, T)$ be the upper and lower games with discrimination and payoff (2.7.1), $V^-(z_0, T)$ and $V^+(z_0, T)$ being the corresponding values of the games. We shall study the game $\Gamma^-(z_0, T)$ in some detail. The game $\Gamma^+(z_0, T)$ is studied in much the same way.

The following two cases (two alternatives) are possible:

1. $V^-(z_0, T) > 0$,

2. $V^-(z_0, T) = 0$.

Consider case 1. Let $K(z_0; u(\cdot), v(\cdot))$ be the payoff function in the game under study. It follows from the definition of $V^-(z,T)$ that for any $\epsilon > 0$ there is such a strategy $v_\epsilon^*(\cdot)$ for Player E that for all strategies $u(\cdot)$

$$K(z_0; u(\cdot), v_\epsilon^*(\cdot)) \geq V^-(z_0, T) - \epsilon.$$

By choosing a sufficiently small ϵ, we may ensure that

$$K(z_0; u(\cdot), v_\epsilon^*(\cdot)) \geq V^-(z_0, T) - \epsilon > 0$$

for all strategies $u(\cdot)$ of Player P.

The form of the function K suggests that, by adopting the strategy $v_\epsilon^*(\cdot)$, Player E can ensure that the inequality

$$\min_{0 \leq t \leq T} \rho(z(t), S) > 0$$

will be satisfied independent of Player P's actions, i.e. in case 1 Player E ensures the avoidance of the point z failing on the set S during the time interval $[0, T]$ independent of Player P's actions.

Consider case 2. Let T_0 be the minimal root of the equation $(T_0 \leq T)$ $V^-(z_0, t) = 0$. The definition of $V^-(z_0, t)$ then suggests that, in the game $\Gamma(z_0, T_0)$ for any $\epsilon > 0$ Player P has the strategy $u_\epsilon^*(\cdot)$ which ensures that the inequality

$$K(z_0; u_\epsilon^*(\cdot), v(\cdot)) \leq V^-(z_0, T) + \epsilon = \epsilon$$

is satisfied for all strategies $v(\cdot)$ of Player E. From the form of the function K it follows that, by adopting the strategy $u_\epsilon^*(\cdot)$. Player P can ensure that the inequality

$$\min_{0 \leq t \leq T} \rho(z(t), S) \leq \epsilon$$

is satisfied independent of Player E's actions. When the strategy $u_\epsilon^*(\cdot)$ is extended to cover the interval $[0, T]$, we see that, in case 2, for any $\epsilon > 0$ Player P can guarantee the entry into the ϵ–neighborhood of the set S in the time T independent of Player E's actions.

We proved the following theorem (about alternative).

Theorem 7 *In the game $\Gamma^-(z, T)$, at least one of the following alternatives holds ($z \in R^n$ and $T > 0$).*

1. *Player E can guarantee the avoidance of the entry into the terminal set S from the initial state during the time T independent of Player P's actions.*

2. *For $\epsilon > 0$, Player P can guarantee the entry into the ϵ–neighborhood of the terminal set S from the initial state z during the time T independent of Player E's actions.*

A similar theorem also holds for the game $\Gamma^+(z, T)$ More general and finer theorems (for the game $\Gamma(z, T)$ were proved by N.N.Krasovsky [12]. The corresponding constructions utilize wide classes of mixed strategies.

The remaining chapters of the book are not dealing with dependent motions.

Chapter 3

Class of pursuit–evasion games with optimal open–loop strategy for evader

3.1 Discrete game with terminal payoff and discrimination for player E

The game $\Gamma_\delta(x_0, y_0, T)$, that will be discussed in this section, differs from the game $\Gamma_\delta(x_0, y_0, T)$ defined in 2, Ch. 2, in the strategy sets P, E for players.

Let $\delta = \{0 = t_0 < t_1 < \ldots < t_k < \ldots < t_n = T\}$ be a fixed partition of the time interval $[0, T]$. The strategy set in the game $\Gamma_\delta(x_0, y_0, T)$ is a subset of the strategy set E in the game $\Gamma(x_0, y_0, T)$ and is defined as follows : $E_\delta = \{(\tau, \beta) : \tau = \delta, (\tau, \beta) \in E\}$, i.e. this is a set of all the pairs $(\tau, \beta) \in E$ for which the partition $\tau = \delta$ is the same.

The strategy space for Player P is composed of pairs $\{\delta, \alpha\}$, where δ is the partition of $[0, T]$ fixed throughout the game (and coinciding with the partition of the same interval for Player E), and α is the map which places an admissible open–loop control $u(t)$ on the interval $[t_k, t_{k+1})$ in correspondence with each point $t_k \in \delta$, the position $x(t_k), y(t_k)$, and the open–loop control $v(t), t \in [t_k, t_{k+1})$.

The strategy spaces for the Player P and E in the game $\Gamma_\delta(x_0, y_0, T)$ are denoted by P_δ, E_δ, respectively.

The construction of strategies for Player P suggests that the information on $x(t_k), y(t_k)$ and the Player E's control $v(t)$ for $t \in [t_k, t_{k+1})$ is supplied at each instant t_k. This may be understood as follows. At the instant t_k Player E is the first to choose $v(t)$ for $t \in [t_k, t_{k+1})$ and reports his choice to Player P. Then Player P chooses $u(t)$ for $t \in [t_k, t_{k+1})$. In this case Player P is in a privileged position, the game is called the game with discrimination for Player E. Note that discrimination for Player E is not assumed in the game $\Gamma(x_0, y_0, T)$.

Denote the strategies in the game $\Gamma_\delta(x_0, y_0, T)$ by $u_\delta(\cdot), v_\delta(\cdot)$. Evidently, all the constructions involving the derivation of a unique solution of the system $\dot{x} = f(x, u_\delta(\cdot)), \dot{y} = g(y, v_\delta(\cdot))$, with the initial conditions and situation $(u_\delta(\cdot), v_\delta(\cdot))$, are made just as in 2, Ch. 2.

Recall that Player E's payoff in the game with terminal payoff is defined as $\rho(x(T), y(T))$, where ρ, is the Euclidean distance between the players' position $x(T), y(T)$ as the game terminates (see 2, Ch. 2).

Denote by $C_P^T(x_0)$ and $C_E^T(y_0)$ the reachability sets for the Player P and E from the initial states x_0, y_0 at the time instant T and $\hat{\rho}(x_0, y_0) = \max_{\eta' \in C_E^T(x_0)} \min_{\xi' \in C_P^T(x_0)} \rho(\xi', \eta')$[1]. Using the designations from 3, Ch. 2, we have

$$P(x_0, 0, T) = C_P^T(x_0), E(y_0, 0, T) = C_E^T(y_0).$$

Let $\hat{\rho}_T(x_0, y_0) = \rho(\xi, \eta)$ ($\hat{\rho}_T$ is achieved at the points $\xi \in C_P^T(x_0), \eta \in C_E^T(y_0)$), where $\rho(\xi, \eta)$ is the Euclidean distance between the points ξ, η. Any trajectory $x^*(t)$ $(0 \le t \le T)$ joining $(x^*(0) = x_0, x^*(T) = \xi)$ points x_0 and ξ is called the conditionally optimal trajectory for Player P and any trajectory $y^*(t)$ $(0 \le t \le T)$ joining $(y^*(0) = y_0,\ y^*(T) = \eta)$ points y_0 and η is called the conditionally optimal trajectory for Player E.

The point M is called the center of pursuit in the game $\Gamma_\delta(x_0, y_0, T)$ if

$$\hat{\rho}_T = \max_{\eta' \in C_E^T(y_0)} \min_{\xi' \in C_P^T(x_0)} \rho(\xi', \eta') = \rho(\xi, M)$$

(ρ is achieved at the points $\xi \in C_P^T(x_0), M \in C_E^T(y_0)$). The center of pursuit may be not unique. To every center of pursuit M correspond points $\xi \in C_P^T(x_0)$ from (3.1.1).

Let $x^*(t)$ be the conditionally optimal trajectory for Player P, leading to ξ, and $y^*(t)$ the conditionally optimal trajectory for Player E, leading to M. We say that the trajectories $x^*(t), y^*(t)$ correspond to one another.

Consider an auxiliary game $\Gamma'_{\delta,\alpha}(x, y, T)$ which is the same as the game $\Gamma_\delta(x, y, T)$ except that here Player E remains at the point y when $t \in [0, \alpha]$, i.e. Player E's motion takes place with the delay α. The motion equations and the payoff of the players are same as in the game $\Gamma_\delta(x, y, T)$.

Let M', $x'^*(t), y'^*(t)$ be the center of pursuit and the relevant conditionally optimal trajectories in the game $\Gamma'_{\delta,\alpha}(x, y, T)$. We say that in the game $\Gamma_\delta(x, y, T)$ there is an invariant center of pursuit if the following conditions are satisfied.

---

[1]In [10, 12] the quantity $\rho_T(x_0, y_0)$ is called the "hypothetic mismatch" between the sets $C_E^T(y_0)$ and $C_P^T(x_0)$.

1. The center of pursuit is the same in all of the games $\Gamma_\delta(x^*(t), y^*(t), T-t)$ from the initial states $x^*(t), y^*(t)$ with the duration $T-t$, where $x^*(t), y^*(t)$ are an arbitrary pair of the relevant conditionally optimal trajectories in the game $\Gamma_\delta(x_0, y_0, T)$.

2. Fix t, $x^*(t)$, $y^*(t)$, and let $x'^*(t+a)$ $(0 \le \alpha \le T-t,\ x'^*(t+a)|_{\alpha=0} = x^*(t))$ be a conditionally optimal trajectory for Player P in the game $\Gamma'_{\delta,\delta}(x'^*(t), y, T-t)$, where $y \in C_E^\delta(y^*(t))$. Consider the current games $\Gamma'_{\delta,\delta-\alpha}(x'^*(t+\alpha), y, T-t-\alpha)$ for $0 \le \alpha \le \delta$ and fixed t, $y \in C_E^\delta(y^*(t))$. Let $M'(\alpha)$ be a center of pursuit in the game $\Gamma'_{\delta,\delta-\alpha}(x'^*(t+\alpha), y, T-t-\alpha)$ for $0 \le \alpha \le \delta$, $y \in C_E^\delta(y^*(t))$. We require the point $M'(\alpha)$ not depend on α for all $0 \le \alpha \le \delta$ (occupies the same position).

Prove the following lemma.

Lemma 1 *Let $x^*(t), y^*(t)$ be a pair of relevant conditionally optimal trajectories in the game $\Gamma_\delta(x, y, T)$, the center of pursuit being invariant. Then there exist the solution $x'(t_0), x'(t_1), \ldots, x'(t_k), \ldots, x'(t_\alpha)$. $t_\alpha = T$, $(x'(t_0) = x_0)$ of the recurrent equations*

$$\hat{\rho}_{T-t_k}(x'(t_k), y^*(t_k)) = \min_{x \in C_P^\delta(x'(t_k))} \hat{\rho}_{T-t_{k+1}}(x, y^*(t_{k+1})), \tag{3.1.2}$$

laying on the conditionally optimal trajectory $x^*(t)$ i.e. $x^*(t_k) = x'(t_k)$.

Proof: From the invariance condition for the center of pursuit M

$$\rho(\xi, M) = \hat{\rho}_{T-t}(x^*(t), y^*(t)), \quad 0 \le t \le T.$$

Along the relevant conditionally optimal trajectories $x^*(t), y^*(t)$ for all $t_k \in \{t_0, \ldots, t_n\}$ we have

$$\hat{\rho}_{T-t_k}(x^*(t_k), y^*(t_k)) = \hat{\rho}_{T-t_{k+1}}(x^*(t_{k+1}), y^*(t_{k+1})). \tag{3.1.3}$$

At the same time, for all $x \in C_P^\delta(x^*(t_k))$ the inequality

$$\hat{\rho}_{T-t_k}(x^*(t_k), y^*(t_k)) \le \hat{\rho}_{T-t_{k+1}}(x, y^*(t_{k+1})).$$

is satisfied. Indeed,

$$\hat{\rho}_{T-t_k}(x^*(t_k), y^*(t_k)) = \max_{y \in C_E^{T-t_k}(y^*(t_k))} \min_{x \in C_P^{T-t_k}(x^*(t_k))} \rho(x, y) =$$

$$\min_{x \in C_P^{T-t_k}(x^*(t_k))} \rho(x', M),$$

$$\hat{\rho}_{T-t_{k+1}}(x, y^*(t_{k+1})) = \max_{y \in C_E^{T-t_{k+1}}(y^*(t_{k+1}))} \min_{x' \in C_P^{T-t_{k+1}}(x)} \rho(x', y) \geq$$

$$\min_{x' \in C_P^{T-t_{k+1}}(x)} \rho(x, M), \quad (x \in C_P^{t_{k+1}-t_k}(x^*(t_k))).$$

However, since $x \in C_P^{t_{k+1}-t_k}(x^*(t_k))$,

$$C_P^{T-t_{k+1}}(x) \subset C_P^{T-t_k}(x^*(t_k)),$$

then

$$\hat{\rho}_{T-t_k}(x^*(t_k), y^*(t_k)) \leq \hat{\rho}_{T-t_{k+1}}(x, y^*(t_{k+1})).$$

The latter inequality, together with Eq. (3.1.3), proves the lemma. ■

Definition 1. The set of points $\{x', y', T'\}$ such that $T' \in [0, T]$, $x' \in C_P^{T-T'}(x)$, $y' \in C_E^{T-T'}(y)$ is called the singular surface in the game $\Gamma_\delta(x, y, T)$ if in the games $\Gamma_\delta(x', y', T')$ there is no invariant center of pursuit. The singular surface is called dispersal if Player E's choice of the control $v(t), t \in [t_k, t_{k+1})$ on the singular surface defines uniquely the invariant center of pursuit M with respect to the reachability sets $C_P^{T-t_k}(x(t_k))$ and $C_E^{T-t_k}(y(t_k), v)$. Here $C_E^{T-t_k}(y(t_k), v)$ is a subset of the set $C_E^{T-t_k}(y(t_k))$, where in Player E may arrive at the instant T from the states $y(t_k)$, with the fixed starting interval of control $v(t), t \in [t_k, t_{k+1})$. This set coincides with $C_E^{T-t_{k+1}}(y(t_{k+1}))$, where $y(t_{k+1})$ is a point wherein Player E arrives from the state $y(t_k)$, when using the control $v(t), t \in [t_k, t_{k+1})$.

Theorem 1 *Suppose that in the game $\Gamma_\delta(x_0, y_0, T)$ there is an invariant center of pursuit $\hat{\rho}_T(x_0, y_0) > 0$. Then there exists an equilibrium point, the conditionally optimal trajectories are optimal, and the value of the game is equal to $V_0(x_0, y_0, T) = \hat{\rho}_T(x_0, y_0)$. Moreover, the optimal strategies for the Players P and E are of the form $v_\delta^*(\cdot) = (\delta, \beta^*)$, where the map β^* associates each point $x(t_k), y(t_k), t_k$ with the control $v^*(t)$, $t \in [t_k, t_{k+1})$ which defines the motion of the point E on the conditionally optimal trajectory $y^*(t)$ in the game $\Gamma_\delta(x_0, y_0, T)$ aiming to the invariant center of pursuit (i.e.,the piecewise open–loop strategy $v_\delta^*(t)$ is time function); $u_\delta^*(\cdot) = (\delta, \alpha^*)$, where α^* places each point $x(t_k), y(t_k), t_k$ and control $v(t), t \in [t_k, t_{k+1})$ of Player E in correspondence with the control $u(t)$, $t \in [t_k, t_k + 1)$ transfering the point x from the state $x(t_k)$ to the state $\bar{x}(t_{k+1})$ such that*

$$\hat{\rho}_{T-t_{k+1}}(\bar{x}(t_{k+1}), y(t_{k+1})) = \min_{x \in C_P^\delta(x(t_k))} \hat{\rho}_{T-t_{k+1}}(x, y(t_{k+1})). \tag{3.1.4}$$

Here $y(t_{k+1})$ is the point wherein Player E arrives when choosing the control $v(t)$, $t \in [t_k, t_{k+1})$ at the time instant t_k.

Proof: We show that

$$K(x_0, y_0; u_\delta^*(\cdot), v_\delta^*(\cdot)) = \hat{\rho}_T(x_0, y_0). \tag{3.1.5}$$

Let $y^*(t)$ be the conditionally optimal strategy for Player E directed towards the invariant center of pursuit M whose existence is assumed in the theorem ($y^*(0) = y_0$, $y^*(T) = M$). In the situation $(x_0, y_0; u^*(\cdot), v^*(\cdot))$ the trajectory $x'(t)$ for Player P at the time instant $t_k, k = 0, 1, \dots, n$ passes through the points $x'(t_k)$ obtained as a solution of the recurrent equation (Lemma 1):

$$\hat{\rho}_{T-t_k}(x'(t_k), y^*(t_k)) = \min_{x'' \in C_P^\delta(x'(t_k))} \hat{\rho}_{T-t_{k+1}}(x'', y^*(t_{k+1})). \tag{3.1.6}$$

For any solution of this system, however, we have

$$\hat{\rho}_T(x_0, y_0) = \hat{\rho}_{T-t_1}(x'(t_1), y^*(t_1)) = \dots$$

$$\dots = \hat{\rho}_{T-t_k}(x'(t_k), y^*_{t_k})) = \dots = \rho(x'(T), M) =$$

$$\rho(x^*(T), M) = K(x_0, y_0; u^*(\cdot), v^*(\cdot)) = const. \tag{3.1.7}$$

We have thus proved the equality (3.1.5).

Show that

$$K(x_0, y_0; u_\delta(\cdot), v_\delta^*(\cdot)) \geq \hat{\rho}_T(x_0, y_0) \geq K(x_0, y_0; u_\delta^*(\cdot), v_\delta(\cdot)). \tag{3.1.8}$$

for all $u_\delta(\cdot) \in P_\delta$ and $v_\delta(\cdot) \in E_\delta$.

First prove the validity of the left–hand inequality. Let $u_\delta(\cdot)$ be a strategy for Player P, and $x(T)$ the point wherein he appears when the game ends in the situation $(u_\delta(\cdot), v_\delta^*(\cdot))$. Since

$$\hat{\rho}_T(x_0, y_0) = \max_{\eta \in C_E^T(y_0)} \min_{\xi \in C_P^T(x_0)} \rho(\xi, M),$$

then, for any point $\xi' \in C_P^T(x_0)$, the inequality $\hat{\rho}_T(x_0, y_0) \leq \rho(\xi', M)$ is satisfied. In particular, $\hat{\rho}_T(x_0, y_0) \leq \rho(x(T), M)$, and since the strategy $v_\delta^*(\cdot)$ transfers y_0 to the point M, then $\rho(x(T), M) = K(x_0, y_0; u_\delta(\cdot), v_\delta^*(\cdot))$, and this proves the validity of the left–hand inequality (3.1.8).

We now prove the validity of the right–hand inequality (3.1.8). In the game $\Gamma_\delta(x_0, y_0, T)$, at each time instant t_k the pursuer P has complete information on the trajectory of Player E for $t \in [t_k, t_{k+1})$. In particular, this means that at each time instant t_k he knows the point $y(t_{k+1})$ which is reached by E in the time $t_{k+1} - t_k$. Now assume that Player E deviates his conditionally optimal trajectory (control $v^*(t)$) during the time $t_{k+1} - t_k$ choosing on the interval $[t_k, t_{k+1})$ the control $v(t)$ different from $v^*(t)$.

Since Player E has held the control $v^*(t)$ up to the instant t_k, then from the definition of the strategy $u^*_\delta(\cdot)$ for Player E and from the invariance of the point M we have

$$\hat{\rho}_T(x_0, y_0) = \hat{\rho}_{T-t_k}(x^*(t_k), y^*(t_k)).$$

Let $D_E^{T-t_k}(y(t_k))$ be a subset of the reachability set $C_E^{T-t_k}(y^*(t_k))$ for Player E, which is obtained from $C_E^{T-t_k}(y(t_k))$ under the additional requirement that, at the instant t_{k+1}, E passes through the point $y(t_{k+1})$. Then

$$\hat{\rho}_T(x_0, y_0) = \hat{\rho}_{T-t_k}(x^*(t_k), y^*(t_k)) \geq$$

$$\geq \max_{\eta \in D_E^{T-t_k}(y^*(t_k))} \min_{\xi \in C_E^{T-t_k}(x^*(t_k))} \rho(\xi, \eta). \tag{3.1.9}$$

Consider the auxiliary game $\Gamma'_{\delta,\delta}(x^*(t_k), y, T - t_k)$ where E remains at the point $y = y(t_{k+1})$ during the time δ.

Let $x'^*(t)$ be a conditionally optimal trajectory for Player P in the game $\Gamma'_{\delta,\delta}(x^*(t_k), y, T - t_k)$. Let $\Gamma'_{\delta,\delta-\alpha}(x'^*(t_k + \alpha), y, T - t_k - \alpha)$ for $0 \leq \alpha \leq \delta$, $y = y(t_k)$, $x'(t_k) = x^*(t_k)$ be the game, where Player E remains at the point $y = y(t_{k+1})$ during the time $t_{k+1} - (t_k + \alpha) = \delta - \alpha$, and $\hat{\rho}'_{T-t_k-\alpha}(x'^*(t_k + \alpha), y)$ be a maximin distance between reachability sets computed in this game. Then

$$\hat{\rho}'_{T-t_k}(x'^*(t_k), y) = \hat{\rho}'_{T-t_k}(x^*(t_k), y(t_{k+1})) =$$

$$\max_{\eta \in D_E^{T-t_k}(y^*(t_k))} \min_{\xi \in C_P^{T-t_k}(x^*(t_k))} \rho(\xi, \eta) \leq \hat{\rho}_{T-t_k}(x^*(t_k), y^*(t_k)). \tag{3.1.10}$$

From the existence of the invariant center of pursuit in the game $\Gamma_\delta(x_0, y_0, T)$ (see condition 2) it follows that, for all $0 \leq \alpha \leq \delta$, the center of pursuit in the games $\Gamma_{\delta,\delta-\alpha}(x'^*(t_k + \alpha), y, T - t_k - \alpha)$ is the same, therefore for all $0 \leq \alpha \leq \delta$.

$$\hat{\rho}'_{T-(t_k+\alpha)}(x'^*(t_k + \alpha), y(t_k)) = const.$$

Hence, in particular, with $\alpha = 0$ and $\alpha = \delta$ we get

$$\hat{\rho}'_{T-t_k}(x'^*(t_k), y) = \hat{\rho}'_{T-t_{k+1}}(x'^*(t_{k+1}), y).$$

However,

$$\hat{\rho}'_{T-t_{k+1}}(x'^*(t_{k+1}), y) = \max_{\eta \in C_E^{T-t_{k+1}}(y(t_{k+1}))} \min_{\xi \in C_P^{T-t_{k+1}}(x(t_{k+1}))} \rho(\xi, \eta) =$$

$$= \hat{\rho}_{T-t_{k+1}}(x'^*(t_{k+1}), y),$$

since, in this case, Player E remains at the point $y(t_{k+1})$ during the time $\delta - \delta = 0$. As a result, we get (see Eq. (3.1.10))

$$\hat{\rho}_{T-t_{k+1}}(x'^*(t_{k+1}), y) \leq \hat{\rho}_{T-t_k}(x^*(t_k), y^*(t_k)), \tag{3.1.11}$$

i.e. there is such point $x'^*(t_{k+1}) \in C_P^\delta(x^*(t_k))$ that (3.1.11) is satisfied. Consequently,

$$\hat{\rho}_{T-t_{k+1}}(\bar{x}, y(t_{k+1})) \leq \hat{\rho}_{T-t_k}(x*(t_k), y^*(t_k)),$$

where $\bar{x}$ is chosen by the strategy $u^*(\cdot)$ from the condition $\min \hat{\rho}_{T-t_k}(x, y(t_{k+1}))$, i.e. in the situation $(u^*(\cdot), v_\epsilon(\cdot))$ we have

$$\hat{\rho}_{T-t_{k+1}}(\bar{x}(t_{k+1}), y(t_{k+1})) \leq \hat{\rho}_{T-t_k}(x^*(t_k), y^*(t_k)) = \hat{\rho}_T(x_0, y_0)$$

if, playing the strategy $v_\delta(\cdot)$, Player E continues to select the controls, which do not coincide with $v^*(t)$, then

$$\hat{\rho}_T(x_0, y_0) = \hat{\rho}_{T-t_k}(x^*(t_k), y^*(t_k)) \geq \hat{\rho}_{T-t_{k+1}}(\bar{x}(t_{k+1}), y(t_{k+1})) \geq$$
$$\geq \hat{\rho}_{T-t^{k+2}}(\bar{x}(t_{k+2}, y(t_{k+2})) \geq \ldots \geq \hat{\rho}_0(\bar{x}(T), y(T)) =$$
$$= \rho(\bar{x}(T), y(T)) = K(x_0, y_0; u_\delta^*(\cdot), v_\delta(\cdot)).$$

This is exactly what proves the validity of the righthand inequality (3.1.8). We have thus proved the theorem. ■

Theorem 1 has a local character and holds in the regions of x_0, y_0, T, where the invariance conditions for the point M are satisfied, i.e. in the space region bounded by the singular surface. Note that $\hat{\rho}_T(x_0, y_0) = 0$ holds for $C_E^T(y_0) \subset C_P^T(x_0)$.

Let $\bar{T} > 0$ be the first time instant when $\hat{\rho}_T(x_0, y_0) = 0$ and the condition $\hat{\rho}_T > 0$ in Theorem 1 is violated explicitly. (Compare this with the first instant of absorption). The following theorem is readily obtained from the continuity of the function $\hat{\rho}_T(x_0, y_0)$ in T.

Theorem 2 *Suppose there is such $\delta_1 > 0$ that in the games $\Gamma_\delta(x_0, y_0; \bar{T} - \delta_2), \delta_2 < \delta_1$, the conditions of Theorem 1 are satisfied (for all $\delta_2 < \delta_1$). Then the value of the game $\Gamma_\delta(x_0, y_0, \bar{T})$ is zero, i.e. for any $\epsilon > 0$ at time T Player P can approach Player E for a distance not exceeding ϵ.*

Proof: The proof of Theorem 2 follows from the fact that, by adopting in the game $\Gamma_\delta(x_0, y_0; \bar{T})$ the optimal strategy $u^*(\cdot)$ in the game $\Gamma_\delta(x_0, y_0; \bar{T} - \delta_2)$, Player P can approach Player E at instant $\bar{T} - \delta_2$ on a distance $\hat{\rho}_{\bar{T}-\delta_2}(x_0, y_0)$, and from the Hausdorff continuity of the reachability sets $C_P^T(x), C_E^T(y)$, at the point $T = 0$. ■

Corollary. If $T \geq \bar{T}$, then the players P and E may be in the states of passive expectation, i.e.they make arbitrary moves along some trajectories $x(t), y(t)$ up to the time instant $\tau = \bar{T}$, where $\bar{T}$ is the minimum root of the equation $\hat{\rho}_{\bar{T}}(x(\tau), y(\tau)) = 0$. For $t \in [\tau, T]$, the players start the pursuit in the game $\Gamma_\delta(x(\tau), y(\tau); T - \tau)$, ensuring a zero value of the game in accordance

with Theorem 2 (provided that the conditions of this theorem are satisfied in the game
$\Gamma_\delta(x(\tau), y(\tau), \bar{T})$).

Consider a global solution of the games. We restrict ourselves to the games of pursuit where all singular surfaces are dispersal.

The theorems 1, 2 make it possible to construct the solution "in small", i.e. with the additional assumption that the players' trajectories do not intersect singular surfaces. For the global solution one chould be able to construct an optimal control on singular surfaces. In the general case, the nonincrease of $\hat{\rho}_T(x, y)$ is to be proved additionally when Player P adopts the strategy $u^*(\cdot)$ (the principle of min ρ), to the point moving on the singular surface (see the examples 4, 5, 7, Ch. 4). In the case of the dispersal surface, the solution is comparatively simple, since Player E's choice of the control $v(t)$ for $t \in [t_k, t_{k+1})$ on a singular surface (which becomes known to the opponent immediately) defines uniquely the center of pursuit M from the states $x(t_k), y(t_k)$ for the sets $C_P^{T-t_k}(x(t_k)), C_E^{T-t_k}(y(t_k))$. The optimal control is intended for aiming at the point M. Thus, the theorems 1, 2 also give a global solution to the games of pursuit where all singular surfaces are dispersal.

3.2 Continuous game without discrimination

Using the continuity of the function $\hat{\rho}_T(x_0, y_0)$ in T, x_0, y_0 we can discard the discrimination requirement.

We define the invariant center of pursuit for the continuous game without discrimination $\Gamma(x, y, T)$. The pursuit center M is called invariant if there is such $\bar{\delta} > 0$ that for all $0 < \delta < \bar{\delta}$ it is invariant in the discrete games $\Gamma_\delta(x, y, T)$ with discrimination for Player E.

Theorem 3 *Suppose in the game $\Gamma(x_0, y_0, T)$ there is an invariant center of pursuit. Then the function $\hat{\rho}_T(x_0, y_0)$ is the value of the game without discrimination (i.e. the game of pursuit where Player P's choice is based only on the information about the state of the system at the time instant t).*

Proof: For the given $\epsilon > 0$, we choose such $\delta_0 > 0$ that

$$|\hat{\rho}_T(x_0, y_0) - \hat{\rho}_T(x', y')| < \frac{\epsilon}{2} \tag{3.2.1}$$

for all x', y' belonging to $C_P^{\delta'}(x_0), C_E^{\delta'}(y_0)$, and for all $\delta' < \delta_0$. Next,let δ_1 be such that for all $x \in C_P^T(x_0)$ and all $\delta' < \delta_1$ the diameter of the set $D(C_P^{\delta'}(x)) < \frac{\epsilon}{2}$ and $\delta = \min\{\delta_0, \delta_1\}$ Player P is prescribed the strategy $u_\epsilon(\cdot)$, under which, on the time interval $[0, \delta]$ he moves arbitrary to a point $x(\delta) \in C_P^\delta(x_0)$. Let

$x(\delta), y(\delta)$ be the states realized by the instant δ. From the state $x' = x(\delta)$, $y(\delta)$ he plays an optimal strategy in the game with discrimination $\Gamma_\delta(x', y_0, T)$ (this is possible, since at each step Player P knows the move made by his opponent). The value of the game $\Gamma_\delta(x', y_0, T)$ is equal to $\hat{\rho}_T(x', y_0)$, but in the game Γ_δ Player P cannot make a final move (having played the strategy $u^*(\cdot)$), since he has lost one (opening) move by choosing an arbitrary motion to the point x' from the point x_0 in the time δ. Therefore, he can guarantee an approach to Player E by the instant T only on a distance $\hat{\rho}_T(x', y_0) + \frac{\epsilon}{2}$ (since the diameter of the reachability set $C_P^\delta(x)$ for one step does not exceed δ for all $x \in C_P^T(x_0)$). Because of Eq. (3.2.1) the strategy $u_\epsilon(\cdot)$ ensures that Player P will aproach Player E by the time instant T on a distance $\hat{\rho}_T(x_0, y_0) + \epsilon$. Since Player E can always provide for himself the payoff $\hat{\rho}_T(x_0, y_0)$, then $u_\epsilon(\cdot)$ is the ϵ–optimal strategy for Player P. This completes the proof of the theorem. ■

Instead of the lower game of pursuit, where Player E is discriminated, we may consider the upper game of pursuit with discrimination of Player P. All previous results are valid if instead of the quantity

$$\check{\rho}_T(x_0, y_0) = \max_{\eta \in C_E^T(y_0)} \min_{\xi \in C_P^T(x_0)} \rho(\xi, \eta)$$

we consider the quantity

$$\rho(x_0, y_0) = \min_{\xi \in C_P^T(x_0)} \min_{\eta \in C_E^T(y_0)} \rho(\xi, \eta).$$

In this case, the definition of the upper center of pursuit and its invariance may be also introduced. When the center of pursuit is invariant the upper center of pursuit is not (this was to be expected, otherwise the quantities $\hat{\rho}, \check{\rho}$ would be equal, which is far from reality). Moreover, we know no case of the invariance of the upper center of the pursuit other than in the game of "simple pursuit" on the sphere. At the same time, the invariance of the center of pursuit is observed in a large class of problems (see [28, 30, 32]). The latter consideration also provides support for the intuitively evident non–simmetry of the players' status in the games of pursuit.

The above–mentioned constructions may also be generalized to the case where the sets $C_P^T(x_0)$ and $C_E^T(y_0)$ are not necessarily closed. Moreover, the ϵ–optimality may be shown for the strategies given above, where under the quantities $\bar{\rho}_T(x, y), \check{\rho}_T(x, y)$ are meant the quantities

$$\hat{\rho}_T(x, y) = \sup_{\eta \in C_E^T(y_0)} \inf_{\xi \in C_P^T(x_0)} \rho(\xi, \eta),$$

$$\check{\rho}(x, y) = \inf_{\xi \in C_P^T(x_0)} \sup_{\eta \in C_E^T(y_0))} \rho(\xi, \eta),$$

and under center of pursuit and the upper center of pursuit are meant the points in the ϵ–neighbourhoods of the points M and M' belonging to the closures of the sets $C_P^T(x_0), C_E^T(y_0)$ and delivering max min and min max in the preceding equalities.

3.3 Arbitrary terminal payoff functions and phase constrains

We now turn to the pursuit–type games of prescribed duration $\Gamma(x_0, y_0, T)$ where the terminal payoff is defined as

$$K(x_0, y_0; u(\cdot), v(\cdot)) = M(x(T), y(T))$$

where $M(x, y)$ is an arbitrary continuous function given on $R^n \times R^n$.

The lower game with discrimination $\Gamma_\delta(x_0, y_0, T)$ is defined as in 2, Ch. 2, except that $M(x(t), y(t))$ is meant everywhere instead of $\rho(x(T), y(T))$.

The function $\bar{M}_T(x, y)$ is the analog of the function $\hat{\rho}(x, y)$ and is defined as follows:

$$\bar{M}_T(x, y) = \max_{\eta \in C_E^T(y)} \min_{\xi \in C_P^T(x)} M(\xi, \eta).$$

Conditionally optimal trajectories are defined as in 1 (for simplicity, the sets $C_E^T(y)$ and $C_P^T(x)$ are assumed to be compact).

The point H is called the center of the game $\Gamma_\delta(x_0, y_0, T)$ if $\bar{M}_T(x_0, y_0) = M(\xi, H)$. (Here ξ, H are the points at which max min of the function $M(\xi, \eta)$ is achieved).

By analogy with 1, we introduce the auxiliary game $\Gamma'_{\delta,\alpha}(x, y, T)$ where Player E remains at the point y during the time α. The center of the auxiliary game is determined as in the game $\Gamma(x, y, T)$ except that the reachabiliy sets, by which max min is taken, are computed for the auxiliary game. As in 1, the notion of the invariant center of the game is introduced with the help of the auxiliary game.

Theorem 1 is fully applicable to the case of the function $\bar{M}_T(x_0, y_0)$.

Theorem 4 *Assume that in the game $\Gamma_\delta(x_0, y_0, T)$ there is an invariant center of the game and $\bar{M}_T(x_0, y_0) > 0$. Then there is an equilibrium point, conditionally optimal trajectories are optimal, the value of the game $V(x_0, y_0, T) = \bar{M}_T(x_0, y_0)$, and optimal strategies for the players P and E are constructed in much the same way as in Theorem 1.*

The proof for Theorem 4 fully coincides with the one for Theorem 1, in 3. The notion of the singular surface is just as in 1.

To generalize Theorem 3 to the games without discrimination, we utilize the continuity of functions $\bar{M}_T(x, y)$ in all of the arguments.

Theorem 5 *Assume that in the game $\Gamma(x_0, y_0, T)$ (without discrimination) there is an invariant center of the game. Then the function $\bar{M}_T(x, y)$ is the value of the game without discrimination.*

Proof: As in Theorem 3, we consider the approximation game $\Gamma_\delta(x_0, y_0, T)$ with descrimination. Here the continuity of the function $\bar{M}_T(x, y)$ is used instead of the continuity of the function $\hat{\rho}(x_0, y_0)$. ■

Theorem 5 is applicable to zero–sum games of pursuit with prescribed duration between the pursuer team acting as one player $P = \{P_1, \dots, P_m\}$ and the evader E.

Suppose the motion equations for the team partners P_i, $i = 1, \dots, m$, and Player E are of the form

$$\dot{x}^{(i)} = f^{(i)}(x^{(i)}, u),\ u \in U^{(i)},\ x^{(i)} \in R^n,\ i = 1, \dots, m,$$

$$\dot{y} = g(y, v),\ v \in V,\ y \in R^n$$

(here $x^{(i)}$ is interpreted as the phase coordinate vector for the team partner P_i). Let $u(\cdot) = \{u_i(\cdot)\}$ be a piecewise open–loop strategy of the team P. The payoff function is defined in the following way $K(x_0^{(i)}, \dots, x_0^{(m)}, y_0, u(\cdot), v(\cdot)) = \min_i \rho(x^{(i)}(T), y(T))$, where $x^i(t)$, $y(t)$, $i = 1, \dots, m-$ are trajectories of the team members P_i and Player E from the initial states x_0^i, y_0, $i = 1, \dots, m$, in the situation $(u(\cdot), v(\cdot))$. In order to have an opportunity to apply Theorem 5, we add to the motion equations for Player E, $n \times (m-1)$ equations of the from

$$\dot{y}_j = 0,\ j = 1, \dots, n \times (m-1),$$

with arbitrary initial conditions. By the state vector of Player E is meant a new extended vector $\bar{y} = (y, y_1, \dots, y_{n(m-1)})$ of dimension $n \times m$. The vectors x and $\bar{y}$ are of the same dimension. We rewrite the payoff in the equivalent form:

$$K(x_0^{(1)}, \dots, x_0^{(m)}, y_0, u(\cdot), v(\cdot)) = M(x(T), \bar{y}(T)) =$$
$$= \min_i \rho(x^{(i)}(T), y(T)).$$

Here by the distance is meant the distance between the vectors $x^{(i)}(T)$ and $y(T)$ (but not $\bar{y}(T)$) in the space R^n. Since the function $M(x, y)$ is continuous, we can apply Theorem 5.

Let a closed convex set S be given in R^n. Consider the pursuit game with prescribed duration and with the additional constraint under which the players P and E are not allowed to leave the set S during the game. Motion equations are of the form Eq.(2.1.1), Eq.(2.1.2), the initial conditions being $(x_0, y_0) \in S$. The payoff is defined as a distance between the players at the time instant T when the play ends. Denote such a game by $\Gamma^{(s)}(x_0, y_0, T)$.

To summarize the previous results, we have to introduce the notion of admissible open–loop controls and admissible strategies.

The open–loop control $u(t)$ for $0 \leq t \leq T$ is called admissible in the game $\Gamma^{(s)}(x_0, y_0, T)$ if under this control the trajectory $x(t)$ corresponding to the given open–loop control belongs to S $(x_0 \in S)$. The admissible open–loop control for Player E is determined in much the same way.

By the reachability set $C_P^T(x_0, S)$ for Player P in the game $\Gamma^{(s)}(x_0, y_0, T)$ is meant a set of points at which Player P can arrive at the time instant T from the point $x_0 \in S$ be employing every admissible open–loop control. The reachability set $C_E^T(y_0, S)$ from Player E is defined in much the same way.

The piecewise open–loop strategy $\bar{u}(\cdot)$ is called admissible if in any situation $(\bar{u}(\cdot), v(\cdot))$, $v(\cdot) \in E$ the trajectory $x(t)$, $x(0) = x_0 \in S$, $0 \leq t \leq T$ belongs to S. An admissible piecewise open–loop strategy for Player E is defined in much the same way.

The set of all admissible piecewise open–loop strategies for the players P and E will be denoted respectively by P^S and E^S. By the equilibrium point and the value of the game are meant those in the class of admissible piecewise open–loop strategies.

It follows from the construction of the sets P^S and E^S that for any $u(\cdot) \in P^S$, $v(\cdot) \in E^S$ and $x_0 \in S$, $y_0 \in S$ the trajectories $x(t), y(t)$, $0 \leq t \leq T$, belong to S.

By analogy with $\hat{\rho}_T(x_0, y_0)$, we define $\hat{\rho}_T(x_0, y_0, S)$ as

$$\hat{\rho}_T(x_0, y_0, S) = \sup_{\eta \in C_E^T(y_0, S)} \inf_{\xi \in C_P^T(x_0, S)} \rho(\xi, \eta).$$

Although this does not play a role in what follows, we assume that the sets $C_E^T(y_0, S)$, $C_P^T(x_0, S)$ are compact and the function $\hat{\rho}_T(x_0, y_0, S)$ is continuous in x_0, y_0, T (in the phase–constrained games this condition may not be fullfilled). Then it is possible to define a center of pursuit and relevant conditionally optimal trajectories $x^*(t)$, $y^*(t)$.

In general, not all of the conditionally optimal trajectories correspond to admissible controls, we agree to consider, without special stipulations, only admissable conditionally optimal trajectories $x^*(t)$, $y^*(t)$ i.e. such that $x^*(t) \in S$, $y^*(t) \in S$ for $0 \leq t \leq T$. From the definition of the sets $C_P^T(x_0, S)$, $C_E^T(y_0, S)$ it follows that such trajectories necessarily exist.

As in 2, we define the discrete game $\Gamma_\delta(x_0, y_0, T)$ with discrimination of Player E. In this game, we find the strategy $u^*(\cdot)$ of Player P from the formula (3.1.4) by replacing therein the set $C_P^\delta(x(t_k))$ by $C_P^\delta(x(t_k), S)$.

By construction, the strategy $u^*(\cdot)$ is admissible. As in 2, we define an invariant center of pursuit. Here by $x^*(t)$, $y^*(t)$ is meant an admissible pair of relevant conditionally optimal trajectories. Theorems 1, 2, 5 hold fully for the case of phase constraints if by the reachability sets are everywhere meant the sets $C_P^T(x,S)$, $C_E^T(y,S)$ both in definition of $\hat{\rho}_T(x,y)$ and in the formulation of an optimal strategy for Player P.

3.4 Optimal time pursuit games

In this section, we will consider the games of pursuit where the payoff function equals the time–to–capture. For the purposes of defininiteness, we assume that the terminal manifold S is the sphere $\rho(x,y) = \ell, \ell > 0$, the function $M(x,y) = +1$ (see in 2, Ch. 2) [28].

Let us assume that the sets $C_P^t(x)$ and $C_E^t(y)$ continuous in the point $t = 0$ uniformly with respect to x and y.

Suppose the following quantity has a meaning

$$\Theta(x,y,\ell) = \max_{v(t)} \min_{u(t)} t_\Pi^\ell(x,y;u(t),v(t)),$$

where $t_\Pi^1(x,y;u(t),v(t))$ is the time of advancing the distance ℓ between the players P and E moving from the initial points x, y under measurable open–loop controls $u(t)$ and $v(t)$, respectively. Also, assume the function $\Theta(x,y,\ell)$ continuous in all of its arguments. Kelendzheridze was the first to introduce this function [9].

Definition 2. Let $(\bar{u}(t),\bar{v}(t))$, $t \in [0,\infty)$ be a pair of open–loop controls, under which $\Theta(x,y,\ell)$ is achieved, and $y^*(t)$, $x^*(t)$ be a corresponding pair of trajectories defined by sistem (2.1.1), (2.1.2) under open–loop controls $\bar{u}(t)$, $\bar{v}(t)$ and initial conditions $x*(0) = x$, $y*(0) = y$.

This pair of trajectories will be referred to as a pair of conditionally optimal trajectories. In what follows, $\Gamma'_\delta(x,y)$ is the game of pursuit which differs from the game $\Gamma(x,y)$ only in that here Player E remains in the initial state y during the time δ. Let $x'^*(t)$ be a conditionally optimal trajectory for Player P in this game.

Theorem 6 *Suppose the following conditions are satisfied*

1. $inf_{x' \in C_P^\delta(x)} \Theta(x',y,\ell)$ *is achieved with* y *fixed for any* x.

2. $\Theta(x^*(t),y^*(t),\ell) + t = const$, $0 \geq t \geq t_\Pi$, *where* $t_\Pi = \min_t\{t : \rho(x^*(t),y^*(t)) \leq \ell\}$.

3. *There is such* $\bar{\delta}$ *that for all* $\delta < \bar{\delta}$ *in the game* $\Gamma'_\delta(x,y)$ *with any fixed* t, $y \in C_E^\delta(y^*(t))$ *for all* $0 \leq \alpha \leq \delta$ *we have*

$$\Theta(x'^*(t+\alpha),y,\ell) + \alpha = const,$$

where $x'*(t) = x^*(t)$ *(here* $\Theta(x'^*(t+\alpha), y, \ell)$ *is computed for the game* $\Gamma'_{\delta-\alpha}(x'^*(t+\alpha), y, \ell)$*) [28].*

Then for any $\ell > 0$, $\epsilon > 0$ *in the game* $\Gamma(x, y)$ *there is the* ϵ*-equilibrium point and the value of the game is equal to* $\Theta(x, y, \ell)$*, i.e. for any* $\epsilon > 0$ *there exist such pair of strategies* $u_\epsilon^*(\cdot)$, $v_\epsilon^*(\cdot)$ *that*

$$\Theta(x, y, \ell) - \epsilon \leq t_\Pi^\ell(x, y; u_\epsilon^*(\cdot), v_\epsilon^*(\cdot)) \leq \Theta(x, y, \ell) + \epsilon, \tag{3.4.1}$$

$$t_\Pi^\ell(x, y; u(\cdot), v_\epsilon^*(\cdot)) + \epsilon \geq t_\Pi^\ell(x, y; u_\epsilon^*(\cdot), v_\epsilon^*(\cdot)) \geq t_\Pi^\ell(x, y, u_\epsilon^*(\cdot), v(\cdot)) - \epsilon, \tag{3.4.2}$$

for all possible strategies $u(\cdot)$ *and* $v(\cdot)$ *of the players* P *and* E*.*

Before passing to the proof of the theorem, we prove some auxiliary statements.

As in 2 of this chapter, we introduce the discrete games $\Gamma_\delta(x, y, \ell)$ with discrimination of Player E. Motion equations for the players are the same, and a partition of the interval $[0, \infty)$ is fixed, i.e. the partition by the set of points $\{t_k\}_{k=0}^{\infty}$ such that $t_{k+1} - t_k = \delta$ for any k. Player E is said to be discriminated, i.e. at the time instant t_k Player P has, apart from information on the time, his own and opponent's position at a given time, information on the control $v(t)$ for $t \in [t_k, t_{k+1})$. Thus, at the instant t_k Player P has information on the opponent's position at the instant t_{k+1}. Let $u_\delta(\cdot)$, $v_\delta(\cdot)$ be strategies for the players P and E respectively in the game $\Gamma_\delta(x, y, \ell)$.

Lemma 2 *Under conditions of Theorem 6, in the game* $\Gamma_\delta(x, y, \ell)$ *for* $\delta < \bar{\delta}$ *there is an equilibrium point in pure strategies and* $\Theta(x, y, \ell)$ *is the value of the game, i.e. there exist a pair of strategies* $u_\delta^*(\cdot)$, $v_\delta^*(\cdot)$ *such that for any strategies* $u_\delta(\cdot)$ *and* $v_\delta(\cdot)$ *of players* P *and* E *in the game* $\Gamma_\delta(x, y, \ell)$ *there is*

$$\begin{aligned} t_\Pi^\ell(x, y, u_\delta(\cdot), v_\delta^*(\cdot)) &\geq t_\Pi^\ell(x, y, u_\delta^*(\cdot), v_\delta(\cdot)) = \\ &= \Theta(x, y, \ell) \geq t_\Pi^\ell(x, y; u_\delta^*(\cdot), v_\delta(\cdot)). \end{aligned} \tag{3.4.3}$$

Proof: Let $\bar{v}(t; x(t_k), y(t_k))$, $\bar{u}(t; x(t_k), y(t_k))$, $t \in [t_k, \infty)$, be a pair of open-loop controls on which $\Theta(x(t_k), y(t_k), \ell)$ is achieved.

Let us define the strategies $u_\delta^*(\cdot)$ and $v_\delta^*(\cdot)$. The strategy $v_\delta^*(\cdot)$ for Player E in the game $\Gamma_\delta(x, y, \ell)$ associates the state $(t_k, x(t_k), y(t_k))$ with the admissible open–loop control $\bar{v}(t, x(t_k), y(t_k))$ for $t \in [t_k, t_{k+1})$. The strategy $u_\delta^*(\cdot)$ for Player P in the game $\Gamma_\delta(x, y, \ell)$ associates the state $(t_k, x(t_k), y(t_{k+1}))$, with the admissible open–loop control $\bar{u}(t; x(t_k), y(t_{k+1}))$, $t \in [t_k, t_{k+1})$ transferring the point $x(t_k)$ to such point $x(t_{k+1}) \in C_P^\delta(x(t_k))$ such that

$$\Theta(x(t_{k+1}), y(t_{k+1}), \ell) = \min_{x' \in C_P^\delta(x(t_k))} \Theta(x', y(t_{k+1}), \ell), \tag{3.4.4}$$

if the ℓ–capture from the points $x(t_k)$, $y(t_k)$ is impossible in the time δ. Otherwise Player P chooses the admissible open–loop control $u(t)$ which allows the ℓ–capture to be realized in the minimum time.

We will show that the situation $(u^*_\delta(\cdot), v^*_\delta(\cdot))$ is the equilibrium point.

Assume that in the situation $(u^*_\epsilon(\cdot), v^*_\epsilon(\cdot))$ for $n \leq k$ the equality holds

$$\Theta(x(t_n), y(t_n), \ell) + t_n = \Theta(x, y, \ell), \tag{3.4.5}$$

where $x(t_n)$, $y(t_n)$ is the position of the players P and E at the instant t_n in the situation $(u^*_\delta(\cdot), v^*_\delta(\cdot))$, establish that Eq.(3.4.5) holds for $n = k+1$.

Let $\bar{v}(t; x(t_k), y(t_k))$, $\bar{u}(t; x(t_k), y(t_k))$, $t \in [t_k, \infty)$ be a pair of open–loop controls on which $\Theta(x(t_k), y(t_k), \ell)$ is achieved, and $x^*(t, x(t_k))$, $y^*(t, y(t_k))$, $t \in [t_k, \infty)$ be a corresponding pair of trajectories defined by the system (2.1.1), (2.1.2) under the initial conditions

$$x^*(t_k, x(t_k)) = x(t_k),$$

$$y^*(t_k, y(t_k)) = y(t_k)$$

and the open–loop controls $\bar{v}(t; x(t_k), y(t_k))$, $\bar{u}(t; x(t_k), y(t_k))$. Introduce the time $\tau = t - t_k$ and denote $x^*(t, x(t_k)) = x(\tau, x(t_k))$, $y^*(t, y(t_k)) = y(\tau, y(t_k))$. From Eq.(3.4.4) we have

$$\Theta(x(t_{k+1}), y(\delta, y(t_k)), \ell) \leq \Theta(x(\delta, x(t_k)), y(\delta, y(t_k)), \ell). \tag{3.4.6}$$

The second condition of Theorem 6 implies

$$\Theta(x(\delta, x(t_k)), y(\delta, y(t_k)), \ell) + \delta = \Theta(x(t_k), y(t_k), \ell). \tag{3.4.7}$$

By the assumption, the condition (3.4.5) is satisfied. Consequently, by introducing the notation $y(\delta, y(t_k)) = y(t_{k+1})$, from (3.4.6) and Eq. (3.4.7) we get

$$\Theta(x(t_{k+1}), y(t_{k+1}), \ell) + t_{k+1} \leq \Theta(x, y, \ell). \tag{3.4.8}$$

At the same time, let x' be an arbitrary point of the set $C^\delta_P(x(t_k))$, and $u'(t)$, $t \in [t_k, t_{k+1})$ be an admissible open–loop control transferring the point $x(t_k)$ to the point x'. Next, let $\bar{\bar{u}}(t)$, $t \in [t_{k+1}, \infty)$ be such admissible open–loop control that

$$t^\ell_\Pi(x', y(t_{k+1}), \bar{\bar{u}}(t), \bar{v}(t, x(t_k), y(t_k))) =$$

$$= \min_{u(t)} t^\ell_\Pi(x', y(t_{k+1}), u(t), \bar{v}(t, x(t_k), y(t_k))) \leq$$

$$\leq \max_{v(t)} \min_{u(t)} t^\ell_\Pi(x', y(t_{k+1}), u(t), v(t)) = \Theta(x', y(t_{k+1}), \ell). \tag{3.4.9}$$

Let D be a set of measurable open–loop controls $u(t)$, $t \in [t_k, \infty)$ that are coincident with the function $u'(t)$ on the interval $[t_k, t_{k+1})$. Denote by $\hat{u}(t)$

the function defined on the interval $[t_k, \infty)$ that is coincident with $u'(t)$ on the interval $[t_k, t_{k+1})$ and with $\bar{\bar{u}}(t)$ on $[t_{k+1}, \infty)$. Then

$$t^{\ell}_{\Pi}(x(t_k), y(t_k), \hat{u}(t), \bar{v}(t, x(t_k), y(t_k))) =$$

$$= \min_{u(t)\in D} t^{\ell}_{\Pi}(x(t_k), y(t_k), u(t), \bar{v}(t, x(t_k), y(t_k)))$$

Next

$$t^{\ell}_{\Pi}(x(t_k), y(t_k), \bar{u}(t, x(t_k), y(t_k)), \bar{v}(t, x(t_k), y(t_k))) =$$

$$= \min_{u(t)} t^{\ell}_{\Pi}(x(t_k), y(t_k), u(t), \bar{v}(t, x(t_k), y(t_k)))$$

(here min is taken for all possible measurable open–loop controls). However, D is a subset of the set of all measurable open–loop controls. Consequently,

$$t^{\ell}_{\Pi}(x(t_k), y(t_k), \bar{u}(t, x(t_k), y(t_k)), \bar{v}(t, x(t_k), y(t_k))) = \Theta(x(t_k), y(t_k), \ell) \leq$$

$$\leq t^{\ell}_{\Pi}(x(t_k), y(t_k), \hat{u}(t), \bar{v}(t, x(t_k), y(t_k))) \leq \delta + \Theta(x', y(t_{k+1}), \ell). \tag{3.4.10}$$

The latter inequality (3.4.10) follows from (3.4.9). Thus,

$$\Theta(x(t_k), y(t_k), \ell) \leq \delta + \Theta(x', y(t_{k+1}), \ell). \tag{3.4.11}$$

The inequality (3.4.11) holds for any point $x' \in C^{\delta}_{P}(x(t_k))$, specifically, for $x(t_{k+1})$. From (3.4.11), (3.4.8), and from the assumption (3.4.5) we have

$$\Theta(x, y, \ell) = \Theta(x(t_{k+1}), y(t_{k+1}), \ell) + t_{k+1}.$$

So (3.4.5) also holds for $n = k+1$. For $n = 0$, Eq.(3.4.5) is obviously satisfied. We have thus proved (3.4.5).

Let N be such that $\Theta(x, y, \ell) = N\delta + \delta'$, where $\delta' < \delta$. Since $N\delta = t_N$, from (3.4.5) we have

$$\Theta(x, y, \ell) = \Theta(x(t_N), y(t_N), \ell) + t_N, \tag{3.4.12}$$

here $\Theta(x(t_N), y(t_N), \ell) = \delta' < \delta$.

From the states $x(t_N)$, $y(t_N)$, having information on Player E's control for the time δ ahead and adopting the strategy $u^*_{\delta}(\cdot)$, Player P may finish the pursuit in the time $\Theta(x(t_N), y(t_N), \ell) = \delta' < \delta$. Thus the hole time of pursuit in the situation $(u^*_{\delta}(\cdot), v^*_{\delta}(\cdot))$ from the initial states x, y along the trajectories $x^*(t)$, $y^*(t)$ is equal to

$$t^{\ell}_{\Pi}(x, y; u^*_{\delta}(\cdot), v^*_{\delta}(\cdot)) = t_N + \delta' = N\delta + \delta' = \Theta(x, y, \ell). \tag{3.4.13}$$

Let us prove the validity of the right–hand side of the inequality (2.4.3). We first show that in the situation $(u_{\delta}(\cdot), v^*_{\delta}(\cdot))$ the following inequality holds

$$\Theta(x(t_k), y(t_k), \ell) \geq \Theta(x(t_{k+1}), y(t_{k+1}), \ell) + \delta, \tag{3.4.14}$$

where $x(t_k)$, $y(t_k)$, $x(t_{k+1})$, $y(t_{k+1})$ are positions of the players P and E at the time instant t_k and t_{k+1} in the situation $(u_\delta(\cdot), v_\delta^*(\cdot))$,

$$\Theta(x(t_k), y(t_k), \ell) = \max_{v(t)} \min_{v(t)} t_\Pi^\ell(x(t_k), y(t_k), u(t), v(t)) =$$

$$= \min_{u(t)} t_\Pi^\ell(x(t_k), y(t_k), u(t), \bar{v}(t, x(t_k), y(t_k))). \tag{3.4.15}$$

Let $\tilde{u}(t)$ $t \in [t_k, t_{k+1})$ be an open–loop control transferring the point $x(t_k)$ to the point $x(t_{k+1})$. The open—loop control transferring the point $y(t_k)$ to the point $y(t_{k+1})$ is the control $\bar{v}(t; x(t_k), y(t_k),\ t \in [t_k, t_{k+1})$. Let D be a set of open–loop controls defined for $t \in [t_k, \infty)$ and coincident with $\tilde{u}(t)$ for $t \in [t_k, t_{k+1})$. Consider the quantity

$$R = \min_{u(t) \in D} t_\Pi^\ell(x(t_k), y(t_k), u(t), \bar{v}(t; x(t_k), y(t_k))).$$

Evidently,

$$\min_{u(t) \in D} t_\Pi^\ell(x(t_k), y(t_k), u(t), \bar{v}(t; x(t_k), y(t_k))) \le$$

$$\le \delta + \Theta(x(t_{k+1}), y(t_{k+1}), \ell). \tag{3.4.16}$$

The set D is a subset of the set of all measurable open–loop controls. Consequently, from (3.4.15) and (3.4.16) it is possible to derive the inequality (3.4.14).

Let $u_\delta^r(\cdot)$ be a strategy of Player P under which, in the situation $(u_\delta^*(\cdot), v_\delta^*(\cdot))$, he makes r times a choice that is different from the choice of $u_\delta^*(\cdot)$. In this notation $u_\delta^*(\cdot) = u_\delta^0(\cdot)$. Assume that for $n \le r$ the inequality

$$t_\Pi^\ell(x, y; v_\delta^*(\cdot), u_\delta^n(\cdot)) \ge \Theta(x, y, \ell). \tag{3.4.17}$$

is satisfied. Prove that (3.4.17) also holds for $n = r + 1$.

Now, suppose that in the game $\Gamma_\delta(x, y, \ell)$ Player P adopts the strategy $u_\delta^{r+1}(\cdot)$, and Player E the strategy $v_\delta^*(\cdot)$. Let t_k be the time instant at which the choice of $u_\delta^{r+1}(\cdot)$ does not coincide with the choice of $u_\delta^*(\cdot)$ for the first time. We introduce the time $\tau = t - t_{k+1}$ and consider the game $\Gamma_\delta(x(t_{k+1}), y(t_{k+1}), \ell)$, where $x(t_{k+1})$, $y(t_{k+1})$ are positions of the players P and E respectively at the instant t_{k+1} in the situation $(u_\delta^{r+1}(\cdot), v_\delta^*(\cdot))$. In this game, Player E adopts the strategy $v_\delta^*(\cdot)$ Player P the strategy under which, in the situation $(u_\delta^r(\cdot), v_\delta^*(\cdot))$, he makes a choice r times different than the choice of $u_\delta^*(\cdot)$, i.e. the strategy $u_\delta^r(\cdot)$. By assumption.

$$t_\Pi^\ell(x(t_{k+1}, y(t_{k+1}), u_\delta^r(\cdot), v_\delta^*(\cdot)) \ge \Theta(x(t_{k+1}), y(t_{k+1}), \ell), \tag{3.4.18}$$

and since

$$t_\Pi^\ell(x, y, v_\delta^*(\cdot), u_\delta^{r+1}(\cdot)) = t_{k+1} + t_\Pi^\ell(x(t_{k+1}), y(t_{k+1}); u_\delta^r(\cdot), v_\delta^*(\cdot)), \tag{3.4.19}$$

then, by employing (3.4.14), we obtain

$$t^{\ell}_{\Pi}(x, y, v^*_\delta(\cdot), u^{r+1}_\delta(\cdot)) \geq t_k + \Theta(x(t_k), y(t_k), \ell). \tag{3.4.20}$$

The choices by the players P and E in the game $\Gamma_\delta(x, y, \ell)$, however, has been coincident with their choices in the situation $(v^*_\delta(\cdot), u^*_\delta(\cdot))$ up to the instant t_k. Consequently (3.4.15) holds for $n \leq k$. Then, from (3.4.20) we find

$$t^{\ell}_{\Pi}(x, y; v^*_\delta(\cdot), u^{r+1}_\delta(\cdot)) \geq \Theta(x, y, \ell) \tag{3.4.21}$$

Thus, (3.4.10) also holds for $n = r+1$, and since this holds obviously for $n = 0$, then it also holds for all $n \leq N$ i.e. for any Player P's strategy in the game $\Gamma_\delta(x, y, \ell)$ we have the inequality.

$$t^{\ell}_{\Pi}(x, y, u_\delta(\cdot), v^*_\delta(\cdot)) \geq \Theta(x, y, \ell) = t^{\ell}_{\Pi}(x, y; u^*_\delta(\cdot), v^*_\delta(\cdot)) \tag{3.4.22}$$

Now prove the validity of the right–hand side of the inequality (3.4.3). First establish the following inequality:

$$\Theta(x(t_k), y(t_k), \ell) \geq \delta + \Theta(x(t_{k+1}), y(t_{k+1}), \ell),$$

where $x(t_k)$, $y(t_k)$, $x(t_{k+1})$, $y(t_{k+1})$ are the points at which the players P and E arrive in the situation $(u^*_\delta(\cdot), v_\delta(\cdot))$ at the instants t_k, t_{k+1}.

Let $v(t)$, $t \in [t_k, t_k + 1)$ be an open–loop control transferring the point $y(t_k)$ to the point $y(t_{k+1})$, and $u(t)$, $t \in [t_k, t_{k+1})$ be an open–loop control transferring the point $x(t_k)$ to the point $x(t_{k+1})$. Let D be a set of measurable open–loop controls that are coincident with the control $v(t)$ on the interval $[t_k, t_{k+1})$, where t_k is the time instant when Player E's choice in the situation $(u^*_\delta(\cdot), v_\delta(\cdot))$ is firstly noncoincident with his choice in the situation $(u^*_\delta(\cdot), v^*_\delta(\cdot))$ Then

$$\Theta(x, y, \ell) = t_k + \Theta(x(t_k), y(t_k), \ell) \geq \max_{v(t) \in D} \min_{u(t)} t^{\ell}_{\Pi}(x(t_k), y(t_k); u(t), v(t)) \tag{3.4.23}$$

Consider an auxiliary game $\Gamma'_{\delta,\delta_1}(x(t_k), y)$, where Player E remains at the point $y = y(t_{k+1})$ during the time δ_1.

Let $x'^*(t)$ be a conditionally optimal trajectory for Player P in the game $\Gamma'_{\delta,\delta_1}(x(t_k), y)$, and $\Gamma'_{\delta,\delta-\alpha}(x'(t_k + \alpha), y)$ for $0 \leq \alpha \leq \delta$, $y = y(t_{k+1})$, $x'^*(t) = x(t_k)$ be the game where Player E remains at the point $y = y(t_{k+1})$ during the time $t_{k+1} - (t_k + \alpha) = \delta - \alpha$. Let $\Theta'(x'^*(t_k + \alpha), y, \ell)$ be the maximin time prior to the ℓ–capture computed in this game. Then

$$\Theta'(x'^*(t_k), y, \ell) = \Theta'(x^*(t_k), y(t_{k+1}), \ell) =$$

$$= \max_{v(t) \in D} \min_{u(t)} t^{\ell}_{\Pi}(x(t_k), y(t_k), u(t), v(t)) \leq \Theta(x(t_k), y(t_k), \ell). \tag{3.4.24}$$

It follows from condition 3 of Theorem 6 that for all $0 < \alpha < \delta$

$$\Theta'(x'^*(t_k+\alpha), y(t_{k+1}), \ell) + \alpha = const. \tag{3.4.25}$$

Hence, in particular, for all $\alpha = 0$ and $\alpha = \delta$ we get

$$\Theta'(x'^*(t_k), y(t_{k+1}), \ell) = \Theta'(x'^*(t_{k+1}), y(t_{k+1}), \ell) + \delta \tag{3.4.26}$$

However

$$\Theta'(x'^*(t_{k+1}), y(t_{k+1}), \ell) = \max_{v(t)} \min_{u(t)} t^{\ell}_{\Pi}(x'^*(t_{k+1}), y(t_{k+1}), u(t), v(t)) =$$

$$= \Theta(x'^*(t_{k+1}), y(t_{k+1}), \ell) \tag{3.4.27}$$

since in this case Player E remains at the point $y(t_{k+1})$ unmoved during the time $\delta = 0$. As a result, we have

$$\Theta(x'^*(t_k), y(t_k), \ell) \geq \Theta(x'^*(t_{k+1}), y(t_{k+1}), \ell) + \delta \tag{3.4.28}$$

Consequently, there exists such $x'^*(t_{k+1}) \in C^{\delta}_{P}(x(t_{k+1}))$ for which (3.4.28) is satisfied. Then the following inequality holds

$$\Theta(x'^*(t_k), y(t_k), \ell) \geq \Theta(\bar{x}, y(t_{k+1}), \ell) + \delta \tag{3.4.29}$$

where $\bar{x}$ is selected under the strategy $u^*(\cdot)$ from the min condition of the function $\Theta(x, y(t_{k+1}), \ell)$ on the set $C^{\delta}_{P}(x(t_k))$ i.e.in the situation $(u^*_{\delta}(\cdot), v_{\delta}(\cdot))$

$$\Theta(x, y, \ell) = \Theta(x(t_k), y(t_k), \ell) + t_k \geq \Theta(x(t_{k+1}), y(t_{k+1}), \ell) + t_{k+1}. \tag{3.4.30}$$

The further prove is in much the same way as the validity of the left–hand side of the inequality Eq.(3.4.3). ∎

We now turn to the proof of Theorem 6.

Proof: By hypothesis, the function $\Theta(x, y, \ell)$ is continuous in all of its arguments. Consequently, for any $\epsilon > 0$ we find such $\delta > 0$. that as soon as

$$\rho((x, y, \ell), (x', y', \ell')) \leq \delta, \tag{3.4.31}$$

$$|Q(x, y, \ell) - Q(x', y', \ell')| \leq \frac{\epsilon}{2}. \tag{3.4.32}$$

The assumption about the continuity of $C^t_P(x)$, $C^t_P(y)$ in $t = 0$ uniformly with respect to x, y suggests that by δ it is possible to find such $\delta_1 > 0$ that for any $x' \in C^{\delta_1}_P(x)$, $y' \in C^{\delta_1}_P(y)$, $\ell' = \ell - \delta_1$ (3.4.31) holds.

Next, by δ_1 it is possible to find such $\sigma_1 > 0$, $\sigma_2 > 0$ that for any $x' \in C^{\sigma_1}_P(x)$, $y' \in C^{\sigma_2}_P(y)$

$$\rho(x, x') \leq \delta_1, \quad \rho(y, y') \leq \delta_1. \tag{3.4.33}$$

Introduce the notation

$$\delta_2 = \min\left\{\delta_1, \delta, \sigma_1, \sigma_2, \frac{\epsilon}{2}\right\}. \tag{3.4.34}$$

Consider the strategy $u_\epsilon^*(\cdot)$ for Player P that is an ordered pair (σ, α), where σ is a partition of the interval $[0, \infty)$ by the points $\{t_k\}_{k=0}^\infty$ with the step equal to $\delta_3 \leq \delta_2$ and α is a map which associates the state t_0, $x(t_0)$, $y(t_0)$ with an arbitrary admissible open–loop control $u(t)$, $t \in [t_0, t_1)$ ($t_0 = 0$, $x(t_0) = x_0$, $y(t_0) = y_0$), and then coincides with the map α' which associates the state $(t_k, x(t_k), y(t_k))$, $k \geq 1$, with an admissible open–loop control $\tilde{u}(t)$, $t \in [t_k, t_{k+1})$ transferring the point $x(t_k)$ to such point $\bar{x} = x(t_{k+1})$ that

$$\Theta(x(t_{k+1}), y(t_k), \ell') = \min_{x' \in C_P^{\delta_3}(x(t_k))} \Theta(x', y(t_k), \ell'), \tag{3.4.35}$$

where $\ell' = \ell - \delta_1$. Then, the map α is an optimal strategy for Player P in the game $\Gamma_{\delta_3}(x(\delta_3), y, \ell')$. Moreover, assume that Player E's strategy $v_\epsilon^*(\cdot)$, under which a partition of the time interval $[0, \infty)$ is chosen with an arbitrary step δ_0 and a map β, coincides with the optimal strategy for Player E in the game $\Gamma_{\delta_0}(x, y, \ell)$.

From the optimality of the strategy $v_\epsilon^*(\cdot)$ in the game $\Gamma_{\delta_3}(x, y, \ell')$. for any strategy $u(\cdot)$ of Player P follows

$$\Theta(x, y, \ell) \leq t_\Pi^\ell(x, y, v_\epsilon^*(\cdot), u(\cdot)). \tag{3.4.36}$$

For $u(\cdot) = u^*(\cdot)$ (3.4.36) takes the form

$$\Theta(x, y, \ell) \leq t_\Pi^\ell(x, y, u_\epsilon^*(\cdot), v_\epsilon^*(\cdot)). \tag{3.4.37}$$

The fact that $u_\epsilon^*(\cdot)$ is an optimal strategy in the game $\Gamma_{\delta_3}(x(\delta_3), y, \ell')$ implies

$$\Theta(x(\delta_3), y, \ell') \geq t_{\delta_3}^{\ell_1}(x, y, u_\epsilon^*(\cdot), v(\cdot)) - \delta_2, \tag{3.4.38}$$

where $t_{\delta_3}^{\ell_1}(x, y, u_\epsilon^*(\cdot), v(\cdot)) = \tau$ is the approaching time for the points $x(t)$, $y(t - \delta_3)$ on a distance ℓ' from the initial positions as x and y in the situation $(u_\epsilon^*(\cdot), v(\cdot))$

Next

$$\rho(x(\tau), y(\tau)) \leq \rho(x(\tau), y(\tau - \delta_3)) + \rho(y(\tau - \delta_3), y(\tau)).$$

However, $\delta_3 \leq \delta_2$ and, based on the choice of δ_2 we get

$$\rho(y(\tau - \delta_3), y(\tau)) \leq \delta_1.$$

Thus,

$$\rho(x(\tau), y(\tau)) \leq \ell' + \delta_1 = \ell. \tag{3.4.39}$$

implies

$$t_\Pi^\ell(x, y; u_\epsilon^*(\cdot), v(\cdot)) \leq t_{\delta_3}^{\ell'}(x, y; u_\epsilon^*(\cdot), v(\cdot)). \tag{3.4.40}$$

From (3.4.32) we have

$$\Theta(x(\delta_3), y, \ell') \leq \Theta(x, y, \ell) + \frac{\epsilon}{2}. \tag{3.4.41}$$

From (3.4.41), (3.4.38) and (3.4.40), and taking in account that $\delta_2 \leq \frac{\epsilon}{2}$, we obtain

$$\Theta(x, y, \ell) \geq t_\Pi^\ell(x, y, u_\epsilon^*(\cdot), v(\cdot)) - \epsilon. \tag{3.4.42}$$

The inequality (3.4.42) holds for any Player E's strategy in the game $\Gamma(x, y, \ell)$. Substituting $v(\cdot) = v_\epsilon^*(\cdot)$ in (3.4.42) we get

$$t_\Pi^\ell(x, y, u_\epsilon^*(\cdot), v_\epsilon^*(\cdot)) - \epsilon \leq \Theta(x, y, \ell). \tag{3.4.43}$$

Using (3.4.43), (3.4.37), (3.4.36) and (3.4.42), we find

$$t_\Pi^\ell(x, y, u_\epsilon^*(\cdot), v(\cdot)) - \epsilon \leq \Theta(x, y, \ell) \leq t_\Pi^\ell(x, y, u_\epsilon^*(\cdot), v_\epsilon^*(\cdot)) \leq$$

$$\leq \Theta(x, y, \ell) + \epsilon \leq t_\Pi^\ell(x, y, u(\cdot), v_\epsilon^*(\cdot)) + \epsilon. \tag{3.4.44}$$

Thus, the situation $(u_\epsilon^*(\cdot), v_\epsilon^*(\cdot))$ is the ϵ–equilibrium point in the game $\Gamma(x,y,\ell)$, and $\Theta(x, y, \ell)$ is the value of the game. ■

Remark. We may abandom the assumption about the existence of

$$\max_{v(t)} \min_{u(t)} t_\Pi^\ell(x, y, u(t), v(t)) \text{ and } \min_{x' \in C_P^\delta(x)} \Theta(x', y, \ell)$$

introducing the quantities

$$\Theta(x, y, \ell) = \sup_{v(t)} \inf_{u(t)} t_\Pi^\ell(x, y, u(t), v(t)),$$

$$\sup_{x' \in C_P^\delta(x)} \Theta(x', y, \ell).$$

In this case, the theorem similar to Theorem 6 is also valid. Because of cumbersome notations, its precise formulation and proof are omitted here.

3.5 Necessary and sufficient conditions for existance of optimal open–loop strategy for player E

In this section, we restrict ourselves to the pursuit games of prescribed duration, although all of its results are extended to cover the optimal time pursuit games.

Let $\Gamma_\delta(x_0, y_0, t)$ be a discrete game of pursuit with the ($\delta = t_{k+1} - t_k$) with prescribed duration T and with discrimination of Player E from the initial states x_0, y_0 Then the following theorem holds

Theorem 7

$$\hat{\rho}_T(x_0, y_0) = Val\Gamma_\delta(x_0, y_0, T), \tag{3.5.1}$$

if and only if for all x_0, $y_0 \in R^n$ *and* $T = \delta k$, $k = 1, 2, \ldots$

$$\hat{\rho}_T(x_0, y_0) = \max_{y \in C_E^\delta(y_0)} \min_{x \in C_P^\delta(x_0)} \hat{\rho}_{T-\delta}(x, y) \tag{3.5.2}$$

($Val\Gamma_\delta(x_0, y_0, T)$ *is the value of the game* $\Gamma_\delta(x_0, y_0, T)$*).*

The proof of the theorem is prefaced with the following lemma.

Lemma 3 *For any* x_0, $y_0 \in R^n$ *and* $T \geq \delta$, *the following inequality is satisfied*

$$\hat{\rho}_T(x_0, y_0) \leq \max_{y \in C_E^\delta(y_0)} \min_{x \in C_P^\delta(x_0)} \hat{\rho}_{T-\delta}(x, y)$$

Proof: By the definition of the function $\hat{\rho}_T(x, y)$, we have

$$\max_{y \in C_E^\delta(y_0)} \min_{x \in C_P^\delta(x_0)} \hat{\rho}_{T-\delta}(x, y) =$$

$$= \max_{y \in C_E^\delta(y_0)} \min_{x \in C_P^\delta(x_0)} \max_{\bar{y} \in C_E^{T-\delta}(y)} \min_{\bar{x} \in C_P^{T-\delta}(x)} \rho(\bar{x}, \bar{y}).$$

For all $\xi \in C_P^\delta(x_0)$ we have the inclusion $C_P^{T-\delta}(x) \subset C_P^T(x_0)$. Consequently, for any $x \in C_P^\delta(x_0)$, $\bar{y} \in C_E^{T-\delta}(y)$

$$\min_{\bar{x} \in C_P^{T-\delta}(x)} \rho(\bar{x}, \bar{y}) \geq \min_{\bar{x} \in C_P^T(x_0)} \rho(\bar{x}, \bar{y}).$$

Then for any $x \in C_P^\delta(x_0)$

$$\max_{\bar{y} \in C_E^{T-\delta}(y)} \min_{\bar{x} \in C_P^{T-\delta}(x)} \rho(\bar{x}, \bar{y}) \geq \max_{\bar{y} \in C_P^{T-\delta}(y_0)} \min_{\bar{x} \in C_P^T(x_0)} \rho(\bar{x}, \bar{y}).$$

Thus

$$\max_{\bar{y}\in C_E^\delta(y_0)} \min_{\bar{x}\in C_P^\delta(x_0)} \hat{\rho}_{T-\delta}(\bar{x},\bar{y}) \geq$$

$$\geq \max_{y\in C_E^\delta(y_0)} \max_{\bar{y}\in C_E^{T-\delta}(y)} \min_{\bar{x}\in C_P^T(x_0)} \rho(\bar{x},\bar{y}) =$$

$$\max_{y\in C_E^T(y_0)} \min_{x\in C_P^T(x_0)} \rho(x,y) = \hat{\rho}_T(x_0,y_0).$$

This completes the proof of the lemma. ■

We now prove the theorem.

Proof: *Necessity.* Suppose the (3.5.1) is satisfied but the (3.5.2) not. Then, by the previous lemma, there exist such x_0, $y_0 \in R^n$, $T_0 = \delta k_0$, $k_0 \geq 1$ that

$$\hat{\rho}_{T_0}(x_0,y_0) < \max_{y\in C_E^\delta(y_0)} \min_{x\in C_P^\delta(x_0)} \hat{\rho}_{T_0-\delta}(x,y). \tag{3.5.3}$$

Denote $x^0(t) = x(t;x_0,u^0(\cdot))$, where $u^0(\cdot)$ is an optimal strategy of Player P in the game $\Gamma_\delta(x_0,y_0,T_0)$. Evidently, there is such point $y^* \in C_E^\delta(y_0)$ for which

$$\min_{x\in C_P^\delta(x_0)} \hat{\rho}_{T_0-\delta}(x,y^*) = \max_{y\in C_E^\delta(y_0)} \min_{x\in C_P^\delta(x_0)} \hat{\rho}_{T_0-\delta}(x,y). \tag{3.5.4}$$

Let $\bar{u}^0(\cdot)$, $\bar{v}^0(\cdot)$ be optimal strategies in the game $\Gamma_\delta(x^0(\delta),y^*,T-\delta)$. We focus on the following strategy $\tilde{v}(\cdot)$ for Player E; at the instant $t=0$ he chooses the function $v \in L([0,\delta],V)$ such that the motion generated by this function transfers Player E from the state y_0 to the state y^* in the time equal to δ, i.e. $y^* = y(\delta;y_0,v)$ and starting from the instant $t=\delta$ Player E plays the strategy $\bar{v}^0(\cdot)$.

Denote by $\hat{u}^0(\cdot)$ a restriction of the strategy $u^0(\cdot)$ to the interval $[\delta,T_0]$. From (3.5.1), (3.5.3), (3.5.4) we find

$$\hat{\rho}_{T_0}(x_0,y_0) \geq K(\bar{u}^0(\cdot),\hat{u}^0(\cdot);x_0,y_0,T_0) =$$

$$= K(\hat{u}^0(\cdot),\bar{v}^0(\cdot);x_0(\delta),y^*,T_0-\delta) \geq \hat{\rho}_{T_0-\delta}(x_0(\delta),y^*) \geq$$

$$\geq \min_{x\in C_P^\delta(x_0)} \hat{\rho}_{T_0-\delta}(x,y^*) = \max_{y\in C_E^\delta(y_0)} \min_{x\in C_P^\delta(x_0)} \hat{\rho}_{T_0}(x_0,y_0) > \hat{\rho}_{T_0}(x_0,y_0).$$

The contradiction obtained proves the necessity of the condition (3.5.2).

Sufficiency. Note that condition (3.5.3) is an equation for the value function of the discrete finitestep game $\Gamma_\delta(x_0,y_0,T)$ (see Theorem 7, Ch. 1, and Theorem 2, Ch. 2). The function $\hat{\rho}_T(x,y)$ satisfies the Eq.(3.5.2); hence it is the value of the game $\Gamma_\delta(x,y,T)$, This proves the theorem. ■

Lemma 4 *If an optimal open–loop strategy (i.e. the strategy which is the function of the time only) for Player E is to exist in the game* $\Gamma(x_0, y_0, T)$, *it is necesary and sufficient that*

$$Val\Gamma(x_0, y_0, T) = \hat{\rho}_T(x_0, y_0).$$

The proof follows from the definition of $\hat{\rho}_T(x_0, y_0)$.

The game $\Gamma(x_0, y_0, T)$ is approximated by the discrete games $\Gamma_\delta(x_0, y_0, T)$.

Theorem 8 *If, for any* x_0, $y_0 \in R^n$, $T > 0$ *in the game* $\Gamma(x_0, y_0, T)$ *Player E is to have an optimal open–loop strategy, it is necessary and sufficient that for any* $\delta > 0$ *and any* $x_0 \in R^n$, $y_0 \in R^n$, $T \geq \delta$

$$\hat{\rho}_T(x_0, y_0) = \max_{y \in C_E^\delta(y_0)} \min_{x \in C_P^\delta(x_0)} \hat{\rho}_{T-\delta}(x, y) \tag{3.5.5}$$

Proof: *Sufficiency.* We show that the condition (3.5.5) implies the equality

$$\hat{\rho}_T(x_0, y_0) = Val\Gamma(x_0, y_0, T) \tag{3.5.6}$$

Evidently, Player E has an open–loop strategy which ensures the payoff the least $\hat{\rho}_T(x_0, y_0)$ since for any $\epsilon > 0$ and any strategy $u(\cdot)$ of Player P the payoff $K(u(\cdot), v^*(t); x_0, y_0, T) \geq \rho_T(x_0, y_0)$, where $v^*(t)$ is open–loop control transferring Player E from the state y_0 to the state $y^* \in C_E^T(y_0)$ (y^* is the center of pursuit) such that

$$\min_{x \in C_P^T(x_0)} \rho(x, y^*) = \max_{y \in C_E^T(y_0)} \min_{x \in C_P^T(x_0)} \rho(x, y) = \hat{\rho}_T(x_0, y_0)$$

Prove that for any $\epsilon > 0$ Player P has a strategy which ensures that he will loose not more than $\hat{\rho}_T(x_0, y_0) + \epsilon$. Fix $\epsilon > 0$. Consider the following strategy $u^\delta(\cdot)$ for Player P: Player P chooses $\delta > 0$ and an arbitrary open–loop control $u_1(t)$, $t \in [0, \delta)$ and, starting from the instant δ, he pursues the point $y(t - \delta)$ in the game $\Gamma_\delta(x(\delta), y_0, T - \delta)$ where $y(t)$ is Player E's motion.

Because of the continuous dependence of the sets $C_P^T(x_0)$, $C_E^T(y_0)$ on x_0, y_0, T, we have that there is $\delta > 0$ such that

$$|\hat{\rho}_{T-\delta_1}(x(\delta_1), y_0) - \hat{\rho}_T(x_0, y_0)| \leq \frac{\epsilon}{2},$$

and there is $\delta_2 > 0$ such that for all $y \in C_E^{T-\delta_2}(y_0)$ and $y' \in C_E^{\delta_2}(y)$ the distance $\rho(y, y') \leq \frac{\epsilon}{2}$. Set $\delta_0 = \min\{\delta_1, \delta_2\}$. Since the strategy $u^{\delta_0}(\cdot)$ is optimal in the game $\Gamma_{\delta_0}(x(\delta_0), y_0, T - \delta_0)$ (see Theorem 7), then it ensures, in this game, an approach to Player E ($y(t - \delta_0)$) on a distance not more than $\hat{\rho}_{T-\delta_1}(x(\delta_1), y_0)$. This implies an approach to Player E on a distance not more than $\hat{\rho}_{T-\delta_0}(x(\delta_0), y_0) +$

$\frac{\epsilon}{2}$, since it follows from the choice of δ_0 that in a time δ_0 he cannot cover a distance more than $\frac{\epsilon}{2}$. However, $|\hat{\rho}_{T-\delta_0}(x(\delta_0), y_0) - \hat{\rho}_T(x_0, y_0)| < \frac{\epsilon}{2}$. Therefore, Player P ensures an approach to Player E on a distance not more than $\hat{\rho}_T(x_0, y_0) + \epsilon$. This completes the proof of the sufficiency.

Necessity. Assume that Player E has an optimal open–loop strategy. Also, suppose that $\delta_0 > 0$ and there are x_0, $y_0 \in R^n$ and T_0 such that

$$\hat{\rho}_{T_0}(x_0, y_0) < \max_{y \in C_E^{\delta_0}(y_0)} \min_{x \in C_P^{\delta_0}(x_0)} \hat{\rho}_{T_0-\delta_0}(x, y) \tag{3.5.7}$$

Because of lemma 3 inequality of opposite sign is not possible. Then Player E has the strategy $v^*(\cdot)$: at the instant $t = 0$ he chooses the open–loop control $v(t)$, $t \in [0, \delta_0]$ transferring y_0 to y^*, where y^* is such that

$$\min_{x \in C_P^{\delta_0}(x_0)} \hat{\rho}_{T_0-\delta_0}(x, y^*) = \max_{y \in C_E^{\delta_0}(y_0)} \min_{x \in C_P^{\delta_0}(x_0)} \hat{\rho}_{T_0-\delta_0}(x, y),$$

and from the instant $t = \delta$ he plays the optimal open–loop strategy $v^*(t)$ in the game $\Gamma(x, y^*, T_0 - \delta_0)$, where x is the state in which Player P appears at the instant $t = \delta_0$. Hence for any strategy $u(\cdot) \in P$

$$\begin{aligned} K(u(\cdot), v^*(\cdot); x_0, y_0, T_0) &= \\ = K(\bar{u}(\cdot), v^*(t); x(\delta_0; x_0, u(\cdot)), y^*, T_0 - \delta_0) &\geq \\ \geq \min_{x \in C_P^{\delta_0}(x_0)} \hat{\rho}_{T_0-\delta_0}(x, y^*) &> \hat{\rho}_{T_0}(x_0, y_0), \end{aligned} \tag{3.5.8}$$

where $\bar{u}(\cdot)$ is a restriction of the strategy $u(\cdot)$ to the interval $[\delta_0, T_0]$. By Lemma 4, the inequality contradicts the optimality of open–loop strategy. This completes the proof of the theorem. ■

By employing Theorem 7, we derive a partial differential equation for the value function of the game. The conditions of Theorem 3 are assumed to be satisfied for the game $\Gamma(x, y, T)$. Then the function $\hat{\rho}_T(x, y)$ is the value of the game $\Gamma(x, y, T)$ with duration T from the initial conditions x, y.

Assume that in some region of the space $R^n \times R^n \times [0, \infty)$ the function $\hat{\rho}_T(x, y)$ has continuous partial derivatives in all its variables. Since $\hat{\rho}_T(x, y)$ is independent of ξ, η then Eq.(3.5.5) may be rewritten as

$$\max_{\xi \in C_E^{\delta}(y)} \min_{\eta \in C_P^{\delta}(x)} [\hat{\rho}_T(x, y) - \hat{\rho}_{T-\delta}(\xi, \eta)] = 0. \tag{3.5.9}$$

Let $\xi \in C_E^{\delta}(y)$, $\eta \in C_P^{\delta}(x)$. Then there is the pair of controls $u(t)$, $v(t)$ which transfers the points y and x, respectively, to the states ξ and η in the time δ. Let $x(t)$, $y(t)$ be relevant trajectories. In this case

$$\xi = \int_0^{\delta} f(x(t), u(t))dt + x,$$

$$\eta = \int_0^\delta g(y(t), v(t))dt + y. \tag{3.5.10}$$

Dividing both sides of the Eq.(3.5.9) by $\delta > 0$, using Eq.(3.5.10) and letting $\delta \to 0$, we obtain

$$\left[\frac{\partial \hat{\rho}}{\partial T} - \max_{v \in V} \sum_{i=1}^{n} \frac{\partial \hat{\rho}}{\partial y_i} g_i(y, v) - \min_{u \in U} \sum_{i=1}^{n} \frac{\partial \hat{\rho}}{\partial x_i} f_i(x, u)\right] = 0, \tag{3.5.11}$$

or

$$\frac{\partial \hat{\rho}}{\partial T} = \max_{v \in V} \sum_{i=1}^{n} \frac{\partial \hat{\rho}}{\partial y_i} g_i(y, v) + \min_{u \in U} \sum_{i=1}^{n} \frac{\partial \hat{\rho}}{\partial x_i} f_i(x, u), \tag{3.5.12}$$

under the initial condition $\hat{\rho}_T(x, y)|_{T=0} = \rho(x, y)$.

Assume that we have somehow succeeded in defining $\bar{u}$, $\bar{v}$, which gives max and min in Eq.(3.5.12), as functions of x, y, T and $\frac{\partial \hat{\rho}}{\partial x}, \frac{\partial \hat{\rho}}{\partial y}$, i.e.

$$\bar{u} = u\left(x, y, \frac{\partial \hat{\rho}}{\partial x}\right), \ \bar{v} = \bar{v}\left(x, y, \frac{\partial \hat{\rho}}{\partial y}\right). \tag{3.5.13}$$

Substituting the (3.5.13) in Eq.(3.5.12), we get

$$\sum_{i=1}^{n} \frac{\partial \hat{\rho}}{\partial y_i} g_i\left(y, \bar{v}\left(x, y, \frac{\partial \hat{\rho}}{\partial y}\right)\right) + \sum_{i=1}^{n} \frac{\partial \hat{\rho}}{\partial x_i} f_i\left(y, \bar{u}\left(x, y, \frac{\partial \hat{\rho}}{\partial x}\right)\right) = \frac{\partial \hat{\rho}}{\partial T} \tag{3.5.14}$$

provided that

$$\hat{\rho}(x, y, T)|_{T=0} = \rho(x, y) \tag{3.5.15}$$

Thus, to define $\hat{\rho}_T(x, y)$, we have the Cauchy problem for the partial differential equation of the first order Eq.(3.5.14) with the initial condition (3.5.15). The Eq.(3.5.14) has been derived independently by various authors (see [1, 33, 65]) and is called the Isaacs equation.

Assume that partial derivatives of the second order of the function $\hat{\rho}$ exist then solution to the Cauchy problem for Eq.(3.5.14) can be obtained by the method of characteristics that are of the form

$$\frac{dx_i}{dt} = f_i\left(x, \bar{u}\left(x, y, \hat{\rho}_x\right)\right), \ i = 1, \ldots, n,$$

$$\frac{d\hat{\rho}_{x_i}}{dt} = -\sum_{j=1}^{n} \frac{\partial f_i(x, \bar{u}(x, y, \hat{\rho}_x))}{\partial x_j} \hat{\rho}_{x_j},$$

$$\frac{dy_i}{dt} = g_i\left(y, \bar{v}\left(x, y, \hat{\rho}_y\right)\right), \ i = 1, \ldots, n, \tag{3.5.16}$$

$$\frac{d\hat{\rho}_{y_i}}{dt} = -\sum_{j=1}^{n} \frac{\partial g_i(y, \bar{v}(x, y, \hat{\rho}_y))}{\partial y_j} \hat{\rho}_{y_j},$$

$$\left(\frac{d\hat{\rho}}{dt}\right)_T = 0.$$

Here dx_j/dt, $d\hat{\rho}_{x_j}/dt$, dy_j/dt, $d\hat{\rho}_{y_j}/dt$ are the time derivatives along the optimal trajectory. The derivation of Eq.(3.5.16) may be found in [1], where the equations of characteristics are shown to determine optimal traectories for players. The initial conditions for integration of equations in (3.5.16) are obtained from (3.5.15).

For the optimal time pursuit game the equation (3.5.15) takes the form

$$\max_{v\in V}\sum_{i=0}^{n}\frac{\partial\Theta}{\partial y_i}g_i(y,v) + \min_{u\in U}\sum_{i=0}^{n}\frac{\partial\Theta}{\partial x_i}f_i(x,u) = -1 \qquad (3.5.17)$$

under the initial condition

$$\Theta(x,y,l)|_{\rho(x,y)=\ell} = 0. \qquad (3.5.18)$$

Here, as in the previous case, the continuous partial derivatives of the first order of the function $\Theta(x,y,l)$ in x, y are assumed to exist.

Assuming that it is possible to define $\bar{u}$, $\bar{v}$, which carry over max and min in Eq.(3.5.6) as functions of x, y, $\frac{\partial\Theta}{\partial x}$, $\frac{\partial\Theta}{\partial y}$, i.e.

$$\bar{u} = u\left(x, y, \frac{\partial\Theta}{\partial x}\right),\ \bar{v} = v\left(x, y, \frac{\partial\Theta}{\partial y}\right),$$

we rewrite equation (3.5.17) in the form

$$\sum_{i=0}^{n}\frac{\partial\Theta}{\partial y_i}g_i\left(y, \bar{v}\left(x, y, \frac{\partial\Theta}{\partial y}\right)\right) + \sum_{i=0}^{n}\frac{\partial\Theta}{\partial x_i}f_i\left(x, \bar{u}\left(x, y, \frac{\partial\Theta}{\partial x}\right)\right) = -1 \qquad (3.5.19)$$

provided that

$$\Theta(x,y,l)|_{\rho(x,y)=\ell} = 0. \qquad (3.5.20)$$

The derivation of equation (3.5.17) is omitted here, since it is similar to the derivation of equation (3.5.14) for the pursuit game with prescribed duration.

Both Cauchy problems (3.5.14), (3.5.15) and (3.5.17), (3.5.18) are nonlinear with respect to partial derivatives, since their solution poses serious problems.

The differentiability of functions $\hat{\rho}$ and Θ presents a very fine and complicated problem. N.N. Krasovsky has shown the differentiability of function $\hat{\rho}$ in the so–called regular case [10].

3.6 Iterative methods for solution of differential game of pursuit

Let $\Gamma_\delta(x,y,T)$ be a discrete form of the differential game $\Gamma(x,y,T)$ of duration $T>0$ terminal payoff $M(x(T),y(T))$ with a fixed step of time–interval partition δ, Player E being discriminated for the time $\delta>0$ ahead. Denote by $V_\delta(x,y,T)$ the value of the game $\Gamma_\delta(x,y,T)$[2]. Then

$$\lim_{\delta\to 0} V_\delta(x,y,T)=V(x,y,T)$$

and optimal strtegies in the game $\Gamma_\delta(x,y,T)$ with sufficiently small δ can be effectively employed to construct ϵ–equilibrium points in the game $\Gamma_\delta(x,y,T)$. We now expound the method.

Zero–order approximation. As a zero–order approximation of the value function of the game $V_\delta(x,y,T)$, we take the function

$$V_\delta^0(x,y,T)=\max_{\eta\in C_E^T(y)}\min_{\xi C_P^T(x)} M(\xi,\eta), \tag{3.6.1}$$

where $C_P^T(x)$, $C_E^T(y)$ are the reachability sets for the players P and E from the initial states x, $y\in R^n$ by the time instant T.

The choice of the function $V_\delta^0(x,y,T)$ as an initial approximation is justified by the fact that in a sufficient broad class of games (a "regular case") it proves to be the value of the game $\Gamma(x,y,T)$ (see in 2, 5 in this chapter).

The following approximation is constructed by this rule

$$V_\delta^1(x,y,T)=\max_{\eta\in C_E^\delta(y)}\min_{\xi\in C_P^\delta(x)} V_\delta^0(\xi,\eta,T-\delta),$$

$$V_\delta^2(x,y,T)=\max_{\eta\in C_E^\delta(y)}\min_{\xi\in C_P^\delta(x)} V_\delta^1(\xi,\eta,T-\delta), \tag{3.6.2}$$

$$\dots$$

$$V_\delta^k(x,y,T)=\max_{\eta\in C_E^\delta(y)}\min_{\xi\in C_P^\delta(x)} V_\delta^{k-1}(\xi,\eta,T-\delta)$$

with $T>\delta$ and $V_\delta^k(x,y,T)=V_\delta^0(x,y,T)$, $k=1,2,\dots$, with $T\le\delta$.

As we may see from the formulas (3.6.2), the operation max min is taken with respect to the reachability sets $C_E^\delta(y)$, $C_P^\delta(x)$ for the time δ, i.e. for one step of the discrete game $\Gamma(x,y,T)$.

Theorem 9 *The sequence of function $\{V_\delta^k(x,y,t)\}$ is monotone nondecreasing.*

---

[2]The terminal payoff is equal to $M(x(T),y(T))$ where $M(x,y)$ is a function continuous on $R^n\times R^n$

Proof: We first prove the inequality[3]

$$V_\delta^1(x,y,T) \geq V_\delta^0(x,y,T)$$

By the definition of the functions $V_\delta^1(x,y,T)$ and $V_\delta^0(x,y,T)$, we have

$$V_\delta^1(x,y,T) = \max_{\eta \in C_E^\delta(y)} \min_{\xi \in C_P^\delta(x)} V_\delta^0(\xi,\eta,T-\delta) =$$

$$= \max_{\eta \in C_E^\delta(y)} \min_{\xi \in C_P^\delta(x)} \left[\max_{\bar\eta \in C_E^{T-\delta}(\eta)} \min_{\bar\xi \in C_P^{T-\delta}(\xi)} M(\bar\xi,\bar\eta) \right].$$

For all $\xi \in C_P^\delta(x)$ there is $C_P^{T-\delta}(\xi) \subset C_P^T(x)$. Consequently, for any $\bar\eta \in C_E^{T-\delta}(\eta)$, $\xi \in C_P^\delta(x)$

$$\min_{\bar\xi \in C_P^{T-\delta}(\xi)} M(\bar\xi,\bar\eta) \geq \min_{\bar\xi \in C_P^T(x)} M(\bar\xi,\bar\eta)$$

Hence we have that for any $\xi \in C_P^\delta(x)$

$$\max_{\bar\eta \in C_E^{T-\delta}(\eta)} \min_{\bar\xi \in C_E^{T-\delta}(\xi)} M(\bar\xi,\bar\eta) \geq \max_{\bar\eta \in C_E^{T-\delta}(\eta)} \min_{\bar\xi \in C_P^T(x)} M(\bar\xi,\bar\eta).$$

Therefore, in particular,

$$\min_{\xi \in C_P^\delta(x)} \left[\max_{\bar\eta \in C_E^{T-\delta}(\eta)} \min_{\bar\xi \in C_P^{T-\delta}(\xi)} M(\bar\xi,\bar\eta) \right] \geq$$

$$\geq \max_{\bar\eta \in C_E^{T-\delta}(\eta)} \min_{\bar\xi \in C_P^T(x)} M(\bar{x}i,\bar\eta).$$

Thus,

$$V_\delta^1(x,y,T) = \max_{\bar\eta \in C_E^\delta(\eta)} \min_{\bar\xi \in C_P^\delta(x)} V_\delta^0(\bar\xi,\bar\eta,T-\delta) \geq$$

$$\geq \max_{\eta \in C_E^\delta(y)} \left[\max_{\bar\eta \in C_E^{T-\delta}(\eta)} \min_{\bar\xi \in C_P^T(x)} M(\bar\xi,\bar\eta) \right] =$$

$$= \max_{\eta \in C_E^T(y)} \min_{\xi \in C_P^T(x)} M(\xi,\eta) = V_\delta^0(x,y,T),$$

i.e.

$$V_\delta^1(x,y,T) \geq V_\delta^0(x,y,T)$$

Assume that for $l \leq k$ the following inequality holds

$$V_\delta^\ell(x,y,T) \geq V_\delta^{\ell-1}(x,y,T). \tag{3.6.3}$$

---

[3]The proof of this inequality is given here only for the purposes of completeness, since it reproduces the one for Lemma 3.

We prove that

$$V_\delta^{k+1}(x,y,T) \geq V_\delta^k(x,y,T). \tag{3.6.4}$$

By definition,

$$V_\delta^{k+1}(x,y,T) = \max_{\eta \in C_E^\delta(y)} \min_{\xi \in C_P^\delta(x)} V_\delta^k(\xi,\eta,T-\delta),$$

$$V_\delta^k(x,y,T) = \max_{\eta \in C_E^\delta(y)} \min_{\xi \in C_P^\delta(x)} V_\delta^{k-1}(\xi,\eta,T-\delta). \tag{3.6.5}$$

From (3.6.3), and using (3.6.5), we obtain the required inequality (3.6.4). We have thus proved that the statement of the theorem holds at each point (x,y,T) in the case $T > \delta$. In the case $T \leq \delta$, however, the statement of the theorem is obviously seen. This complete the proof of theorem. ■

Theorem 10 *The sequence* $\{V_\delta^k(x,y,T)\}$ *converges in a finite number of steps* N, *with the estimate* $N \leq \left[\frac{T}{\delta}\right] + 1$, *where the brackets stand for the greatest integer.*

Proof: Let $N \leq \left[\frac{T}{\delta}\right] + 1$. Show that

$$V_\delta^N(x,y,T) = V_\delta^{N+1}(x,y,T). \tag{3.6.6}$$

The expression (3.6.6) is obtained from the construction of the sequence $\{V_0^k(x,y,T)\}$. Indeed,

$$V_\delta^N(x,y,T) = \max_{\eta^1 \in C_E^\delta(y)} \min_{\xi^1 \in C_P^\delta(x)} V_\delta^{N-1}(\xi^1,\eta^1,T-\delta) =$$

$$= \max_{\eta^1 \in C_E^\delta(y)} \min_{\xi^1 \in C_P^\delta(x)} \max_{\eta^2 \in C_E^\delta(\eta^1)} \cdots$$

$$\cdots \max_{\eta^{N-1} \in C_E^\delta(\eta^{N-2})} \min_{\xi^{N-1} \in C_P^\delta(\xi^{N-2})} V_\delta^1(\xi^{N-1},\eta^{N-1},T-(N-1)\delta).$$

Similarly, we get

$$V_\delta^{N+1}(x,y,T) = \max_{\eta^1 \in C_E^\delta(y)} \min_{\xi^1 \in C_P^\delta(x)} \max_{\eta^2 \in C_E^\delta(\eta^1)} \cdots$$

$$\cdots \max_{\eta^{N-1} \in C_E^\delta(\eta^{N-2})} \min_{\xi^{N-1} \in C_P^\delta(\xi^{N-2})} V_\delta^2(\xi^{N-1},\eta^{N-1},T-(N-1)\delta).$$

However, $T-(N-1)\delta = \alpha < \delta$, therefore

$$V_\delta^1(\xi^{N-1},\eta^{N-1},T-(N-1)\delta) =$$

$$= V_\delta^2(\xi^{N-1},\eta^{N-1},T-(N-1)\delta) = V_\delta^0(\xi^{N-1},\eta^{N-1},\alpha),$$

whence follows (3.6.6).

The coincidence of members of the sequence Eq. (3.6.2) for $k \geq N$ is obtained from Eq.(3.6.6) by induction. This proves the theorem. ■

Theorem 11 *The limit of the sequence* $\{V_\delta^K(x,y,T)\}$ *coincides with the value of the game* $\Gamma_\delta(x,y,T)$.

Proof: Basically, Theorem 11 is the corollary of Theorem 10. Indeed, let

$$V_\delta(x,y,T) = \lim_{k\to\infty} V_\delta^k(x,y,T). \tag{3.6.7}$$

Since the convergence takes place in a finite number of steps not exceeding $N = \left[\frac{T}{\delta}\right] + 1$, then in the recurrent Eq.(3.6.2) it is possible to convert to the limit as $k \to \infty$. The limiting function $V_\delta(x,y,T)$ satisfies the equation

$$V_\delta(x,y,T) = \max_{\eta\in C_E^\delta(y)} \min_{\xi\in C_P^\delta(x)} V_\delta(\xi,\eta,T-\delta) \tag{3.6.8}$$

under the initial condition

$$V_\delta(x,y,T)\,|_{0\le T\le\delta} = \max_{\eta\in C_E^\delta(y)} \min_{\xi\in C_P^T(x)} M(\xi,\eta),$$

which is a sufficient condition for the function $V_\delta(x,y,T)$ to be the value of the game $\Gamma(x,y,T)$.

With knowledge of the function $V_\delta(x,y,T)$, and using the Eq.(3.6.8), we can construct optimal piecewise open–loop strategies in the game $\Gamma_\delta(x,y,T)$. The ϵ–optimal strategies in the basic game $\Gamma(x,y,T)$ are constructed with the help of strategies that are optimal in the game $\Gamma_\delta(x,y,T)$ (see 2, 5 in this chapter).

As is clear from (3.6.2), the coincidence of two successive approximation at the steps k and $k+1$ means that the corresponding approximation is the value of the game $\Gamma_\delta(x,y,T)$, since in this case all subsequent approximations coincide with the k–th approximation. Such a coincidence is a criterion for termination of computations. We have every reason to say that in the broad class of problems the convergence occurs faster than in the time mentioned in Theorem 11.In particular, the computation in the "regular case" terminates at the first step after the function $V_\delta^1(x,y,T)$ has been computed (at the same time, this provides a "regularity" criterion).

We will modify the above method of successive approximations.

As an initial approximation, we take the function $\tilde V_\delta^0(x,y,T) = V_\delta^0(x,y,T)$, where $V_\delta^0(x,y,T)$ is defined by (3.6.1). The following approximation is constructed by the rule

$$\tilde V_\delta^{k+1}(x,y,T) = \max_{i\in[1:N]} \max_{\eta\in C_E^{i\delta}(y)} \min_{\xi\in C_P^{i\delta}(x)} \tilde V_\delta^k(\xi,\eta,T-i\delta) \tag{3.6.9}$$

for $T > \delta$, where $N = \frac{T}{\delta}$, and $\tilde V_\delta^{k+1}(x,y,T) = V_\delta^0(x,y,T)$ for $T \le \delta$.

The statements of the theorems 9–11 holds for the sequence of functions $\{\tilde{V}_\delta^k(x,y,T)\}$, as well as for sequence of functions $\{V_\delta^k(x,y,T)\}$. These statements for the sequence of functions $\{\tilde{V}_\delta^k(x,y,T)\}$ are proved in much the same way as similar reasonings for the sequence of functions $\{V_\delta^k(x,y,T)\}$. The functional equation for the value function of the game $\Gamma_\delta(x,y,T)$ in the region $\{(x,y,T)|T > \delta\}$ is of the form

$$V(x,y,T) = \max_{i\in[1:N]} \max_{\eta\in C_E^{i\delta}(y)} \min_{\xi\in C_P^{i\delta}(x)} V(\xi,\eta,T-i\delta), \tag{3.6.10}$$

where $N = \frac{T}{\delta}$, and the initial condition remains the same,i.e.

$$V(x,y,T)\,|_{T\leq\delta} = \max_{\eta\in C_E^T(y)} \min_{\xi\in C_P^T(x)} M(\xi,\eta)$$

We prove the equivalence of Eq.(3.6.8) and Eq.(3.6.10). ∎

Theorem 12 *Eq.(3.6.8) and Eq.(3.6.10) with the initial condition*

$$V(x,y,T)\,|_{T\leq\delta} = \min_{\eta\in C_E^T(y)} \min_{\xi\in C_P^T(x)} M(\xi,\eta) \tag{3.6.11}$$

are equivalent.

Proof: Suppose the function $V_\delta(x,y,T)$ satisfies Eq. (3.6.8) and initial condition (3.6.11). We show that it satisfies Eq. (3.6.10) in the region $\{(x,y,T)|T > \delta\}$

Indeed, since $V_\delta(x,y,T)$ satisfies (3.6.8), then the relationships

$$V_\delta(x,y,T) = \max_{\eta\in C_E^\delta(y)} \min_{xi\in C_P^\delta(x)} V_\delta(\xi,\eta,T-\delta) =$$

$$= \max_{\eta\in C_E^\delta(y)} \min_{\xi\in C_P^\delta(x)} \max_{\bar\eta\in C_E^\delta(\eta)} \min_{\bar\xi\in C_P^\delta(\xi)} V_\delta(\bar\xi,\bar\eta,T-2\delta) \geq$$

$$\geq \max_{\eta\in C_E^\delta(y)} \max_{\bar\eta\in C_E^\delta(\eta)} \min_{\xi\in C_P^\delta(x)} \min_{\bar\xi\in C_P^\delta(\xi)} V_\delta(\bar\xi,\bar\eta,T-2\delta) =$$

$$= \max_{\eta\in C_E^{2\delta}(y)} \min_{\xi\in C_P^{2\delta}(x)} V_\delta(\xi,\eta,T-2\delta) \geq \ldots$$

$$\ldots \geq \max_{\eta\in C_E^{i\delta}(y)} \min_{\xi\in C_P^{i\delta}(x)} V_\delta(\xi,\eta,T-i\delta) \geq \ldots$$

holds in the region $\{(x,y,T)|T > \delta\}$. Since the equality

$$V_\delta(x,y,T) = \max_{\eta\in C_E^\delta(y)} \min_{\xi\in C_P^\delta(x)} V(\xi,\eta,T-\delta),$$

is satisfied for $i = 1$, then the equality

$$V_\delta(x, y, T) = \max_{i \in [1:N]} \max_{\eta \in C_E^{i\delta}(y)} \min_{\xi \in C_P^{i\delta}(x)} V(\xi, \eta, T - \delta),$$

where $N = \frac{T}{\delta}$, holds, which proves the required statement.

Now let the function $V_\delta(x, y, T)$ in the region $\{(x, y, T) | T > \delta\}$ satisfy Eq. (3.6.10) and initial condition (3.6.11). We show that it also satisfies Eq.(3.6.8). Suppose the opposite is true. Then from (3.6.10) we find

$$V_\delta(x, y, T) > \max_{\eta \in C_E^\delta(y)} \min_{\xi \in C_P^\delta(x)} V_\delta(\xi, \eta, T - \delta). \tag{3.6.12}$$

In the region $\{(x, y, T) | T > \delta\}$. However,

$$\max_{\eta \in C_E^\delta(y)} \min_{\xi \in C_P^\delta(x)} V_\delta(\xi, \eta, T - \delta) =$$

$$= \max_{\eta \in C_E^\delta(y)} \min_{\xi \in C_P^\delta(x)} \max_{i \in [1:N-1]} \max_{\bar{\eta} \in C_E^{i\delta}(\eta)} \min_{\bar{\xi} \in C_P^\delta(\xi)} V_\delta(\bar{\xi}, \bar{\eta}, T - (i+1)\delta) \geq$$

$$\geq \max_{\eta \in C_E^\delta(y)} \max_{i \in [1:N-1]} \max_{\bar{\eta} \in C_E^{i\delta}(\eta)} \min_{\xi \in C_P^\delta(x)} \min_{\bar{\xi} \in C_P^{i\delta}(\xi)} V_\delta(\bar{\xi}, \bar{\eta}, T - (i+1)\delta) =$$

$$= \max_{i \in [1:N-1]} \max_{\eta \in C_E^{i\delta}(y)} \max_{\bar{\eta} \in C_E^{i\delta}(\eta)} \min_{\xi \in C_P^\delta(x)} \min_{\bar{\xi} \in C_P^{i\delta}(\xi)} V_\delta(\bar{\xi}, \bar{\eta}, T - (i+1)\delta) =$$

$$= \max_{i \in [2:N]} \max_{\eta \in C_E^{i\delta}(y)} \min_{\xi \in C_P^{i\delta}(x)} V_\delta(\xi, \eta, T - i\delta) = V_\delta(x, y, T),$$

since, by (3.6.12), the rigorous inequality holds for $i = 1$. As a result, we get a contradiction which completes the proof of the theorem. ■

Consider now the optimal time pursuit game $\Gamma(x, y)$. As is seen from 4, Ch. 2, the existence of the ϵ, δ–equilibrium points in the game $\Gamma^\ell(x, y)$ for any $\epsilon > 0$, $\delta > 0$ is connected with the solution of the game $\Gamma(x, y, T)$ with prescribed duration T, with the payoff equal to $w = \min_{0 \leq t \leq T} \rho(x(t), y(t))$. In the game $\Gamma(x, y, T)$ with perscribed dupation, by the solution is commonly meant the ϵ–equilibrium point. For any $\epsilon > 0$,the existence of the ϵ–equilibrium point and the existence of the game value for the games with perscribed duration is proved under sufficiently general assumptions in 3, Ch. 2.

Let $V(x, y, T)$ be the value of such a game. Fix the initial state x, $y \in R^n$ and consider the function

$$w(t) = V(x, y, t)$$

where $w(t)$ is a continuous non–increasing function with $t \in [0, \infty)$ (see 4, Ch. 2). Let $\Gamma_\delta(x, y, T)$ be the game with prescribed duration T with the payoff w and the value $V_\delta(x, y, T)$, where Player E is discriminated for the time $\delta > 0$.

Based on 3, Ch. 2,

$$w_\delta(t) = V_\delta(x, y, t) \leq V(x, y, t) = w(t),$$

$$\lim_{\delta\to 0} V_\delta(x, y, t) = V(x, y, t).$$

Denote by $\Gamma_\delta^\ell(x, y)$ (see 4 in this chapter) the optimal time game up to the ℓ–capture instant, where the evader is discriminated for the time δ ahead. Assume that the values of the games $\Gamma_\delta^\ell(x, y)$, and $\Gamma^\ell(x, y)$ exist are finite and are equal, respectively to V_ℓ^δ and $V^\ell(x, y)$. Then the following theorem holds.

Theorem 13 $\lim_{\delta\to 0} V_\delta^\ell(x, y) = V_\delta^\ell(x, y)$ *at each point* $(x, y) \in R^n \times R^n$.

Proof: As is evident from Theorem 3, Ch. 2, $V^\ell(x, y)$ and $V_\delta^\ell(x, y)$ are the minimum roots of equations

$$V(x, y, T) = \ell, \; V_\delta(x, y, T) = \ell. \tag{3.6.13}$$

for fixed x, y

$$w(T) = V(x, y, T) \geq V_\delta(x, y, T) = w_\delta(T) \tag{3.6.14}$$

(3.6.14) and the functions $w(T)$, $w_\delta(T)$ are monotonically nonincreasing. Since $V^\ell(x, y)$ and $V_\delta^\ell(x, y)$ are the minimum roots of (3.6.13), then there are such $\alpha > 0$, $\alpha_1 > 0$ that on the intervals $[V^\ell(x, y) - \alpha, V^\ell(x, y)]$, $[V_\delta^\ell(x, y) - \alpha_1, V_\delta^\ell(x, y)]$ the functions $w(T)$ and $w_\delta(T)$ are strictly descreasing. Then we have

$$\lim_{\delta\to 0} w_\delta(T) = w(T).$$

Therefore, for any $\alpha > 0$ there are such $\delta_1 > 0$ an $T' \in [V^\ell(x, y) - \alpha, V^\ell(x, y)]$ that

$$w_{\delta_1}(T') > w(V^\ell(x, y)) = \ell,$$

and since $w_{\delta_1}(V^\ell(x, y)) \leq w(V^\ell(x, y)) = \ell$, then the minimum root of equation $w_{\delta_1}(T) = \ell$ belongs to the interval $[V^\ell(x, y) - \alpha, V^\ell(x, y)]$. Consequently, $V_{\delta_1}^\ell(x, y)$ also belongs to this interval. Since α may be chosen arbitrary small, and the sequence $\{V_\delta^\ell(x, y)\}$ is monotonically non decreasing in δ, then

$$\lim_{\delta\to 0} V_\delta^\ell(x, y) = V^\ell(x, y).$$

This completes the proof of the theorem. ■

Theorem 13 suggests that the sequence $\{V_\delta^\ell(x, y)\}$ can be used for approximation of $V^\ell(x, y)$.

Fix a particular $\delta > 0$. The computation of $V_\delta^\ell(x, y)$ is a sufficiently tedious operation, since it involves the solution of discrete Isaacs equation. We suggest the recurrent procedure of finding $V_\delta^\ell(x, y)$

Let $t_{\Pi}^{\ell}(x, y; u(t), v(t))$ be the functional, that is defined for any admissable open–loop control $u(t)$ and $v(t)$ and is equal to the time to the ℓ–capture instant, from the initial states x, $y \in R^n$, if the ℓ–capture occurs when the controls $u(t)$ and $v(t)$ are used, otherwise it equals ∞. Next, let

$$\Theta^{\ell}(x, y) = \sup_{v(t)} \inf_{u(t)} t_{\Pi}^{\ell}(x, y; u(t), v(t)).$$

As a zero–order approximation, we take

$$V_{\delta,0}^{\ell}(x, y) = \Theta^{\ell}(x, y).$$

The choice of zero–order approximation is justified by the fact that in some cases this provides the value of the game (see 4 in this chapter). The second approximation is constructed by the rule

$$V_{\delta,1}^{\ell}(x, y) = \sup_{\eta \in C_E^{\delta}(y)} \inf_{\xi \in C_P^{\delta}(x)} V_{\delta,0}^{\ell}(\xi, \eta) + \delta \tag{3.6.15}$$

for $(x, y) \notin L_0$ and $V_{\delta,1}^{\ell}(x, y) = V_{\delta,0}^{\ell}(x, y)$ for $(x, y) \in L_0$, where $L_0 = \{(x, y) \in R^n \times R^n | V_{\delta,0}^{\ell}(x, y) \leq \delta\}$.

As in the proof of a similar statement in 5 of this chapter (see Lemma 3), we may show that for any states x, y, and for any ℓ and $\delta \in (0, \Theta^{\ell}(x, y))$, the following inequality holds

$$\Theta^{\ell}(x, y) \leq \sup_{\eta \in C_E^{\delta}(y)} \inf_{\xi \in C_P^{\delta}(x)} \Theta^{\ell}(\xi, \eta) + \delta.$$

Hence

$$V_{\delta,1}^{\ell}(x, y) \leq V_{\delta,0}^{\ell}(x, y).$$

The approximation $k + 1$ is constructed by the rule

$$V_{\delta,k+1}^{\ell}(x, y) = \sup_{\eta \in C_E^{\delta}(y)} \inf_{\xi \in C_P^{\delta}(x)} V_{\delta,k}^{\ell}(\xi, \pi) + \delta \tag{3.6.16}$$

for $(x, y) \in L_k$ and $V_{\delta,k+1}^{\ell}(x, y) = V_{\delta,k}^{\ell}(x, y)$ for $(x, y) \in L_k$, where $L_k = \{(x, y) \in R^n \times R^n | V_{\delta,k}^{\ell}(x, y) \leq (k + 1)\delta\}$. The construction of the sequence $\{V_{\delta,k}^{\ell}(x, y)\}$ implies $V_{\delta,k+1}^{1}(x, y) \geq V_{\delta,k}^{1}(x, y)$. Evidently, the sequence $\{V_{\delta,k}^{1}(x, y)\}$ for $(x, y) \in L_i$ converges in not more than i steps, and the i–th approximation coincides with the value of the game $\Gamma_{\delta}^{1}(x, y)$.

Indeed, in the passage in Eq.(3.6.14) to the limit as $k \to \infty$ at the point $(x, y) \in L_i$ (the latter is possible because of the convergence of the process in a finite number of steps), we get

$$V_{\delta}^{\ell}(x, y) = \sup_{\eta \in C_E^{\delta}(y)} \inf_{\xi \in C_P^{\delta}(x)} V_{\delta}^{\ell}(\xi, \eta) + \delta, \tag{3.6.17}$$

here for $(x, y) \in L_0$

$$V_\delta^\ell(x, y) = \Theta^\ell(x, y)$$

However, Eq.(3.6.17) with a suitable initial condition on the set L_0 provides a sufficient condition for $V_\delta^\ell(x, y)$ to be the value of the game (x, y). Thus, the following theorem holds.

Theorem 14 *Let $(x, y) \in L_i$ and k be the first step at which $V_{\delta,k+1}^\ell(x, y) = V_{\delta,k}^\ell(x, y)$. Then for all $i \geq k$*

$$V_{\delta,i}^\ell(x, y) = V_{\delta,k}^\ell(x, y) = V_\delta^\ell(x, y),$$

where $V_\delta^1(x, y)$ is the value of the game.

Knowing $V_\delta^\ell(x, y)$ and using Eq.(3.6.17), we may find the ϵ, δ–optimal strategies for the players P and E, as it has been done in 4 of this chapter.

Chapter 4

Examples of differential games of pursuit

4.1 Games with prescribed duration without phase constraints

This chapter studies differential games of pursuit whose solution is based on the results presented in Ch. 3.

Example 1. Let the reachability sets for the Player P and E be such that

$$\max_{\eta \in C_E^T(y)} \min_{\xi \in C_P^T(x)} M(\xi, \eta) = \min_{\xi \in C_P^T(x)} \max_{\eta \in C_E^T(y)} M(\xi, \eta) \tag{4.1.1}$$

Let ξ_1, H be an arbitrary pair of points at which max min is achieved in (4.1.1). Then by Theorem 1, Ch. 1. min max is also achieved in the same points. In the latter case, the point H satisfies the first condition of invariance. Indeed, for all $0 \leq t \leq T$ the point, ξ_1 belongs to the set $C_P^{T-t}(x^*(t))$, and the point H belongs to the set $C_E^{T-t}(y^*(t))$. Moreover, from Theorem 1, Ch. 1, for all $\eta \in C_E^T(y)$, $\xi \in C_P^T(x)$ we have

$$M(\xi_1, \eta) \leq M(\xi_1, H) \leq M(\xi, H).$$

Since for $t \in [0, T]$ $C_E^{T-t}(y^*(t)) \subset C_E^T(y)$, $C_P^{T-t}(x^*(t)) \subset C_P^T(x)$, then the above inequality also holds for all $\eta \in C_E^{T-t}(y^*(t))$, $\xi \in C_P^{T-t}(x^*(t))$. By applying again Theorem 1, Ch. 1, we obtain

$$\max_{\eta \in C_E^{T-t}(y^*(t))} \min_{\xi \in C_P^{T-t}(x^*(t))} M(\xi, \eta) = \min_{\xi \in C_P^{T-t}(x^*(t))} \max_{\eta \in C_E^{T-t}(y^*(t))} M(\xi, \eta) = M(\xi_1, H).$$

The latter relationship shows that the first condition for invariance of the point H is satisfied.

When the (4.1.1) is satisfied, the differential multistage game $\Gamma(x, y, T)$ converts to a simultaneous single–stage game and it follows from (4.1.1) that such a game has an equilibrium point in pure strategies. In this case, the optimal strategy for Player P is the choice of the point ξ_1, and the one for Player E is the choice of the point H. From this it follows that in the corresponding differential game (under condition (4.1.1)) the optimal strategies for the players employ only the information on the players' positions at the initial instant of the game and on the game duration T. The continuous "tracking" of the opponent is unwarranted.

Example 2. (simple pursuit). The pursuit occurs over the entire Euclidean space. Motion equations are of the form

$$\dot{x}_i = \alpha u_i,\ i = 1, \ldots, n,\ \sum_{i=1}^{n} u_i^2 \leq 1,$$

$$\dot{y}_i = \beta v_i,\ i = 1, \ldots, n,\ \sum_{i=1}^{n} v_i^2 \leq 1,$$

where $\alpha = const$, $\beta = const$, $\alpha > \beta$.

The set $C_P^T(x)$ is a sphere with the radius αT and its center at the point x. The set $C_E^T(y)$ is a sphere with the radius βT and its center at the point y. The payoff function is defined as follows:

$$H(x(T), y(T)) = \rho(x(T), y(T)) = \sqrt{\sum_{i=1}^{n}(x_i(T) - y_i(T))^2}.$$

Geometrically, the invariance of the center of pursuit (the center of the game) is straightforward, and Theorem 3, Ch. 3, is applicable here. Indeed (see the Example 1 from 2, Ch. 1),

$$\hat{\rho}_T(x, y) = \rho(x, y) - (\alpha - \beta)T \quad (\rho(x, y) > 0).$$

The quantity $\hat{\rho}_T(x, y)$ is achieved at the points M_1, M (see Fig.5–7, Ch. 1), where M is the intersection point for the line of the centers $xy(OO_1)$, that is farthest from $x(O)$, and the point M is the point of the set $C_P^T(x_1)$ (S_1), that is nearest to the point M. The conditionally optimal trajectories for the Players P and E are the straight line segments xM_1, (OM_1) and yM (O_1M). The first condition of invariance follows from the fact that for any $0 \leq t \leq T$ the point $M(t)$, that is the center of pursuit with respect to the states $x^*(t), y^*(t), T - t$ remains on the center line $xy(OO_1)$, i.e. it coincides with M.

We now assume that at some instant $0 \leq t \leq T$ Player E deviates from the conditionally optimal trajeciory during the time δ, and makes a transition to the point $y' \in C_E^{\delta}(y^*(t))$. Then the center of pursuit with respect to the sets $C_P^{T-t}(x^*(t))$ and $C_E^{T-t-\delta}(y')$, i.e. the center of pursuit M' in the game $\Gamma'_{\delta,\delta}(x^*(t), y', T-t)$, is on the center line $x^*(t)$, y'. The conditionally optimal trajectory $x'^*(t)$ for Player P in this game is a straight line segment of lenght $\alpha(T-t)$ focussed on M'.

Consider the games $\Gamma'_{\delta,\alpha}(x'^*(t+\alpha), y', T-t-\alpha)$ for $0 \leq \alpha \leq \delta$. In these games, the center of pursuit $M'(\alpha)$ is on the center line $x'^*(t+\alpha)$, y', and since the trajectory $x'^*(t)$ is focussed on $M' = M'(\alpha)$, then for all $0 \leq \alpha \leq \delta$ it coincides with $M' = M'(0) = M'(\alpha)$, $0 \leq \alpha \leq \delta$. We have thus proved the second condition of invariance.

Consequently, the value of the game

$$\hat{\rho}_T(x,y) = \rho(x,y) - (\alpha - \beta)T.$$

The optimal strategy for Player E is open–loop and provides an open–loop trnsition to the center of pursuit M. Using the ϵ–optimal strategy for Player P at each instant of time is minimizing the function $\hat{\rho}$ with respect to the realized state of the game. Geometrically, this involves the motion towards the realized point y. In particular, when the equilibrium point is concerned, this leads to the motion along the conditionally optimal straight trajectory focussed on the point M (see Example 10, Ch 2).

If $R_1 = |xM_1| \leq |xM| - R_2 = |xM| - |yM|$, then based on Example from Chapter 1 (see 2, case 2)

$$\max_{\eta \in C_E^T(y)} \min_{\xi \in C_P^T(x)} \rho(\xi,\eta) = \min_{\xi \in C_P^T(x)} \max_{\eta \in C_E^T(y)} \rho(\xi,\eta) = \hat{\rho}_T(x,y)$$

the condition of Example 1 are satisfied.

Now write formally the Isaacs equation for the "simple pursuit" game. It is of the form

$$\frac{\partial \hat{\rho}}{\partial T} = \max_{|v| \leq \beta} \sum_{i=1}^{2} \frac{\partial \hat{\rho}}{\partial y_i} v_i + \min_{|u| \leq \alpha} \sum_{i=1}^{2} \frac{\partial \hat{\rho}}{\partial x_i} u_i \tag{4.1.2}$$

with the initial condition

$$\hat{\rho}_T(x,y)|_{T=0} = \sqrt{(x_1-y_1)^2 + (x_2-y_2)^2}. \tag{4.1.3}$$

In (4.1.2), max and min are achieved for

$$\bar{u}_1(x,y,\hat{\rho}_x) = \alpha \frac{-\frac{\partial \hat{\rho}}{\partial x_1}}{\sqrt{\left(\frac{\partial \hat{\rho}}{\partial x_2}\right)^2 + \left(\frac{\partial \hat{\rho}}{\partial x_2}\right)^2}},$$

$$\bar{u}_2(x,y,\hat{\rho}_x) = \alpha \frac{-\frac{\partial \hat{\rho}}{\partial x_2}}{\sqrt{\left(\frac{\partial \hat{\rho}}{\partial x_1}\right)^2 + \left(\frac{\partial \hat{\rho}}{\partial x_2}\right)^2}},$$

$$\bar{v}_1(x, y, \hat{\rho}_y) = \beta \frac{\frac{\partial \hat{\rho}}{\partial y_1}}{\sqrt{\left(\frac{\partial \hat{\rho}}{\partial y_1}\right)^2 + \left(\frac{\partial \hat{\rho}}{\partial y_2}\right)^2}},$$

$$\bar{v}_2(x, y, \hat{\rho}_y) = \beta \frac{\frac{\partial \hat{\rho}}{\partial y_2}}{\sqrt{\left(\frac{\partial \hat{r}ho_y}{\partial y_2}\right)^2 + \left(\frac{\partial \hat{\rho}}{\partial y_2}\right)^2}},$$

and (4.1.2) takes the form

$$-\frac{\partial \hat{\rho}}{\partial T} = \alpha \left[\left(\frac{\partial \hat{\rho}}{\partial x_1}\right)^2 + \left(\frac{\partial \hat{\rho}}{\partial x_2}\right)^2\right]^{\frac{1}{2}} - \beta \left[\left(\frac{\partial \hat{\rho}}{\partial y_1}\right)^2 + \left(\frac{\partial \hat{\rho}}{\partial y_2}\right)^2\right]^{\frac{1}{2}} \tag{4.1.4}$$

with the initial condition

$$\hat{\rho}_T(x, y)|_{T=0} = \sqrt{(x_1 - y_1)^2 + (x_2 - y_2)^2} \tag{4.1.5}$$

We now check the continuous differentiability of the function $\hat{\rho}_T(x, y)$:

$$\frac{\partial \hat{\rho}}{\partial T} = -(\alpha - \beta),$$

$$\frac{\partial \hat{\rho}}{\partial x_i} = \frac{x_i - y_i}{\sqrt{(x_1 - y_1)^2 + (x_2 - y_2)^2}},$$

$$-\frac{\partial \hat{\rho}}{\partial y_i} = \frac{x_i - y_i}{\sqrt{(x_1 - y_1)^2 + (x_2 - y_2)^2}}. \tag{4.1.6}$$

It follows from (4.1.6) that function $\hat{\rho}_T(x, y)$ has continuous derivatives everywhere except of the surface

$$\sqrt{(x_1 - y_1)^2 + (x_2 - y_2)^2} = 0. \tag{4.1.7}$$

The surface (4.1.7) corresponds to the case $\hat{\rho}_T(x, y) = 0$, since in this case $x = y$ and $C_P^T(x) \supset C_E^T(y)$ $(\beta \leq \alpha)$. Thus, the function $\hat{\rho}_T(x, y)$ turns out to be a solution to the Cauchy problem for (4.1.4) in the space region which does not contain the surface the (4.1.7), the latter being singular for the Isaacs (4.1.4).

Note that this singularity has been overcome in the procedure developed in Ch. 3 (theorem 2,3).

Example 3. (dynamic game of pursuit with frictional forces). The pursuit takes place over the entire plane. Motion equations are of the form

$$\dot{q}_i = p_i, \quad |u| \leq 1,$$

$$\dot{r}_i = s_i, \quad |v| \le 1,$$
$$\dot{p}_i = \alpha u_i - k_P p_i, \quad i = 1, 2.$$
$$\dot{s}_i = \beta v_i - k_E s_i,$$

Here q and r are respectively geometrical coordinates of the Players P and E, s (p) is a momentum of Player E (P), and k_P, k_E are some constants interpreted as friction coefficients. The payoff function is defined as follows:

$$H(q(T), r(T)) = \rho(q(T), r(T)) = \sqrt{\sum_{i=1}^{2}(q_i(T) - r_i(T))^2}.$$

First consider the case $k_p \ne 0$, $k_E \ne 0$. The set $C_P^T(q)$ (in the space q) is the circle with the radius

$$R_P(T) = \alpha \frac{e^{-k_P T} + k_P T - 1}{k_P^2}$$

and its center at the point

$$\bar{q} = q + p\frac{1 - e^{-k_P T}}{k_P},$$

where q, p are the starting position and initial momentum for Player P. The set $C_E^T(r)$ (in the space) is the circle with the radius

$$R_E(T) = \beta \frac{e^{-k_E T} + k_E T - 1}{k_E^2}$$

and its center at the point

$$\bar{r} = r + s\frac{1 - e^{k_E T}}{k_E}$$

where r, s are the starting position and initial momentum for Player E. Let us make the following assumption:

$$\alpha > \beta, \quad \frac{\alpha}{k_P} > \frac{\beta}{k_P}. \tag{4.1.8}$$

The quantity $\hat{\rho}_T(q, r)$ is equal to the distance $\rho(M', M)$ between the point M (the center of pursuit), lying on the center line and being farthest from the center of the circle $C_P^T(q)$, and the point M' being nearest to the point M in the circle $C_P^T(q)$, and is computed from the formula

$$\hat{\rho}_T(q, r) = \rho(O, O') + R_E - R_P$$

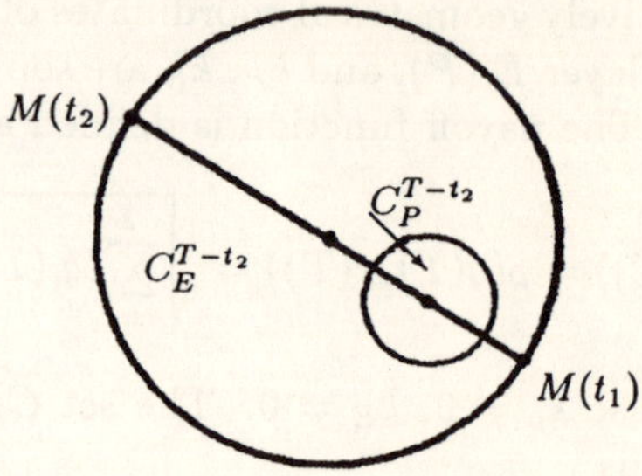

Fig. 14.

(in this example, the centers of circles $C_P^T(q)$, $C_E^T(r)$ do not coincide with the points q, r), whence we have that

$$\hat{\rho}_T(q,r) = \sqrt{\sum_{i=1}^{2}\left(q_i - r_i + p_i\frac{1-e^{-k_P T}}{k_P} - s_i\frac{1-e^{-k_E T}}{k_E}\right)^2} -$$

$$-\left(\alpha\frac{e^{-k_P T} + k_P T - 1}{k_P^2} - \beta\frac{e^{-k_E T} + k_E T - 1}{k_E^2}\right).$$

Let $q^*(t)$, $r^*(t)$ be conditionally optimal trajectories. Then, to satisfy the first condition of invariance, it is sufficient that along the conditionally optimal trajectories

$$R_P(T-t) > R_E(T-t) \tag{4.1.9}$$

for all $0 \leq t < T$, where $R_P(T-t)$ and $R_E(T-t)$ are respectively the radius of the circles $C_P^{T-t}(q^*(t))$ and $C_E^{T-t}(r^*(t))$ (note that the condition (4.1.8) is also satisfied for the simple pursuit, since in this case $R_P(T-t) = \alpha(T-t)$, $R_E(T-t) = \beta(T-t)$, $\alpha > \beta$). Indeed, suppose there are such instants t_1 and t_2 that

$$R_P(T-t_1) > R_E(T-t_1),$$

$$R_P(T-t_2) < R_E(T-t_2).$$

Then the situation shown in Fig.14 becomes possible. Here the center of pursuit has "jumped" from the point $M(t_1)$ to the diametrically opposite point $M(t_2)$. The inequality (4.1.9) excludes such a possibility. We now obtain a sufficient condition for the coefficients α, β, k_P, k_E for which (4.1.9) takes place. Denote $T - t = \tau$.

Let us compute R'_P, R'_E:

$$R'_P(\tau) = \alpha\frac{1-e^{-k_P\tau}}{k_P}, \quad R'_E(\tau) = \beta\frac{1-e^{-k_E\tau}}{k_E},$$

$$R'_P(0) = 0, \quad R'_E(0) = 0.$$

We find second derivatives:

$$R''_P(\tau) = \alpha e^{-k_P\tau}, \quad R''_E(\tau) = \beta e^{-k_E\tau}.$$

Moreover,

$$R''_P(\tau)|_{\tau=0} = \alpha, \quad R''_E(\tau)|_{\tau=0} = \beta.$$

From (4.1.8) we obtain

$$R''_P(0) > R''_E(0).$$

Consequently, in some neighbourhood of zero

$$R''_P(\tau) > R''_E(\tau),$$

and since $R'_P(0) = R'_E(0) = 0$, then in the same neighbourhood

$$R'_P(\tau) > R'_E(\tau).$$

Hence $R_E(0) = R_P(0) = 0$ and for small τ we get

$$R_P(\tau) > R_E(\tau).$$

Evidently, if $\alpha > \beta$ and $k_P \leq k_E$, then condition (4.1.8) is satisfied, and $R_P(\tau) > R_E(\tau)$ for all $\tau > 0$. Therefore, of interest is proof of (4.1.9) for $\alpha > \beta$ and $k_P > k_E$ ($\frac{\alpha}{k_P} > \frac{\beta}{k_E}$).

We will show that in this case there is always

$$R'_P(\tau) > R'_E(\tau). \tag{4.1.10}$$

Indeed, suppose the opposite is true. Then there is such instant τ^1 that

$$R'_P(\tau^1) = \frac{\beta}{k_E}\left(1 - e^{-k_p\tau^1}\right) = \frac{\alpha}{k_P}\left(1 - e^{-k_p\tau^1}\right) = R'_E(\tau^1) \tag{4.1.11}$$

But $\frac{\beta}{k_E} < \frac{\alpha}{k_P}$ (see condition (4.1.8)).

From $k_E < k_P$ follows $k_E\tau^1 < k_P\tau^1$, or $e^{k_E\tau^1} < e^{k_P\tau^1}$, whence

$$\frac{\beta}{k_E}\left(1 - e^{-k_E\tau^1}\right) < \frac{\alpha}{k_P}\left(1 - e^{-k_P\tau^1}\right).$$

But this contradicts (4.1.11), i.e. (4.1.10) holds for all $\tau > 0$. Since in the neighbourhood of zero $R_P(\tau) > R_E(\tau)$, then (4.1.10) implies $R_P(\tau) > R_E(\tau)$ for all $\tau > 0$.

We have thus proved (4.1.9) under condition (4.1.8) for all $\tau > 0$. The second condition of invariance is satisfied and proved as in Example 2.

We now construct the equation with partial derivatives of the first order (see (3.5.12)) for the game involved. It is of the form

$$\frac{\partial\hat{\rho}}{\partial T} = \sum_{i=1}^{2}\left(\frac{\partial\hat{\rho}}{\partial q_i}p_i + \frac{\partial\hat{\rho}}{\partial r_i}s_i\right) - \sum_{i=1}^{2}\left(\frac{\partial\hat{\rho}}{\partial q_i}k_P p_i + \frac{\partial\hat{\rho}}{\partial p_i}k_E s_i\right) +$$

$$+\beta \max_{|v|\le 1}\sum_{i=1}^{2}\frac{\partial\hat{\rho}}{\partial s_i}v_i + \alpha \min_{|u|\le 1}\sum_{i=1}^{2}\frac{\partial\hat{\rho}}{\partial p_i}u_i \tag{4.1.12}$$

With the initial condition

$$\hat{\rho}_T(q,r)|_{T=0} = \sqrt{\sum_{i=1}^{2}(q_i - r_i)^2}. \tag{4.1.13}$$

Controls $\bar{u}$, $\bar{v}$ from (3.5.13), on which min and max are achieved in (4.1.12), are determined by the formulas

$$\bar{u}_i = \frac{-\frac{\partial\hat{\rho}}{\partial p_i}}{\sqrt{\left(\frac{\partial\hat{\rho}}{\partial p_1}\right)^2 + \left(\frac{\partial\hat{\rho}}{\partial p_2}\right)^2}}.$$

$$\bar{v}_i = \frac{\frac{\partial\hat{\rho}}{\partial s_i}}{\sqrt{\left(\frac{\partial\hat{\rho}}{\partial s_1}\right)^2 + \left(\frac{\partial\hat{\rho}}{\partial s_2}\right)^2}}, \quad i = 1,2.$$

Substituting $\bar{u}_i$, $\bar{v}_i$ into (4.1.12), we obtain the Cauchy problem for equation with partial derivatives of the first order

$$\frac{\partial\hat{\rho}}{\partial T} = \sum_{i=1}^{2}\left(\frac{\partial\hat{\rho}}{\partial q_i}p_i + \frac{\partial\hat{\rho}}{\partial r_i}s_i\right) - \sum_{i=1}^{2}\left(\frac{\partial\hat{\rho}}{\partial p_i}k_P p_i + \frac{\partial\hat{\rho}}{\partial s_i}k_E s_i\right) -$$

$$-\alpha\left[\sum_{i=1}^{2}\left(\frac{\partial\hat{\rho}}{\partial s_i}\right)^2\right]^{\frac{1}{2}} + \beta\left[\sum_{i=1}^{2}\left(\frac{\partial\hat{\rho}}{\partial p_i}\right)^2\right]^{\frac{1}{2}} \tag{4.1.14}$$

with the initial condition

$$\hat{\rho}_T(q,r) = \left[\sum_{i=1}^{2}(q_i - r_i)^2\right]^{\frac{1}{2}}. \tag{4.1.15}$$

Introduce the notations

$$\frac{\partial \hat\rho}{\partial q_i} = \hat\rho_{q_i}, \ \frac{\partial \hat\rho}{\partial p_i} = \hat\rho_{p_i}, \ \frac{\partial \hat\rho}{\partial s_i} = \hat\rho_{s_i},$$

$$\frac{\partial \hat\rho}{\partial r_i} = \hat\rho_{r_i}, \ \frac{\partial \hat\rho}{\partial T} = \hat\rho_T, \ i = 1,2.$$

Characteristics of (4.1.14) are optimal trajectories in the game of pursuit [1]. Equations of Characteristics are of the form (see (3.5.18))

$$\dot q_i = p_i, \ \dot p_i = \alpha \frac{-\hat\rho_{p_i}}{\sqrt{\hat\rho_{p_1}^2 + \hat\rho_{p_2}^2}}, \ \dot r_i = s_i,$$

$$\dot s_i = \beta \frac{\hat\rho_{s_i}}{\sqrt{\hat\rho_{s_1}^2 + \hat\rho_{s_2}^2}}, \ \dot{\hat\rho}_{q_i} = 0, \ \dot{\hat\rho}_{p_i} = -\hat\rho_{q_i} - k_P \hat\rho_{p_i},$$

$$\dot{\hat\rho}_{r_i} = 0, \ \dot{\hat\rho}_{s_i} = -\hat\rho_{r_i} - k_E \hat\rho_{s_i}, \ \dot{\hat\rho}_T = 0, \ i = 1,2.$$

When solving the equation (4.1.14) with the initial condition (4.1.15) by the method of characteristics, we obtain the previously derived expression

$$\hat\rho_T(q,r) = \sqrt{\sum_{i=1}^{2} \left(q_i - r_i + p_i \frac{e^{-k_P T} - 1}{k_P} - s_i \frac{e^{-k_E T} - 1}{k_E} \right)^2} - \\ - \left(\alpha \frac{e^{-k_P T} + k_P T - 1}{k_P^2} - \beta \frac{e^{-k_E T} + k_E T - 1}{k_E^2} \right). \qquad (4.1.16)$$

The function $\hat\rho$ from (4.1.16) has continuous partial derivatives over the entire region of space except the points lying on the surface

$$\sqrt{\sum_{i=1}^{2} \left(q_i - r_i + p_i \frac{e^{-k_P T} - 1}{k_P} - s_i \frac{e^{-k_E T} - 1}{k_E} \right)^2} = 0, \qquad (4.1.17)$$

where the partial derivatives $\hat\rho_{p_i}$, $\hat\rho_{s_i}$, $i = 1,2$ are not defined. The surface (4.1.2) corresponds to the initial states, for which the distance $\rho(O, O')$ between the centers of reachability circles is zero, since the expression on the left side of (4.1.12) is exactly the distance between the centers of reachability circles. Thus, the function $\hat\rho_T$ proves to be a solution to the Cauchy problem for the equation (4.1.14) with the initial condition (4.1.15) in the space region free from singularities, at these points $C_P^T(q) \supset C_E^T(r)$ and $\hat\rho_T(q,r) = 0$. When the condition $R_P(\tau) > R_E(\tau)$ ($\alpha > \beta$, $\frac{\alpha}{k_P} > \frac{\beta}{k_E}$) is satisfied, the results of Chapter 1 (see the Theorem 2, 3) makes it possible to overcome this singularity [27] just as in Example 1.

Consider the case $k_P = k_E = 0$ (the absence of friction) [35] in some detail. The set $C_P^T(q)$ is a circle with the radius $R_P(T) = \alpha\frac{T^2}{2}$ and its center at the point $\bar{q} = q + pT$, where q, p are the starting position and initial momentum for Player P. The set $C_E^T(r)$ (in the space q, r) is a circle with the radius $R_E(T) = \beta\frac{T^2}{2}$ and its center at the point $\bar{\rho} = r + sT$, where r, s are the starting position and initial momentum for Player E. The condition (4.1.9) converts to the inequality $\alpha\frac{T^2}{2} > \beta\frac{T^2}{2}$, and the existence condition for the invariant center of pursuit (4.1.9) converts to the inequality

$$\alpha > \beta \tag{4.1.18}$$

Physically, this means that the maximum force applied by Player P exceeds the force applied by Player E.

When (4.1.18) is satisfied, the value of the game is equal to $\hat{\rho}_T(q,r)$. It is determined from the formula

$$\hat{\rho}_T(q,r) = \sqrt{\sum_{i=1}^{2}(q_i - r_i + T(p_i - s_i))^2 - (\alpha - \beta)\frac{T^2}{2}}. \tag{4.1.19}$$

The equation with partial derivatives of the first order for the function $\hat{\rho}_T(q,r)$ is derived from (4.1.14) by substituting $k_P = k_E = 0$

Example 3a. The pursuit occurs over the entire plane. Motion equations are of the form

$$\begin{aligned} \dot{q}_i &= p_i, \\ \dot{p}_i &= \alpha u_i - k_P p_i, \quad |u| \le 1, \quad i = 1,2 \\ \dot{y}_i &= \beta v_i, \quad\quad\quad\quad |v| \le 1, \quad i = 1,2, \end{aligned}$$

Here q and y are respectively the geometrical positions of the players P and E, and p is Player P's momentum. Thus, in the case at hand, Player E moves in accordance with the "simple motion", and Player P representing a material point of unit mass, moving under the action of the friction force and the force α.

Player E's payoff is defined as a distance between the players' geometrical locations when the play terminates:

$$H(q(T), y(T)) = \rho(q(T), y(T)) = \sqrt{\sum_{i=1}^{2}(q_i(T) - y_i(T))^2}.$$

First consider the case $k_P \neq 0$. The set $C_P^T(q)$(in the space q) is a circle with the radius

$$R_P(T) = \alpha\frac{e^{-k_P T} + k_P T - 1}{k_P^2}$$

and its center at the point

$$\bar{q} = q + p\frac{1 - e^{k_P T}}{k_P^2},$$

where q, p are the starting position and initial momentum for Player P. The set $C_E^T(y)$ is a circle with the radius $R_E(T) = \beta T$ and its center at the point y. As before, the quantity $\hat{\rho}_T(q, y)$ is computed by the formula

$$\hat{\rho}_T(q, y) = \rho(q, y) + R_E - R_P,$$

$$\hat{\rho}_T(q, y) = \sqrt{\sum_{i=1}^{2}\left(q_i - y_i + p_i\frac{1 - e^{-k_P T}}{k_P}\right)^2} -$$

$$-\alpha\frac{e^{-k_P T} + k_P T - 1}{k_P^2} + \beta T.$$

As shown in the previous example, to satisfy the invariance conditions for the center of pursuit M it is sufficient that along the conditionally optimal trajectories $q^*(t)$, $y^*(t)$ the inequality

$$R_P(T - t) > R_E(T - t) \tag{4.1.20}$$

is satisfied for all $0 \leq t \leq T$, where $R_P(T - t)$, $R_E(T - t)$ are the radius of reachability circles $C_P^{T-t}(q^*(t))$, $C_E^{T-t}(y^*(t))$. In what follows, $T - t$ is denoted by τ.

We find the condition (4.1.20) to be satisfied for the case under study. Compute R'_P and R'_E:

$$R'_P(\tau) = \alpha\frac{1 - e^{-k_P\tau}}{k_P}, \quad R'_E(\tau) = \beta.$$

Since $\beta > 0$ and $R'_P(0) = 0$, there is such neighbourhood of zero that $R'_P(\tau) < R'_E(\tau) = \beta$. Since $R_P(0) = R_E(0) = 0$, then $R_P(\tau) < R_E(\tau)$, $\tau \neq 0$ for τ from this neighbourhood. Thus, the sufficient condition of invariance (4.1.20) is violated. At the same time, under some initial conditions the center of pursuit is found to be invariant. Let us deduce these conditions.

Consider the difference

$$\varphi(\tau) = R_P(\tau) - R_E(\tau) = \alpha\frac{e^{-k_P\tau} + k_P\tau - 1}{k_P^2} - \beta\tau.$$

Show that $\varphi(\tau)$ goes to zero at a unique point which is a monotonicity point of the function $\varphi(\tau)$. Compute $\varphi'(\tau)$ and $\varphi''(\tau)$:

$$\varphi'(\tau) = \alpha\frac{1 - e^{k_P\tau}}{k_P} - \beta, \quad \varphi''(\tau) = \alpha e^{k_P\tau}.$$

Since $\varphi''(\tau) > 0$, then $\varphi'(\tau)$ increases monotonically. However, $\varphi'(0) = -\beta < 0$.

Let $\beta < \frac{\alpha}{k_P}$. Then there is such τ_0 that $\varphi'(\tau_0) = 0$, or

$$\tau_0 = \frac{1}{k_P} \ln \frac{1}{1 - \frac{\beta}{\alpha} k_P}. \tag{4.1.21}$$

For $\tau > \tau_0$ the function $\varphi'(\tau) > 0$, i.e. the function $\varphi(\tau)$ on the interval $[\tau_0, \infty)$ is strictly increasing. Since $\varphi'(0) < 0$, $\varphi(0) < 0$ for $0 < \tau < \tau_0$, then $\varphi(\tau_0) < 0$ and $\varphi(\tau)$ goes to zero for some $\bar{\tau}_0 > \tau_0$. In the same time

$$\lim_{\tau \to \infty} \varphi(\tau) = \lim_{\tau \to \infty} \left[\frac{\alpha}{k_P^2} + \left(\frac{\alpha}{k_P} - \beta \right) \tau \right] = +\infty, \quad \left(\frac{\alpha}{k_P} > \beta \right).$$

Suppose the following condition is satisfied

$$\hat{\rho}_T(q, y) > R_E(\tau_0) - R_P(\tau_0). \tag{4.1.22}$$

Then the point M is invariant. In fact, the invariance of the point M may be violated (see Fig.14) only if during the motion a long conditionally optimal trajectories the point $\bar{M}$, that is diametrically opposite to the point M,in the set $C_E^T(y)$ is more removed from $C_P^T(q)$ than the point M (or it is at the same distance from $C_P^T(q)$ as the point M). Geometrically, it is evident that under condition (4.1.22) this is not possible. When the condition (4.1.22) is satisfied, for all τ we have

$$R_E(\tau) - R_P(\tau) \leq \hat{\rho}_T(q, y). \tag{4.1.23}$$

In the case $k_P = 0$ the inequality (4.1.23) will have the form

$$\beta\tau - \frac{\alpha\tau^2}{2} \leq \rho_T(q_0, y_0), \tag{4.1.24}$$

or, computing a maximum of the function $\beta\tau - \frac{\alpha\tau^2}{2}$

$$\beta - \alpha\bar{T}_{\max} = 0, \ \bar{T}_{\max} = \frac{\beta}{\alpha}, \tag{4.1.25}$$

we rewrite (4.1.24) as

$$\frac{\beta^2}{2\alpha} \leq \hat{\rho}_T(q_0, y_0). \tag{4.1.26}$$

The function $\hat{\rho}_T(q, y)$ is continuously differentiable everywhere except the surface

$$\sqrt{\sum_{i=1}^{2} \left(q_i - y_i + p_i \frac{1 - e^{-k_P T}}{k_P} \right)^2} = 0, \tag{4.1.27}$$

corresponding to the initial states for which the center of circles $C_P^T(q)$ and $C_E^T(y)$ coincide. This function is a solution to the Cauchy problem everywhere off the set (4.1.27) for the equation with partial derivatives of the first order

$$\frac{\partial \rho}{\partial T} = \sum_{i=1}^{2} \left(\frac{\partial \hat{\rho}}{\partial q_i} p_i - \frac{\partial \hat{\rho}}{\partial p_i} k_P p_i \right) - \alpha \left[\sum_{i=1}^{2} \left(\frac{\partial \hat{\rho}}{\partial p_i} \right)^2 \right]^{\frac{1}{2}} + \beta \left[\sum_{i=1}^{2} \left(\frac{\partial \hat{\rho}}{\partial y_i} \right)^2 \right]^{\frac{1}{2}}$$

with the initial condition

$$\hat{\rho}_T|_{T=0} = \sqrt{(q_i - y_i)^2 + (q_2 - y_2)^2}.$$

When the conditions (4.1.21), (4.1.22), (4.1.26) are not satisfied, the singular surface (4.1.27) may not be dispersal and the optimal strategy for Player E may not be open–loop.

4.2 Phase–constrained "simple pursuit" games

Example 4. ("simple pursuit" game on the half–plane). The players move in the half–plane S (they cannot leave S until the play terminates) in accordance with the motion equations

$$\begin{aligned} \dot{x}_i &= \alpha u_i, \quad i = 1,2, \quad |u| \leq 1, \\ \dot{y}_i &= \beta v_i, \quad i = 1,2, \quad |v| \leq 1. \end{aligned} \tag{4.2.1}$$

where $\alpha = const$, $\beta = const$, $\alpha > \beta$ (see Example 2 in this chapter). The game has the prescribed duration T, and Player E's payoff is equal to a distance between him and Player P at the game termination instant T.

Denote by A_1 a set of initial positions, where center of pursuit M belongs to a boundary of the set S. Also, denote by A_2 a set of positions, for which this condition is not satisfied. The solution of the game from the initial positions belonging to the set A_2 coincides with the solution to the "simple pursuit" game without phase constraints (see Example 2 in this chapter).

Of greater interest is the case where x, y, $T \in A_1$. Let us choose a coordinate system in such a way that the abscissa axis passes along the boundary of the half–plane S (Fig.15).

Let $S_R(x)$ be a minimum sphere with its center at x such that $S_R(x) \bigcap S \supset C_E^T(y) \bigcap S$. Let $\bar{S}_R(x) = \overline{S_R(x) \bigcap S}$ be a boundary of the set $S_R(x) \bigcap S$. Then the center of pursuit in the game $\Gamma(x, y, T)$ for x, y, $T \in A_1$ is the intersection point of the sphere $S_{\beta T}(y)$ and the sphere $\bar{S}_R(x)$, liying in S. For $\hat{\rho}_T(x, y) > 0$ the point M is invariant. Indeed, the conditionally optimal trajectory $\bar{y}(t)$ of Player E prescribes the motion along the line $[y, M]$, and the conditionally

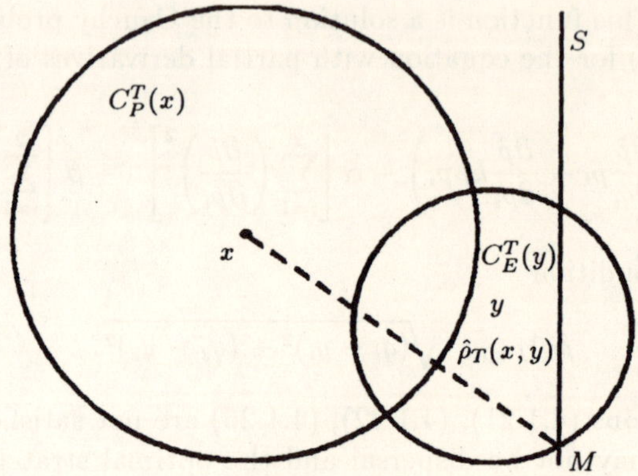

Fig. 15.

optimal trajectory $\bar{x}(t)$ of Player P prescribes the motion along the straight line from x towards M.

Geometrically, the set $S_R(\bar{x}(t)) \supset C_E^T(\bar{y}(t)) \bigcap S$ for all $0 \leq t \leq T$ and, consequently, the point remains invariant in all of the games $\Gamma(\bar{x}(t), \bar{y}(t), T-t)$. Therefore, using Theorem 3, Ch. 3 we may write immediately

$$\hat{\rho}_T(x,y) = Val\Gamma(x,y,T),$$

which gives

$$Val\Gamma(x,y,T) = -\alpha T + \sqrt{\left[x_1 - y_1 + \sqrt{(\beta T)^2 - y_1^2}\right]^2 + x_2^2}, \qquad (4.2.2)$$

if $x_1 - y_1 > 0$, or

$$Val\Gamma(x,y,T) = -\alpha T + \sqrt{\left[x_1 - y_1 + \sqrt{(\beta T)^2 - y_2^2}\right]^2 + x_2^2}, \qquad (4.2.3)$$

if $x_1 - y_1 < 0$.

Recall the equation with partial derivatives of the first order for the value function of the "simple pursuit" game:

$$-\frac{\partial V}{\partial T} = \alpha\left[\left(\frac{\partial V}{\partial x_1}\right)^2 + \left(\frac{\partial V}{\partial x_2}\right)^2\right]^{\frac{1}{2}} - \beta\left[\left(\frac{\partial V}{\partial y_1}\right)^2 + \left(\frac{\partial V}{\partial y_2}\right)^2\right]^{\frac{1}{2}}. \qquad (4.2.4)$$

The function Eq.(4.2.2), Eq.(4.2.3) satisfy the Eq.(4.2.4) in the region $x_2 > 0$, $y_2 > 0$ with the initial condition

$$V(x,y,T)|_{y_2=0} = -\alpha T + \sqrt{(x_1 - y_1 - \beta T)^2 + x_2^2} \qquad (4.2.5)$$

(if $x_1 - y_1 > 0$), that is the basic equation for the game of simple pursuit.

Singular surface. A singular surface is coming into play when the positions of the Players P and E are on a perpendicular to the boundary of the set S, i.e. a singular surface is of the form $x_1 = y_1$. We show that for all points x, y on the singular surface the value of the game is equal to $\hat{\rho}_T(x, y)$ and the optimal strategy $u^*(\cdot)$ for Player P satisfies the condition of Theorem 1, Ch. 3, i.e. Player P chooses the segment of his open–loop control from the condition $\min \hat{\rho}$.

Consider two cases.

1. The point E lies at the foot of a perpendicular dropped from the point P to the boundary of the set S. In this case:
   a) the reachability set $C_P^T(x) \bigcap S$ of Player P does not contains the initial position E (Fig.16);
   b) the reachability set $C_P^T(x) \bigcap S$ of Player P contains the initial position E (Fig.17);
2. The point E lies on a perpendicular dropped from the point P to the boundary of the set S, but it does not belong to the boundary of S.

Consider case 1a. To prove the optimality of the strategy $u^*(\cdot)$ for Player P, it suffices to show that, when this strategy is adopted, $\hat{\rho}$ decreases if Player E remains on the singular surface (in this case, he remains at the foot of the perpendicular dropped from the point P to the boundary of the set S).

Consider Fig.16. Let the point E be at y_0 during time t. Since the length of the interval $[N(T), y_0]$ is strictly positive here, then as the point x moves in accordance with the strategy $u^*(\cdot)$ (this strategy prescribes the motion along the perpendicular x_0, y_0 since if the point y_0 is in the state of rest, then such a motion gives $\min \hat{\rho}$) the angle $[x(\tau), N(\tau), y_0]$ increases for $0 \leq \tau \leq t$. From $x(\tau)$ and parallel to x_0M we draw the line $x(\tau)M'$, on which from the point $x(\tau)$ we lay off the line segment $x(\tau)N'$ of lenght $\alpha(T - \tau)$. From the similarity of triangles we have that the points N, N', $N(T)$ are lying on the same straight line. Hence we get

$$\hat{\rho}_T(x_0, y_0) = \rho(N, M) > \rho(N', M') > \rho(x(\tau), M(\tau)) - \alpha(T - \tau) =$$

$$= \rho(N(\tau), M(\tau)) = \hat{\rho}_{T-\tau}(x(\tau), y(\tau)).$$

We have thus proved the optimality of the strategy $u^*(\cdot)$ for the case under study.

Consider case 1b (see Fig.17). As in the previous case, assume that the point E is at y_0 during the time t. Let $\bar{u}(\cdot)$ be an auxiliary strategy for Player

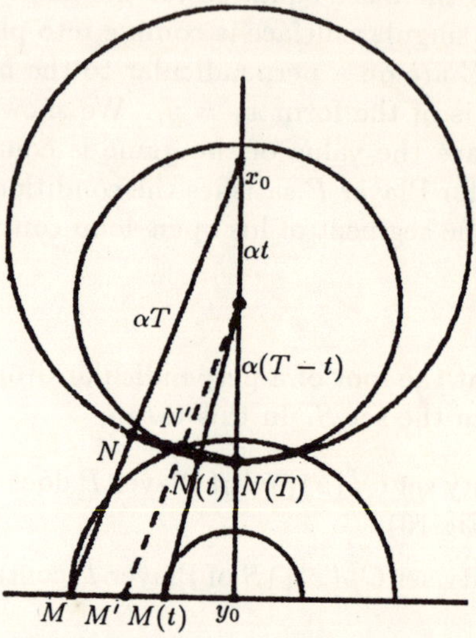

Fig. 16.

P, under which he moves during the time $[0, t]$ towards the point y_0, so that the line segment $x(\tau)M(\tau)$ for $0 \le \tau \le t$ moves parallel to itself.

We prescribe this strategy to Player P up to the instant $\bar{t}$ until case 1b converts to case 1a provided, of course, that $\bar{t} \le t$ i.e. until the circle $C_P^{T-t}(x_0)$ touches the boundary of S at the point y_0. The latter is allways possible, since under the strategy $\bar{u}(\cdot)$ the motion does not occur at a maximum speed (to maintain the parallelism in approach, Player P is to decrease his speed).

As before, note that the points N, $N(\tau)$, y_0 are lying on the same straight line. Hence

$$\hat{\rho}_T(x_0, y_0) = \rho(N, M) > \rho(N(\tau), M(\tau)) = \hat{\rho}_{T-t}(x(t), y_0).$$

In this manner $\hat{\rho}$ decreases up to the instant $\bar{t}$. Case 1b then converts to case 1a. This completes the proof of the optimality of the strategy $u^*(\cdot)$, since although the strategy $\bar{u}(\cdot)$ does not carry ever the point P to the point min $\hat{\rho}$, the quantity $\hat{\rho}$ decreases, hence the strategy $u^*(\cdot)$ transferring the point P to the point min $\hat{\rho}$ ensures a decrease in $\hat{\rho}$

We have thus shown that, in case 1, $\hat{\rho}_T(x_0, y_0)$ is the value of the game and $u^*(\cdot)$ is the optimal strategy for Player P.

Consider case 2 (Fig.18). We show that, when the strategy $u^*(\cdot)$ is adopted by Player P, the quantity $\hat{\rho}$ decreases if the point E remains on the singular surface. Note that Player E is concerned with the motion for which the velocity of the current center of pursuit $M(t)$ towards the point $\bar{M}$ is minimum. Such

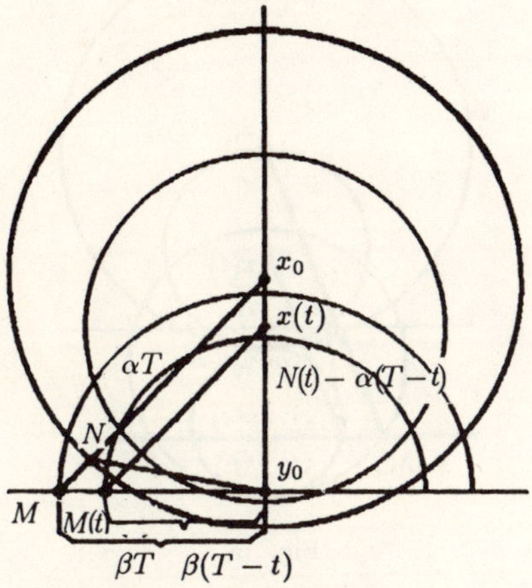

Fig. 17.

a motion corresponds to the case where the point E moves from the position y_0 towards the point $\bar{M}$ at a maximum velocity β.

We draw the straight line L through the point N parallel to the boundary of the set S. When the point x moves towards the point $\bar{K}$ at a maximum velocity α, the intersection point $K(t)$ of the boundary of the set $C_P^{T-t}(x(t))$ and the straight line L moves from N to $\bar{K}$. Since $|\frac{x_0 N}{y_0 M}| < |\frac{x_0 \bar{K}}{y_0 \bar{M}}|$, then the velocity of the point $K(t)$ at each instant t is less then the velocity of the point M' (here we consider the time t that is less than the time during which the point E reaches the point M). This means that the angle $K(t)M'\bar{M}$ increases continuously. But the angle $NM\bar{M}$ is acute and, therefore, continues to be acute during some time $[0, \delta]$. This enables us conclude that on the interval $[0, \delta]$.

$$\hat{\rho}_T(x_0, y_0) = NM > KM' > N'M' = \hat{\rho}_{T-t}(x(t), y(t)),$$

i.e. $\hat{\rho}$ decreases. If there is an tnstant $\bar{t}$, at which the angle $KM'\bar{M}$ turns out to be right, then we draw a new chord L' from the point N' and consider the intersection point K'' for the boundary of $C_P^{T-t}(x(t))$ and the chord L' on some interval $[\bar{t}, \bar{t} + \delta_1]$ where the angle $K''M'\bar{M}$ is acute. As before, we prove that $\hat{\rho}$ decreases for $t \in [\bar{t}, \bar{t} + \delta]$. This process may be continued up to the instant $\bar{\bar{t}}$ until the point E occupies the position $\bar{M}$ since the time $\bar{\bar{t}}$, during which the point E arrives at the point $\bar{M}$ is strictly less than the time, during which the point P arrives at the point of $\bar{K}$. In this manner we prove that there is the strategy $\bar{u}(\cdot)$ of Player P (i.e. he moves on the singular surface at a maximum velocity) under which $\hat{\rho}$ decreases on the interval $[0, \bar{\bar{t}}]$.

The further decrease in $\bar{\rho}$ follows from case 1. Since $\hat{u}(\cdot)$ is a strategy of

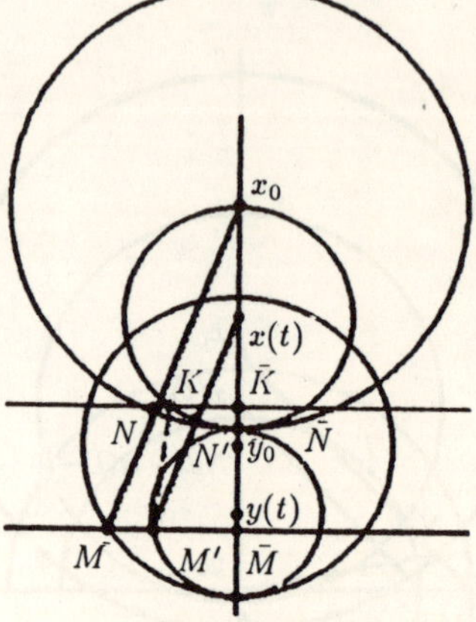

Fig. 18.

Player P, then the strategy $u^*(\cdot)$ transferring at each step the point x to the point min $\hat{\rho}$ ensures a decrease in $\hat{\rho}$. We have thus obtained a complete solution to the "simple pursuit" game in the half-plane.

Theorem 1 *In the game of "simple pursuit " in the half-plane, an optimal strategy for Player P is the strategy $u^*(\cdot)$ and the value of the game is equal to $\hat{\rho}_T(x_0, y_0)$.*

Example 5. ("simple pursuit" within an acute angle. The case of prescribed duration and terminal payoff). Let S be the interior of the acute angle together with the boundary. We restrict ourselves to the case where the intersection $C_E^T(y_0) \bigcap S$ does not include the vertice of the angle 0 (x_0, $y_0 \in S$), (Fig.19).

There are not more than four points of intersection of the boundary of the set $C_E^T(y_0)$ with the sides of the angle. Denote them by N, M_1, M_2, M_3, M_4, respectively. Let N be the distant point of intersection of the straight line x_0, y_0 with the boundary of the set $C_E^T(y_0)$. Then for any arrangement of the points x_0, $y_0 \in S$ the center of pursuit is among the points N, M_1, M_2, M_3, M_4. When the center of pursuit coincides with the point N, the game involved is equivalent to the "simple pursuit" game free of phase constraints and reduces to the above-mentioned case (see Example 1 in this chapter).

Since $\alpha > \beta$ for all x_0, $y_0 \in S$ then no more than two centers of pursuit may exist. Moreover, such a pair may be formed only by the points M_1, M_2 or M_3, M_4 lying pairwise on one of the angle sides. The singular surface in the game exactly corresponds to the set of initial states for which there are to centers of pursuit. Geometrically, it corresponds to the case where the points x, y belong to one of the perpendiculars L_1 or L_2 erected from the middle portions of the line segments M_1M_2 and M_3M_4 see (Fig.19).

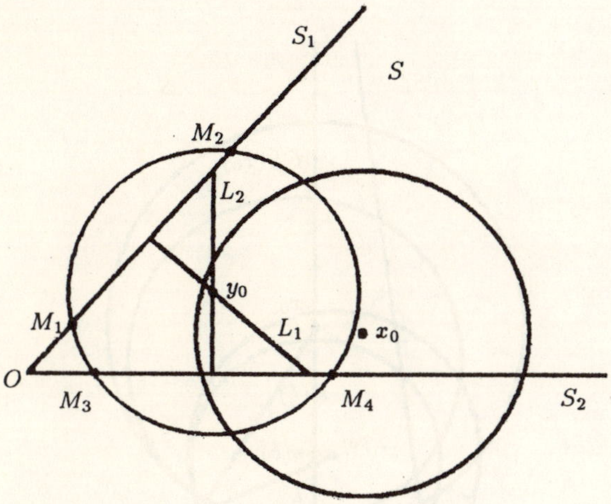

Fig. 19.

When dealing with the region bounded by the singular surface, we find that in such regions there is the invariant center of pursuit and the value of the game is equal to $\hat{\rho}_T(x_0, y_0)$, since in each such case the game is equivalent to the game of pursuit on the half-plane, where the boundary is one of the angle sides.

For the global soluion of the game, we have to study the players' behavior on the singular surface. For the purposes of definiteness, assume that the point x, y is on the perpendicular L_1 erected from the middle of the line segment M_1M_2 lying on the side S_1 of the angle. When dealing with the auxiliary game of "simple pursuit" on the half-plane with the boundary S_1 and prescribing an optimal strategy in this game to Player P (see the previous example), we find that, as shown above, the quantity $\hat{\rho}$ only decreases when the points x, y are on L. This means that the strategy $u^*(\cdot)$ for Player P ensures the constancy of $\hat{\rho}$ in the region bounded by the singular surface, and its decrease on this surface. So the value of game is $\hat{\rho}_T(x_0, y_0)$ and an optimal strategy of Player P coincides to the strategy $u^*(\cdot)$ from Theorem 1, Ch. 3.

Example 6. ("simple pursuit", where $\hat{\rho}_T(x_0, y_0)$ is not the value of the game). Consider the previous example where the vertex of angle O is the interior point of the set $C_E^T(y)$ (Fig.20). Let L be a perpendicular erected from the middle of the line segment OM_1. Assume that the point $x \in L$. At this point there are two centers of pursuit, and with all $y' \in S$ for which $O \in C_E^T(y')$ the condition of Theorem 1, Ch. 3, are violated at this points. Thus, we are facing here the whole region on singularity rather than a singular surface. We will show that for such points x, y there is no strategy $u^*(\cdot)$ which ensures a nonincrease in

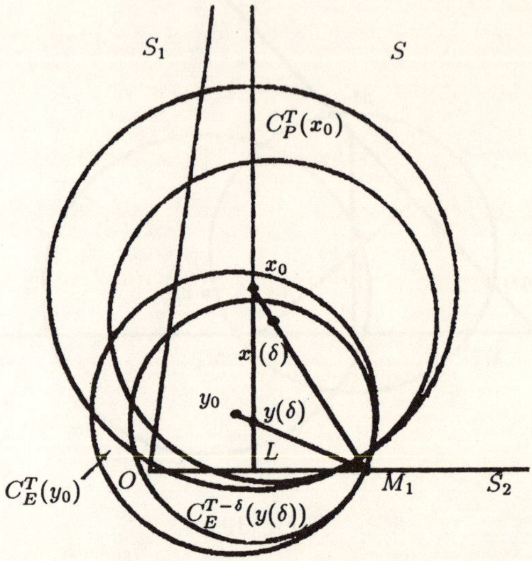

Fig. 20.

$\hat{\rho}_T(x, y)$.

Examine Fig.20 (consider the case $C_P^T(x) \supset C_E^T(y) \bigcap S$). Let the time interval $[0, \delta)$ be such that for all $y' \in C_E^\delta(y)$ the set $C_E^{T-\delta}(y')$ contains the point O. Suppose the point E moves towards the point M_1. The new reachability set $C_E^{T-\delta}(y(\delta))$ contains both the point M_1 and point O. To ensure the distance from $C_P^{T-\delta}(x(\delta))$ to M_1 equal to $\hat{\rho}_T(x, y)$ Player P must move towards the point M_1. But this involves an increase in the distance from $C_P^{T-\delta}(x(\delta))$ to the point O, i.e. an increase in $\hat{\rho}$. The latter means that for any strategy of Player P there is Player E's strategy (viz., the strategy which focussed the motion of the point E on the point M_1), for which $\hat{\rho}_{T-\delta}(x(\delta), y(\delta)) > \hat{\rho}_T(x, y)$.

A more complex case of "simple pursuit" in the circle is discussed in [15]. But there only the game of kind is presented, which substantially simplifies the problem.

4.3 "Simple pursuit" game with two pursuers and one evader

Example 7. Consider the case where several pursuers and one evader participate in the game, i.e. the team of pursuers P consists of $P = \{P_1, P_2\}$ players and all the participants of the game have a simple motion. Thus, motion equations are of the form:

$$\begin{array}{lll} \text{for } P_i & \dot{x}_j^{(i)} = u_j^{(i)}, & |u^i| \le \alpha^{(i)}, \\ \text{for } E & \dot{y}_j = v_j, & |v| \le \beta, \end{array}$$

$$i = 1,2, \quad j = 1,2, \quad \beta < \min_i \alpha^{(i)}.$$

The game has a prescribed duration T, and the payoff of Player E is equal to $\min_i \rho(x^{(i)}(T), y(T))$

Consider the properties of the center of pursuit M. By definition,

$$\hat{\rho}_T(x^{(1)}, x^{(2)}, y) = \max_{\eta \in C_E^T(y)} \min_{\xi_1^{(1)} \in C_{P_1}^T(x^{(1)}), \xi^{(2)} \in C_{P_2}^T(x^{(2)})} \left[\min_i \rho(\xi^{(i)}, \eta) \right].$$

The point M is a point from $C_E^T(y)$ which is farthest from the union of the sets $C_{P_1}^T(x^{(1)})$ and $C_{P_2}^T(x^{(2)})$.

The following two cases are possible:

1. A minimum in the previous formula is achieved at one point (for the purposes of definiteness, let it be the point $\xi_1 \in C_{P_1}^T(x^{(1)})$).

2. A minimum is achieved simultaneously at two points: $\xi_1 \in C_{P_1}^T(x^{(1)})$, $\xi_2 \in C_{P_2}^T(x^{(2)})$. Then $\rho(\xi_1, M) = \rho(\rho_2, M)$.

Consider the first case. The following lemma holds.

Lemma 1 *In the first case, the point M is on the center line of reachability regions $C_{P_1}^T(x^{(1)})$ and $C_E^T(y)$ coincides with the center of pursuit M for the "simple pursuit" game with the Players P_1, E, and the game itself is equivalent to this game.*

Proof: Since $\hat{\rho}_T$ is achieved at one point ξ_1, then

$$\min_{\xi \in C_{P_2}^T(x^{(2)})} \rho(\xi, M) = \rho(\xi_2, M) > \rho(\xi_1, M) = \min_{\xi \in C_{P_1}^T(x^{(1)})} \rho(\xi, M).$$

Suppose the statement of the lemma is not true and $\bar{M} \neq M$. Then M is not the point in the set $C_E^T(y)$ which is farthest from the set $C_{P_1}^T(x^{(1)})$. Therefore, for any $\epsilon > 0$ in its ϵ–neighbourhood there is the point $M'_\epsilon \in C_E^T(y)$ which is farther from $C_{P_1}^T(x^{(1)})$ then the point M, i.e.

$$\min_{\xi \in C_{P_1}^T(x^{(1)})} \rho(\xi, M'_\epsilon) > \min_{\xi \in C_{P_1}^T(x^{(1)})} \rho(\xi, M).$$

Let $\rho(\xi_2, M) - \rho(\xi_1, M) = \delta$. Also, let

$$\min_{\xi \in C_{P_1}^T(x^{(1)})} \rho(\xi, M'_\epsilon) = \rho(\xi'_1, M'_\epsilon),$$

$$\min_{\xi \in C^T_{P_2}(x^{(2)})} \rho(\xi, M'_\epsilon) = \rho(\xi'_2, M'_\epsilon).$$

We have

$$\rho(\xi'_1, M'_\epsilon) > \rho(\xi_1, M).$$

Choose $\epsilon > 0$ in such a way that

$$|\rho(\xi'_2, M'_\epsilon) - \rho(\xi_2, M)| < \frac{\delta}{3},$$

$$|\rho(\xi'_1, M'_\epsilon) - \rho(\xi_1, M)| < \frac{\delta}{3}.$$

Then

$$\rho(\xi'_2, M'_\epsilon) - \rho(\xi'_1, M'_\epsilon) > \frac{\delta}{3}.$$

Hence the point M, is farther from the union of the sets $C^T_{P_1}(x^{(1)}) \bigcup C^T_{P_2}(x^{(2)})$ than the point M on $\frac{\delta}{3}$. This contradicts to the fact that the point M is the center of pursuit. This completes the proof of the first part of the lemma. Its second part is the consequence of the first one. ■

It follows from the lemma that for the complete solution of the problem it suffices to consider case 2.

Let $\rho(\xi_1, M) = \ell_1$, $\rho(\xi_2, M) = \ell_2$. In case 2, $\ell_1 = \ell_2$. Denote the L-neighbourhood of the set $C^T_{P_i}(x_{(i)})$ by $U^T_L(x^{(i)})$. The set $U^T_L(x^{(i)})$ is a circle with $\alpha^{(i)}T + \ell$ as a radius, and its center at the point $x^{(i)}$. The definition of the quantity $\hat{\rho}_T(x^{(i)}, x^{(2)}, y)$ implies

$$C^T_E(y) \subset U^T_\ell(x^{(1)}) \bigcup U^T_\ell(x^{(2)}).$$

Prove that, in the case involved, the quantity $\hat{\rho}_T(x^{(1)}, x^{(2)}, y)$ is the value of the game (in case 1, this result is obtained as a consequence of the solution of the "simple pursuit" game). To this end, we shall check the invariance condition and study the singular surfaces.

First consider the case where the points $x^{(1)}$, $x^{(2)}$, y are not lying on the same straight line. We prove that the first condition of invariance is satisfied. Let $y^*(t)$, $x^{(1)*}(t)$, $x^{(2)*}(t)$ be conditionally optimal trajectories (in this case, $u^*(t)$ is a straight trajectory from y to M, and $x^{(i)*}(t)$ a straight trajectory from $x^{(i)}$ to ξ_i).

The first condition of invariance may be written as

$$C^{T-t}_E(y^*(t)) \subset U^{T-t}_i(x^{(1)*}(t)) \bigcup U^{T-t}_i(x^{(2)*}(t)), 0 \le t \le T \qquad (*)$$

The boundary of the set $U^{T-t}_\ell(x^{(1)*}(t)) \bigcap U^{T-t}_\ell(x^{(2)*}(t))$ is composed of two line segments $C^{T-t}_\ell(x^{(1)*}(t))$ and $C^{T-t}_\ell(x^{(2)*}(t))$, each being a portion of the circle

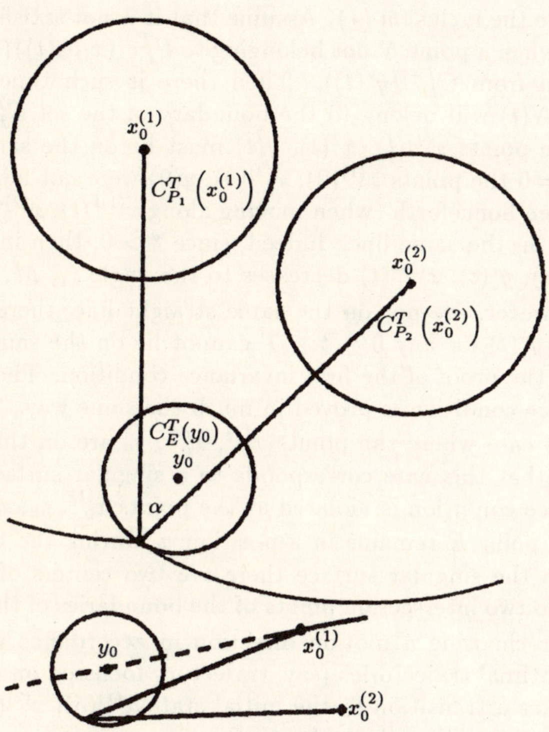

Fig. 21.

with $\alpha^i(T-t)+\ell$ (Fig.21) as a radius and its center at the point $x^{(i)*}(t)$. One of the intersection points of these circles, M, is stationary (i.e. it does not vary for $0 \le t \le T$), and another the point, $N(t)$, is dependent on the parameter t. Denote by α the angle $x_0^{(1)} M x_0^2$ not exceeding π.

Lemma 2 *Suppose that the point M does not coincide with the point of intersection $M^{(i)}$ of the center line $x_0^{(i)}$, y_0 and the boundary of $C_E^T(y_0)$. Then M lies within the angle α.*

Proof: Suppose the opposite is true, and the point y_0 lies out of the angle α (see Fig.21). Then, of two line segments $x_0^{(i)} M$ there is the one which does not intersect with the line segment $x_0^{(3-i)} M^{(i)}$. Let it be the segment $x_0^{(1)} M^{(1)}$. It can be readily seen that the point $M^{(1)}$ is farther from the points $x_0^{(1)}$ and $x_0^{(2)}$ than the point M. Consequently, it lies farther from the sets $C_{P_1}^T(x_0^{(1)})$ and $C_{P_2}^T(x_0^{(2)})$ since these are the circles with centers at the points $x_0^{(1)}$, $x_0^{(2)}$. This, however, is inconsistent with M being the center of pursuit. This completes the proof of the lemma. ■

We now prove the inclusion (∗). Assume that it is not satisfied. Then there is the instant t when a point N not belonging to $U_\ell^{T-t}(x_{(1)*}(t)) \bigcap U_\ell^{T-t}(x_{(2)*}(t))$ may be choosen from $C_E^{T-t}(y^*(t))$. Then there is such time instant $\bar{t} < t$ that the point $N(\bar{t})$ will belong to the boundary of the set $C_E^{T-\bar{t}}(y(\bar{t}))$. That implies that the points $x^{(1)}(\bar{t})$, $x^2(\bar{t})$, $y(\bar{t})$ must lie on the same line. But if at the instant t=0 the points $x^{(1)}(0)$, $x^{(2)}(0)$, $y(0)$ were not lying on the same straight line, then honceforth (when moving along $x^{(1)*}(t)$, $x^{(2)*}(t)$, $y^*(t)$) they would never lie on the same line. Indeed, since $\ell > 0$, then in course of time the angle $x^{(1)*}(t)$, $y^*(t)$, $x^{(2)*}(t)$ decreases to the angle ξ_1, M, ξ_2. The points $x_0^{(1)}$, y, $x_0^{(2)}$, however, are not on the same straight line, therefore the points $x^{(1)*}(t), x^{(2)*}(t), y^*(t)$ for any $0 \le t \le T$ cannot lie on the same straight line. This completes the proof of the first invariance condition. The validity of the second invariance condition is proved in much the same way.

Consider the case where the points $x_0^{(1)}$, $x_0^{(2)}$, y_0 are on the same straight line. We show that this case corresponds to a singular surface. Indeed, the second invariance condition is violated at the points $x_0^{(1)}$, y_0, $x_0^{(2)}$.

Suppose the point E remains in a position y_0 during the time δ. For the initial states on the singular surface there are two centers of pursuit M, N corresponding to two intersection points of the boundaries of the sets $U_\ell^T(x_0^{(1)})$, $U_\ell^T(x_0^{(2)})$. After choosing a motion direction in accordance with one of the conditionally optimal trajectories (say, trajectory focussed on M), in the time δ Player P makes a transition to the initial state $x^{(1)}(\delta)$, $x^{(2)}(\delta)$ (Player E is resting at the point y_0) for which the center of pursuit is now the point N.

The singular surface under study is not dispersal, since there is a control for Player E (viz., the control $v \equiv 0$) which, for $t \in [0, \delta]$, retains the points $x^{(1)}(t), x^{(2)}(t)$, $y(t)$ on S provided that Player P chooses the control $u(\cdot)$ in accordance with Theorem 3, Ch. 3, i.e. from the condition $\min \hat{\rho}$. Indeed, the motion under the strategy u^* (condition $\min \hat{\rho}$) focusses the points P_1, P_2 on the point E, and since the latter is in the state of rest, then the game does not leave S (Fig.22).

We show that for such a behavior of Player E the function $\hat{\rho}$ decreases, which means that the value of the game is equal to $\hat{\rho}$. Now, assume that Player E employs the control $v \equiv 0$ during the time $[0, \delta]$. Then the point $x^{(i)}(t)$ moves towards the point y_0 on a distance $\alpha_i\delta$ ($\alpha_i > \beta$) in a straight line, which is perpendicular to the straight line My_0, and the point M occupies the position $M(\delta)$ on the straight line. Here we are facing the situation which is similar to the behaviour on the singular surface in case 1, Example 4 ("simple pursuit" in the half-plane). Reproducing the reasonings given there, we may show that the length of the segment $[\xi_i(0), M]$ is stricly greater than the length of the segment $[\xi_i(\delta), M(\delta)]$. However, the length of the first segment $\ell = \hat{\rho}$ and the lenght of the second is $\hat{\rho}_{T-\delta}(x^{(1)}(\delta), x^{(2)}(\delta), y_0)$. This proves the decrease

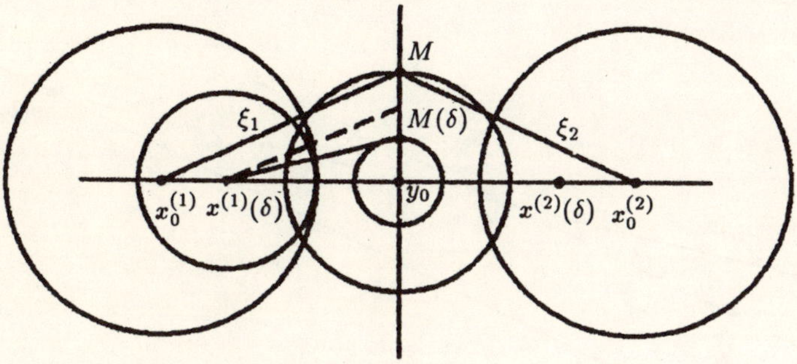

Fig. 22.

of $\hat{\rho}$ and implies the following theorem (see (Fig.22).

Theorem 2 *In the game of "simple pursuit" of Player E by the team* $P = \{P_1, P_2\}$, *the value of the game is* $\hat{\rho}_T(x_0^{(1)}, x_0^{(2)}, y_0)$, *and an optimal strategy for Player P is constructed according to Theorem 1, Ch. 3.*

4.4 Relations between maximin time of pursuit and time of absorption

Let $L_P^T(x)$ be the ℓ–neighbourhood of the set $C_P^T(x)$. Denote by $\bar{T}$ the first time instant when

$$L_P^{\bar{T}}(x) \supset C_E^{\bar{T}}(y). \tag{4.4.1}$$

N.N Krasovsky was the first to introduce the quantity $\bar{T}$ and called it the absorption time.

Under conditions of Theorem 3, Ch. 3, the value of the pursuit game of prescribe duration $\bar{T}$ is precisely equal to ℓ (the terminal payoff).

Evidently, the following inequality always holds

$$\Theta(x, y, \ell) \leq \bar{T}, \tag{4.4.2}$$

were $\Theta(x, y, \ell)$ is the maximin time up to the ℓ–capture instant (until the distance ℓ is approached), and $\bar{T}$ is the absorption time (ℓ–absorption). Indeed, moving in the game $\Gamma(x, y, \bar{T})$ (along the conditionally optimal trajectory $y'^*(t)$), in the optimal time game of pursuit $\Gamma(x, y, \ell)$ Player E can be sure that Player P cannot approach him on a distance not exceeding ℓ up to the instant Θ. Since the value of the game $\Gamma(x, y, \bar{T})$ is equal to ℓ, then by the instant T Player P ensures an approach to Player E on the distance ℓ, whenever the motion of the latter may be. Consequently,$\Theta \leq \bar{T}$ (see 4, Ch. 3).

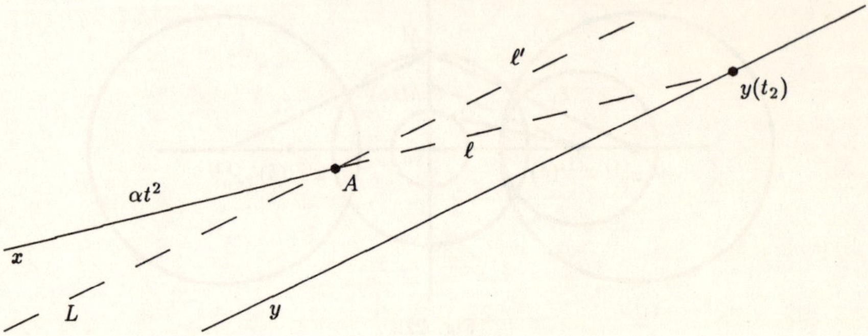

Fig. 23.

The inequality (4.2.2) may also be interpreted somewhat differently. In the game $\Gamma(x, y, \ell)$ Player P ensures an approach to Player E on a distance ℓ in the time $\Theta(x, y, \ell)$. Here he also reckons the cases of approaching Player E on the distance ℓ in the time less than Θ. At the same time, that the ℓ equals the value of the game $\Gamma(x, y, \bar{T})$ merely signifies that Player P can ensure an approach to Player E for a distance not exceeding ℓ exactly at the instant $\bar{T}$. The latter provides a more complex problem (for Player P), since the approach to Player E on the distance ℓ before the instant $\bar{T}$ is not taken into consideration.

Consider the game of "simple pursuit" in some closed convex set S on the plane. This means that there are phase constraints: the players P and E are not allowed to leave the set S as the game proceeds). We prove the following obvious lemma.

Lemma 3 *Let the point E move along the ray from the point y at a constant velocity β. Then for $\alpha > \beta$, $\rho(x, y) > \ell$, the equation*

$$\rho(x, y(t)) - \alpha t = \ell \quad (\ell \geq 0) \tag{4.4.3}$$

has a unique solution.

Proof: Suppose there are two solutions $t_1 > t_2 \geq 0$ (the equation (4.4.3) is a quadratic equation). Examine Fig.23. Here $|xA| = \alpha t_2$, $|Ay(t_2)| = \ell$. From the point A the straight line L is drawn parallel to $y(t)$. Prescribe Player P a motion from the point X to the point A with velocity α and further along L with a velocity β. The velocity $\beta < \alpha$ and therefore such a motion is possible up to the instant $t_1 > t_2$.

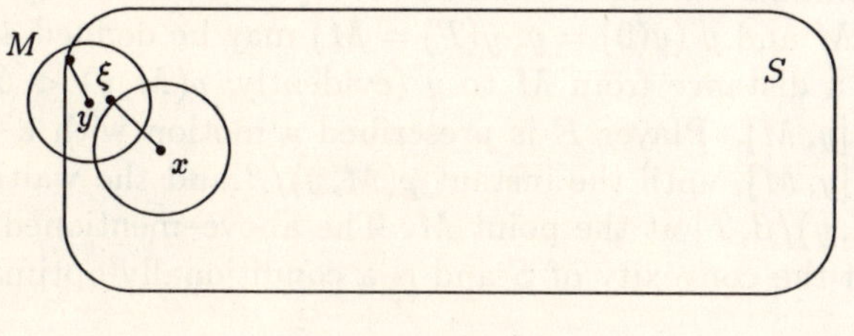

Fig. 24.

Let $\bar{x}(t)$ be a corresponding trajectory. By construction, $\rho(\bar{x}(t_1), y(t_1)) = \ell$. However, since the trajectory $\bar{x}(t)$ is a polygonal line, then it is evident that by employing a straight–line motion from x to $y(t_1)$, Player P will aproach Player E by the instant t_1 on a distance less than ℓ, i.e. $\hat{\rho}(x, y(t_1)) - \alpha t_1 < \ell$. This is inconsistent with the definition of t_1. We have proved the lemma. ∎

Theorem 3 *Let the conditions of Theorem 3, Ch. 3, be satisfied in the "simple pursuit" game of prescribed duration T in S. Then*

$$Val\Gamma(x, y, T) = \hat{\rho}_T(x, y) = \ell,$$

$$Val\Gamma(x, y, \ell) = \Theta(x, y, \ell) = T$$

(here $\Gamma(x, y, \ell)$ is the optimal time pursuit game until a distance $\ell > 0$ is reached).

Proof: The equality $Val\Gamma(x, y, T) = \hat{\rho}_T(x, y) = \ell$ follows from Theorem 3, Ch. 3. Denote by $M \in S$ a center of pursuit, i.e. the point for which

$$\ell = \rho(\bar{\xi}, M) = \max_{\eta \in C_E^T(y)} \max_{\xi \in C_P^T(x)} \rho(\xi, \eta) = \hat{\rho}_T(x, y).$$

The set $C_E^T(y)$ is an intersection of the sphere $S_y(\beta T)$, with radius βT ($v_1^2 + v_2^2 \leq \beta^2$) and its center at y, and the set S. Similarly, the set $C_P^T(x)$ is an intersection of the sphere $S_x(\alpha T)$, with radius αT ($u_1^2 + u_2^2 \leq \alpha^2$) and its center at x, and the set S. The point $\bar{\xi} \in S$ lies on the boundary of the sphere $S_x(\alpha T)$, since $M \in S$, $x \in S$ and because of the convexity of S the hole segment $[x, M] \subset S$ and $\bar{\xi} \in [x, M]$ because it is a point from $C_P^T(x)$ that is nearest to the point M (Fig.24).

Thus, the conditionally optimal trajectory for Player P is a straight line.

We show that among the conditionally optimal trajectories for Player E there is also the straight one. Indeed, if M is not on the boundary of S, then the point M belongs to the boundary of $S_y(\beta T)$, and Player E's trajectory is straight. If the point M belongs to the boundary of S, then any trajectory joining the points M and y ($y(0) = y$, $y(T) = M$) may be defined as follows.

Let $\rho(M, y)$ be a distance from M to y (evidently, $\rho(M, y) < \beta T$). Consider the segment $[y, M]$. Player E is prescribed a motion with a velocity β along the interval $[y, M]$, until the instant $\rho(M, y)/\beta$ and the waiting on the time interval $[\rho(M, y)/\beta, T]$ at the point M. The above–mentioned motion is possible because of the convexity of S and is a conditionally optimal straight trajectory.

As shown before, according to (4.2.2) $\Theta(x, y, \ell) \leq T$. Assume that the inequality $\Theta(x, y, \ell) < T$ is true.

Let $\bar{x}(t)$, $\bar{y}(t)$ be the corresponding conditionally–optimal trajectories in the optimal time pursuit game $\Gamma(x, y, \ell)$. As before, we may show that $\bar{x}(t)$, $\bar{y}(t)$ are the straight line trajectories emanating from the point x, y; here $\bar{x}(t)$ is focussed on the point $\bar{y}(\Theta(x, y, \ell))$ (the "capture point"). Since $\Theta(x, y, \ell) < T$ and the game is simple pursuit, the point $\bar{y}(\Theta(x, y, \ell)) \in C_E^T(y)$ (since E can realize the following motion, he moves in the a straight line towards the point $\bar{y}(\Theta(x, y, \ell))$ until the instant Θ and then stays at the point $\bar{y}(\Theta(x, y, \ell))$. on the time interval $[\Theta, T]$).

Now, let, $\bar{x}(t)$, $\bar{y}(t)$ be a pair of conditionally optimal trajectories in the game $\Gamma(x, y, \ell)$. Player E is prescribed a motion from the points y to the point M along a straight trajectory with velocity β and then a waiting at the point M until the instant $\Theta(x, y, \ell)$. The Player P cannot aproach him on a distance ℓ, since $\Theta < T$ (see Lemma 3) and

$$\rho(\bar{x}(\Theta), \bar{y}(\Theta)) > \rho(x(T), y(T)) = \rho(\xi, M) = \hat{\rho}_T(x, y) = \ell.$$

The contradiction obtained proves the theorem. ■

Theorem 3 contains the method of finding solution of the optimal time pursuit games in the "simple motion" case. The solution of the optimal time pursuit game (in the case if the conditions of Theorem 3 are fullfilled) is constructed as follows. For the initial conditions of the game $\Gamma(x, y, \ell)$ (the optimal time pursuit game from the initial condition x, y until the capture at the distance ℓ), the game $\Gamma(x, y, \bar{T})$ of prescribed duration $\bar{T}$ is constructed. Moreover, $\bar{T}$ is chosen from the condition $\bar{T} = \min\{T : \ Val\Gamma(x, y, T) = \ell\}$, i.e. $\bar{T}$ is the minimum duration of the game $\Gamma(x, y, T)$ for which its value is equal to ℓ (the capture radius in the game $\Gamma(x, y, \ell)$). If $\bar{T} = \infty$, then the ℓ–approach cannot be ensured by Player P in a finite time.

The examples of simple pursuit 2, 4, 5 show that the conditions of Theorem 3, Ch. 3 are satisfied. Consequently, Theorem 3 is applicable to them. In Example 2, the set S coincides with the entire plane, in Example 4 with the half plane and in Example 5 with the interior of the angle.

In Example 2, the function $\Theta(x, y, \ell)$ is computed in a simple way and is the time to the ℓ–capture instant provided that Player E moves from Player P in a straight line with the velocity β, and Player P moving towards Player E in a straight line at the velocity α. This gives us the following formula

$$\Theta(x, y, \ell) = \frac{\sqrt{(x_1 - y_1)^2 + (x_2 - y_2)^2} - \ell}{\alpha - \beta}. \tag{4.4.4}$$

The function $\Theta(x, y, \ell)$ has continuous partial derivatives everywhere except the surface

$$\sqrt{(x_1 - y_1)^2 + (x_2 - y_2)^2} = 0 \tag{4.4.5}$$

which corresponds to a coincidence of the positions of players P and E.

If the distance between the players at the initial instant is assumed to be strictly greater than zero and $\ell > 0$ then the players' trajectories do not penetrate the surface (4.4.5) during the game, and hence the function $\Theta(x, y, \ell)$ satisfies the equation

$$\max_{v \in V} \sum_{i=1}^{2} \frac{\partial \Theta}{\partial y_i} v_i + \min_{u \in U} \sum_{i=1}^{2} \frac{\partial \Theta}{\partial x_i} u_i = -1 \tag{4.4.6}$$

with the initial condition

$$\Theta(x, y, \ell)|_{\rho(x,y)=\ell} = 0.$$

Substituting the controls $\bar{v}$, $\bar{u}$ into (4.4.6) (see Example 2 in this chapter), on which max and min are realized in this equation, we obtain the equation with partial derivatives of the first order

$$-\alpha \left[\left(\frac{\partial \Theta}{\partial x_1} \right)^2 + \left(\frac{\partial \Theta}{\partial x_2} \right)^2 \right] + \beta \left[\left(\frac{\partial \Theta}{\partial y_1} \right)^2 + \left(\frac{\partial \Theta}{\partial y_2} \right)^2 \right] = -1 \tag{4.4.7}$$

with the initial condition

$$\Theta(x, y, \ell)|_{\rho(x,y)=\ell} = 0. \tag{4.4.8}$$

For the game of "simple pursuit" in the half–plane (see Example 4 in this chapter) and "simple pursuit" within the angle (see Example 5), the function $\Theta(x, y, \ell)$ cannot be computed explicitly, simple (see Example 2 in this chapter) it turns out to be the root of the following equation (see (4.2.2), (4.2.3)):

$$-\alpha\Theta + \sqrt{\left(x_1 - y_1 + \sqrt{(\beta\Theta)^2 - y_2^2}\right)^2 + x_2^2} = \ell.$$

As noted before, (see (4.2.2)), $\Theta(x, y, \ell) \leq \bar{T}$,i.e. the maximin time of pursuit until the ℓ–capture instant does not exceed the absorption time (ℓ–absorption). One obvious sufficient condition may be given for the equality in (4.2.2) to be satisfied.

Let $y^*(t)$ be an optimal trajectory and $v^*(t)$ be an optimal open–loop control for Player E in the game $\Gamma(x, y, \bar{T})$. and let $u^*(t)$ be an optimal open–loop control for Player P realizing the ℓ–capture of Player E in a minimum time provided that he employs the open–loop control $v^*(t)$ previously reported to Player P.

Denote by $t_\Pi(x, y; u^*(t), v^*(t))$ the corresponding time until the ℓ–capture instant. Then if

$$t_\Pi(x, y; u^*(t), v^*(t)) = \bar{T}, \tag{4.4.9}$$

then also

$$\Theta(x, y, \ell) = \bar{T}. \tag{4.4.10}$$

Indeed, since $\Theta(x, y, \ell) = \max_{v(t)} \min_{u(t)} t_\Pi(x, y, u(t), v(t))$, then from the definition of $u^*(t)$ we have

$$\Theta(x, y, \ell) \geq t_\Pi(x, y; u^*(t), v^*(t)) = \bar{T},$$

and, using (4.4.2), obtain (4.4.10).

Chapter 5

"Life line" game of pursuit

5.1 Definition of "life line" game

This chapter deals with a family of zero–sum two–person games, each being a model of pursuit in a closed convex set $S \subset R^2$. Two points – Pursuer P and Evader E – have modulus – constrained velocities and move within the set S with the possibility to switch the direction of motion (simple motion) at each instant of time. Evader E is said to be captured as soon as the distance between him and Pursuer P reaches the value less than or equal to $\ell > 0$. The number ℓ is called the capture radius. Player P seeks to approach Player E before the latter intersects the "life line", i.e. the boundary of the set S. Player P must stay in S throughout the duration of the game[1].

Motion equations are of the form

$$\dot{x}_i = u_i,$$
$$\dot{y}_i = v_i, \quad i = 1, 2,$$

$$u_1^2 + u_2^2 \leq \alpha^2,\ \alpha = const,\ \beta = const,\ v_1^2 + v_2^2 \leq \beta^2,\ \alpha > \beta \tag{5.1.1}$$

The game is studied in the class of piecewise open–loop strategies (see in 2, Ch. 2). The payoff function is defined in the following way.

Let $x(t)$, $y(t)$ be a solution of the system of differential equations (5.1.1) in the situation $(u(\cdot), v(\cdot))$ with the initial positions $x_0, y_0 \in S$ and let

$$t_{s_P} = \inf\{t : x(t) \in S\},$$

$$t_{s_E} = \inf\{t : y(t) \in S\}.$$

[1]The problem for the half–plane was first stated in [1]. The solution for $\ell = 0$ was obtained in [31], for $\ell \geq 0$ in [34]. Another method of solving the problem for the half–plane case is given in [43].

Then we require that $x(t) \notin S$ for all $t > t_{S_P}$ and $y(t) \notin S$ for all $t > t_{S_E}$ (the latter condition restricts the strategy classes for the players).

Next, let $t_P = \min\{t : \rho(x(t), y(t)) = \ell\}$ (if there are no such t_P, then t_P is assumed to be equal to ∞).

Payoff function. Let $x(t)$, $y(t)$ be the trajectories for the players P and E corresponding to the initial conditions $x_0, y_0 \in S$ in the situation $(u(\cdot), v(\cdot))$. Then the payoff function (Player P's payoff) is defined as follows:

$$K(x_0, y_0; u(\cdot), v(\cdot)) = \begin{cases} +1, & \text{if } t_P \leq t_{S_E}, \quad t_P < \infty \\ 0, & \text{if } t_P = t_{S_E} = \infty, \\ -1, & \text{if } t_P > t_{S_E}, \quad t_{S_E} < \infty \end{cases}$$

With the players' strategy sets and the payoff function defined, for each pair of initial positions $x_0, y_0 \in S$ the game is denoted by $\Gamma(x_0, y_0)$.

From the form of the payoff function it follows that the game $\Gamma(x_0, y_0)$ is the game of kind where Player P seeks to approach Player E before the latter intersects the boundary of the set S. Assume that the game is of zero-sum two-person type, i.e. Player E's payoff is Player P's payoff with opposite sign.

We first prove basic theorems for the games with Player E discriminated for the time $\delta > 0$ ahead and then show that these theorems also hold for the game $\Gamma(x_0, y_0)$ (without discrimination for Player E).

5.2 Discrete game

Let δ be a fixed partition of the interval $[0, \infty)$ with a constant step $\Delta > 0$. The game $\Gamma_\Delta(x_0, y_0)$ is the game with discrimination of Player E over the class of strategies P and E defined in 2, Ch. 2.

Let $\bar{u}_\delta(\cdot)$ be a fixed strategy for Player P which can, in any situation $(\bar{u}_\delta(\cdot), v_\delta(\cdot))$, ensure the ℓ-capture of Player E in the game $\Gamma_\Delta(x_0, y_0)$. Let $C_{\bar{u}_0}(x_0, y_0)$ be a set of all possible locations of Player E's position at the ℓ-capture instant in the situation $\bar{u}_\delta(\cdot), v_\delta(\cdot)$ for different $v_\delta(\cdot) \in E$ (in what follows, Player E's position at the ℓ-capture instant is also called the "capture point"). Evidently, if the set $C_{\bar{u}_\delta(\cdot)}(x_0, y_0)$ has a nonempty intersection with a complement of the set S Player P can not guarantee the ℓ-capture of E in the set S using $\bar{u}_\delta(\cdot)$. Consequently, in this case Player E can always choose a strategy under which the ℓ-capture in the situation $(\bar{u}_\delta(\cdot), v_\delta^*(\cdot))$ occurs in the complement of the set S. There may exist a whole set of such strategies.

Theorem 1 *Suppose the following conditions are satisfied:*

1. *the intersection of $C_{\bar{u}_\delta(\cdot)}(x_0, y_0)$ with the complement of the set S is nonempty;*

2. *there exist the strategy* $v_\delta^*(\cdot)$ *under which, in the situation* $(\bar{u}_\delta(\cdot), v_\delta^*(\cdot))$, *Player P speeds up to the "capture point" from the initial point* x_0, *and the "capture point" (at the instant of* ℓ *capture) does not belong to* S.

The speed up to "capture point" is interpreted as follows.

Let $\bar{x}(t)$, $y^*(t)$ *be the trajectories of the players P and E in the situation* $(\bar{u}_\delta(\cdot), v_\delta^*(\cdot))$ *from the initial states* x_0, y_0. *Then we say that the strategy* $\bar{u}_\delta(\cdot)$ *in the situation* $(\bar{u}_\delta(\cdot), v_\delta^*(\cdot))$ *speeds up to the "capture point" if, in his motion along the trajectory* $\bar{x}(t)$, *Player P, realizes the* ℓ*-capture of Player E as fast as possible (provided that Player E moves along the trajectory* $y^*(t)$ *and his trajectory is previously reported to Player P).*

The solution $(\bar{u}_\delta(\cdot), v_\delta^*(\cdot))$ *constitutes an equilibrium point in the game* $\Gamma_\delta(x_0, y_0)$, *and the value of the game is* -1 *(i.e. under conditions of the theorem, none of the strategies of Pursuer P can ensure the* ℓ*-capture of Player E in S).*

Proof: Let $x^*(t)$, $y_\delta^*(\cdot)$ be trajectories of the players P and E in the situation $(\bar{u}_\delta(\cdot), v_\delta^*(\cdot))$ in the game $\Gamma_\delta(x_0, y_0)$. Denote by $\tilde{v}_\delta^*$ a strategy for Player E under which he chooses at each time instant a direction of motion along the trajectory $y^*(t)$ irrespective of Player P's actions. Whenever the strategy $u_\delta(\cdot) \in P$ may be, the ℓ–capture in the situation $(u_\delta(\cdot), \tilde{v}_\delta^*(\cdot))$ cannot take place before the instant of time

$$t_P(\bar{u}_\delta(\cdot), \tilde{v}_\delta^*(\cdot)) = t_P = t_P(\bar{u}_\delta(\cdot), v_\delta^*(\cdot)),$$

because in the situation $(\bar{u}_\delta(\cdot), \tilde{v}_\delta(\cdot))$ or $(u_\delta(\cdot), v_\delta^*(\cdot))$ Player P speeds up to the "capture point". This means that

$$t_P(u_\delta(\cdot), \tilde{v}_\delta^*(\cdot)) \geq t_P(\bar{u}_\delta(\cdot), \tilde{v}_\delta^*(\cdot)) > t_{S_E}$$

for all strategies $u_\delta(\cdot)$ for Player P. We have thus proved the theorem. ■

When the set $C_{\bar{u}_\delta(\cdot)}(x_0, y_0)$ is contained in S, the ℓ–capture of Player E in S is always possible if the strategy $\bar{u}_\delta(\cdot)$ is adopted by Player P. In this case, the strategy $\bar{u}_\delta(\cdot)$ turns out to be optimal for Player P (it may not be unique), and the value of the game is $+1$. Optimal for Player E is any strategy $v_\delta(\cdot)$.

Now, to solve the game it suffices to establish the existence of strategy $v_\delta(\cdot)$ for Player E satisfying the conditions of Theorem 1, or to establish the existence of strategy $\bar{u}_\delta(\cdot)$ under which the set $C_{\bar{u}_\delta(\cdot)}(x_0, y_0)$ is contained in S.

We now turn to construction of the pair of strategies $\bar{u}_\delta^*(\cdot)$, $\bar{v}_\delta^*(\cdot)$ possessing the property mentioned above.

Suppose that from the initial position y_0 Player E chooses a constant control $v = const$ ($|v| = \beta$) (i.e. he moves in a straight line with a velocity β along some ray y_0A). For each such motion, Player P has a unique constant control $\bar{u} = const$ which guarantees him the ℓ–capture of Player E from the position

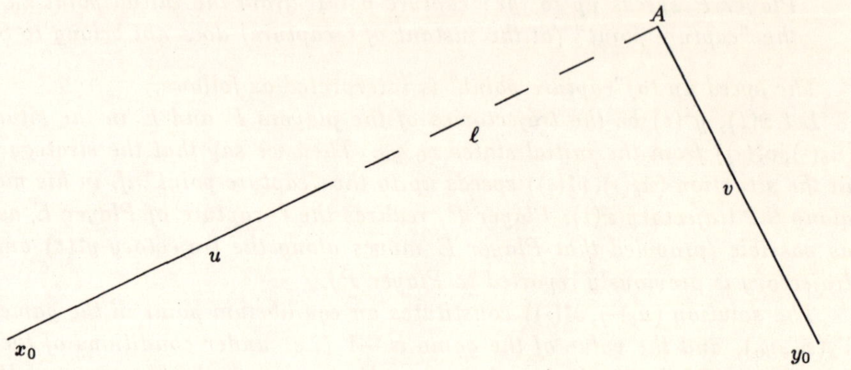

Fig. 25.

x_0 in a minimum time. This control prescribes him the motion, which is called the speed up to – "capture point", along the ray x_0B focussed on the "capture point" (Fig. 25).

Definition 1. Each open–loop control $v(t)$ transferring $y(t_k)$ to $y(t_{k+1})$ is associated with the open–loop control $\bar{v}(t)$, $|\bar{v}(t)| = \beta$ transferring $y(t_k)$ to $y(t_{k+1})$ along a single–vertex polygonal trajectory $\bar{y}(t)$ with its vertex at the point $\bar{y}(\bar{t}_k) = \bar{y}\left(\frac{t_{k+1}-t_k}{2}\right)$ (Player E's speed being maximal along $\bar{y}(t)$). Π–strategy $u_\delta^\Pi(\cdot)$ places each state $x(t_k)$, $y(t_k)$ and the interval of the open–loop control $u(t)$ for $t \in [t_k, t_{k+1})$, i.e. the point $y(t_{k+1})$, in correspondence with the interval of the open–loop control $u(t)$ for $t \in [t_k, t_{k+1})$ which, for $t \in [t_k, \bar{t}_k)$ coincides with the open–loop control realizing the speed up to "capture point" motion provided that from the time t_k, and throughout the duration of the game, Player E moves at a maximum speed along the straight trajectory $y(t_k)$, $\bar{y}(\bar{t}_k)$ towards the point $\bar{y}(\bar{t}_k)$ and for $t \in [\bar{t}_k, t_{k+1})$ it coincides with the open–loop control realizing the speed up to – "capture point" motion provided that from the time $\bar{t}_k$ and throughout the duration of the game, Player E moves at a maximum speed along the straight trajectory $\bar{y}(\bar{t}_k)$, $y(t_{k+1})$ towards the point $y(t_{k+1})$. Moreover, if at an instant $t' \in [t_k, t_{k+1})$ it turns out that $\rho(x(t'), y(t')) = \ell$, then under the strategy $u_\delta^\Pi(\cdot)$ Player P chooses $u(t)$ on the time interval $[t', t_{k+1})$ in such a way that $\rho(x(t), y(t)) = \ell$ for all $t \in [t', t_{k+1})$ [31][2].

Let us determine the structure of the set $C_{u_\delta^\Pi(\cdot)}(x_0, y_0)$. Consider the subset

[2] Π–strategy has the following important property. If in the situation $(u_\delta^\Pi(\cdot), v(\cdot))$ the "capture point" $y(t_P) \in S$, then for all $0 \le t \le t_P$ the trajectory $x(t) \in S$,i.e. for the capture of Player E, Player P does not leave the set S.

$\bar{C}_{u_\delta^\Pi(\cdot)}(x_0, y_0)$ which is obtained from the latter provided that Player E adopts constant strategies $v_\delta(\cdot)$, $|v_\delta(\cdot)| = \beta$ only (i.e. Player E moves only at a constant speed β along the rays emanating from the point y_0.

Let $\bar{\bar{C}}_{u_\delta^\Pi(\cdot)}(x_0, y_0)$ be a convex hull of the set $\bar{C}_{u_\delta^\Pi(\cdot)}(x_0, y_0)$ We show that

$$\bar{\bar{C}}_{u_\delta^\Pi(\cdot)}(x_0, y_0) = C_{u_\delta^\Pi(\cdot)}(x_0, y_0)$$

and the set $\bar{C}_{u_\delta^\Pi(\cdot)}(x_0, y_0)$ is the boundary of the set $\bar{\bar{C}}_{u_\delta^\Pi(\cdot)}(x_0, y_0)$ (see in 3).

Theorem 2 *Assume that the set $\bar{C}_{u_\delta^\Pi(\cdot)}(x_0, y_0)$ intersects the complement of the set S. Then, for Π–strategy to satisfy the condition of Theorem 1, it is sufficient that there be the strategy $v^*(\cdot)$ for Player E, under which he moves along some ray at the speed β, and such that the "capture point" in the situation $(u_\delta^\Pi(\cdot), v^*(\cdot))$ belongs to the complement of the set S.*

Proof: This theorem follows from that in the situation $(u_\delta^\Pi(\cdot), v^*(\cdot))$ (see the definition of the Π–strategy) Player P also moves along some ray passing through x_0 to the "capture point", i.e. he realizes the speed–up to the "capture point". ■

Now we show that under the condition $\bar{C}_{u_\delta^\Pi(\cdot)}(x_0, y_0) \bigcap S \neq \Lambda$ such a strategy exists (see Lemma 1). This is a ray passing through y_0 to the "capture point". Now we have the theorem that follows.

Theorem 3 *To avoid the ℓ–capture of Player E in S, it is necessary and sufficient that the set $C_{u_\delta^\Pi(\cdot)}(x_0, y_0)$ have a nonempty intersection with the complement of the set S.*

5.3 Proof of one geometric lemma

Let Cartesian coordinates x, y be introduced on the plane. Assume that at $t = 0$ the pursuer P is situated in the point $(0, 0)$, and the Evader E in $(a, 0)$, $a > \ell$. For the simplicity of notation, we set $a = 1$, $\beta = \lambda$, $\lambda < 1$. Denote the position of P at the time instant t by $P(t) = (x_P^t, y_P^t)$, and Player E by $E(t) = (x_E^t, y_E^t)$[3].

The players P and E are assumed to move in a straight lines with maximal speeds along the rays L_P and L_E "meet" at the time instant t at the point w, $w \in L_P \bigcap L_E$. The coordinates of the point $w(x, y)$ then satisfy the equations

$$x^2 + y^2 = (t + \ell)^2, \; t > 0$$

[3]Here, for convenience of investigation of the geometric properties of capture curves, we sightly depart from the adopted notation $P - (x_1, x_2), E - (y_1, y_2)$.

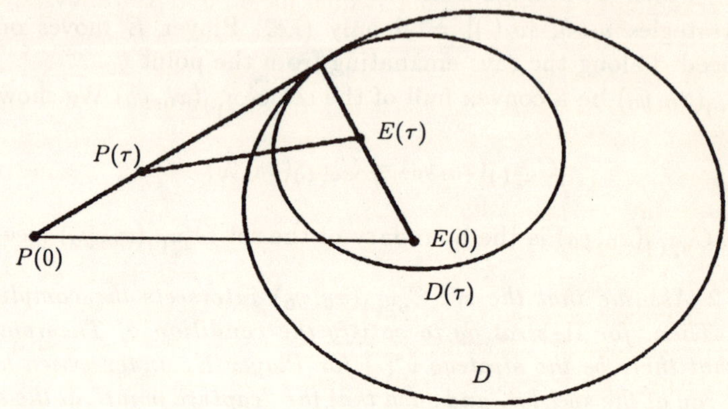

Fig. 26.

$$(x-a)^2+y^2=(\lambda t)^2, \tag{5.3.1}$$

Equations (5.3.1) on the plane x, y define the curve called the Cartesian oval [44]. Since $t>0$ then we get a section of the Cartesian oval. Denote it by D.

Assume that at the time instant τ the players P and E change the direction of the motion (Player P changes this direction so as to realize the ℓ–capture). Then we have a new "capture" curve $D(\tau)$ given by the equations

$$(x-x_P^\tau)^2+(y-y_P^\tau)^2=(t-\tau+\ell)^2,\ \tau<t$$

$$(x-x_E^\tau)^2+(y-y_E^\tau)^2=[\lambda(t-\tau)]^2 \tag{5.3.2}$$

(t is the "capture" time for the players P and E when they are moving along L_P and L_E, respectively). Note that with $\tau=0$ the equations (5.3.2) convert to (5.3.1).

We have the following basic lemma [34].

Lemma 1 *$D(\tau)$ is closed, bounded convex curve for $\lambda<1$, $a>\ell$, $\tau<t$. Moreover, with $\tau>0$ the curve $D(\tau)$ lies in a closed region bounded by the curve $D=D(0)$ (Fig.26).*

Before proving the lemma, three simple remarks will be given without proof.

Remark 1. If any straight line intersects the curve K at no more than two points, then this curve is strictly convex.

Remark 2. Let

$$y^2-x^2=m^2,$$

$$(y-y_0)^2-\mu^2(x-x_0)^2=n^2,$$

$$\mu>1,\ y_0\geq 0,\ |x_0|\geq y_0.$$

be upward branches of two hyperbolas which have no more than two common points.

Remark 3. By the homothety of the branch γ of the hyperbola with the homothetic center at the point $M \in \gamma$ this branch changes to the branch γ' of the hyperbola, γ and γ' being tangent at the point $M \in \gamma$. If the homothety coefficient $0 < k < 1$, then the convex region G bounded by γ contains γ'. If, however,$k > 1$, then γ' is outside G.

Proof:(of the lemma). In Euclidean space with coordinates (x, y, t), the equations from (5.3.2) give the right circular cones with the axes parallel to the axis t. Under the condition $\tau < t$ each of the equations (5.3.2) give the upper (i.e., directed to the increase of t) sections of these cones $K(P, \tau)$, $K(E, \tau)$ and the curve $D(\tau)$ is a projection of the intersection line of these cones onto the plane x, y.

We show that $D(\tau)$ is convex. To this end, (see Remark 1), is satisfies to check whether the straight line intersects $D(\tau)$ at no more than two points. Let p be a straight line in the plane x, y and π the plane passing through p and being parallel to the axis t. Here $\pi \bigcap K(P, 0)$ and $\pi \bigcap K(E, 0)$ are the upper branches of two hyperbolas having, by Remark 2, no more than two common points. We have thus proved the convexity.

We prove that $D(\tau)$ with $\tau > 0$ lies in a closed (convex) region which is bounded by $D = D(0)$. To do this, it suffices to establish that, for any straight line passing through the point $w = D(\tau) \bigcap D$, the chord in $D(\tau)$ is the portion of the chord of the same straight line which is cut out in D.

Suppose that the players P and E are moving along the straight lines L_P and L_E, respectively. At the instant of time t we associate them with the points $\tilde{P}(t) \in K(P, 0)$ and $\tilde{E}(t) \in K(E, 0)$ whose projections are respectively $P(t)$ and $E(t)$. The points $\tilde{P}(t)$ and $\tilde{E}(t)$ move along the rays $\tilde{L}_P \subset K(P, 0)$ and $\tilde{L}_E \subset K(E, 0)$ intersecting at the point $\tilde{w}$ (on the plane x, y the point $\tilde{w}$ is projected into the point w). Since the directions of motions of players P and E change at the instant $t = \tau$, then the points $P(t)$ and $E(t)$ with $t > \tau$ are associated with the points $\tilde{P}(t) \in K(P, \tau)$ and $\tilde{E}(t) \in K(E, \tau)$ in much the same way (Fig.27).

Let p be the straight line on the plane passing through the point w, and π the plane passing through the straight line p and parallel to the axis t. Introduce the notation $\gamma(P, \tau) = \pi \bigcap K(P, \tau)$, $\gamma(E, \tau) = \pi \bigcap K(E, \tau)$.

The curves $\gamma(P, 0)$, $\gamma(E, 0)$, $\gamma(P, \tau)$, $\gamma(E, \tau)$ are the branches of hyperbolas lying in the plane π (see the Fig.27). All of them are passing thought the point $\tilde{w}$.

The real axes of these hyperbolas are parallel to the axis t. The center of hyperbola $\gamma(P, 0)$ is at the point A, that is an ortogonal projection onto π of the point $(0, 0, -\ell)$- the vertex of the cone $K(P, 0)$. In the same manner

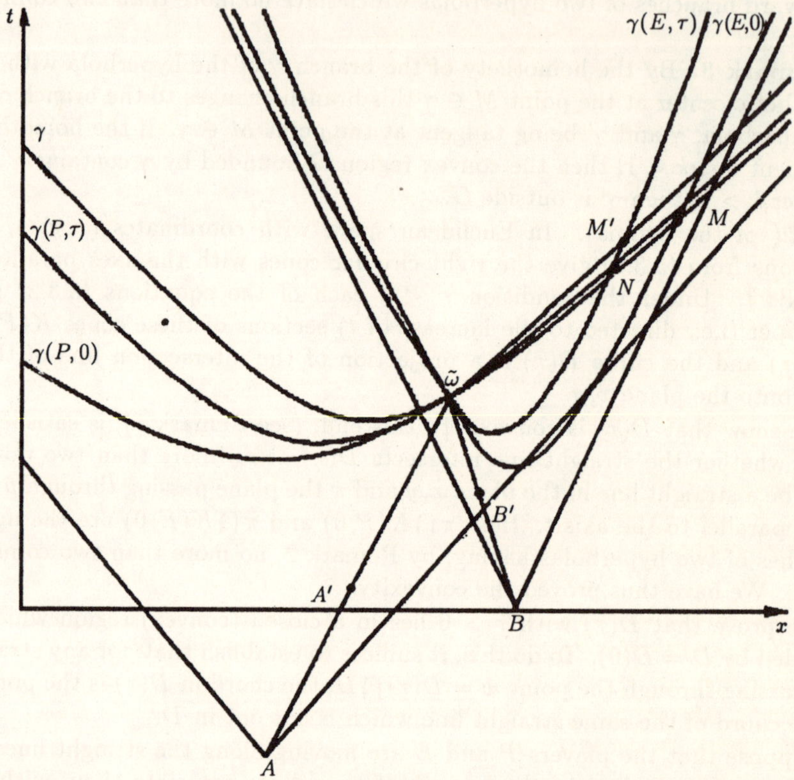

Fig. 27.

the center of the hyperbola $\gamma(P,\tau)$ is the projection onto the same plane π of the vertex of the cone $k(P,\tau)$ so that the points A, A' and $\tilde{w}$ lie on the same straight line, that is a projection of the straight line $P(0)w$ onto π. From this it follows that $\gamma(P,\tau)$ is obtained from $\gamma(P,0)$ by the homothety with homothetic coefficient

$$k_P = \frac{\tilde{w}A'}{\tilde{w}A} = \frac{\tilde{t} - \tau + \ell}{\tilde{t} + \ell},$$

where $\tilde{t}$ is the applicate of the point $\tilde{w}$. Similarly let B and B' be centers of hyperbolas $\gamma(E,0)$ and $\gamma(E,\tau)$. The points B, B' and $\tilde{w}$ lie on the same straight line, so that $\gamma(E,\tau)$ is obtained from $\gamma(E,0)$ by the homothety with the homothetic coefficient

$$k_E = \frac{\tilde{w}B'}{\tilde{w}B} = \frac{\tilde{t} - \tau}{\tilde{t}}.$$

Clearly, $k_E < k_P < 1$.

Let M be the second intersection point of $\gamma(P,0)$ and $\gamma(E,0)$, and N be the second intersection point of the hyperbolas $\gamma(P,\tau)$ and $\gamma(E,\tau)$. We will show that the projection of the segment $\tilde{w}M$ onto the plane x,y covers the projection of $\tilde{w}N$. Indeed, for the homothety with the coefficient k_E and its center at $\tilde{w}$, the hyperbola $\gamma(E,0)$ converts to the hyperbola $\gamma(E,\tau)$, and the hyperbola $\gamma(P,0)$ to the branch γ, where the second intersection point M' lies on the segment $\tilde{w}M$. The curve $\gamma(P,\tau)$ is obtained from γ by the homothety, with the coefficient $k = \frac{k_P}{k_E} > 1$, so that N (see Remark 3) lies on the arc $\tilde{w}M$ of the curve $\gamma(E,\tau)$. Since $\gamma(E,\tau)$ is uniquely projected onto the plane x,y, then the projection of N lies on the projection of the segment $\tilde{w}M$. We have thus proved the lemma. ■

5.4 Basic theorem

By employing Lemma 1, we now show that D is oval, that is a set of "captures points" for the straight–line motion of Player E, and it coincides with the boundary of the set $\bar{C}_{u_\delta^\Pi(\cdot)}(x,y)$.

Let $\bar{D}(x,y)$ be the region that is bounded by oval D constructed for the initial states x,y of the players P and E. We choose δ in such a way that the diameter of the set $C_E^\delta(y)$ is less than ϵ. Then the "capture point" in the situation $(x,y;u_\delta^\pi(\cdot),v_\delta(\cdot))$ under any strategy $v_\delta(\cdot)$ belongs to the ϵ– neighborhoods of the set $\bar{D}(x,y)$. Indeed, let $x(t)$, $y(t)$ be the motion trajectories for the players in the situation $(x,y;u_\delta^\pi(\cdot),v_\delta(\cdot))$, and t_0, $t_1,\ldots$, $t_n,\ldots$ be the break points of the time interval $[0,\infty)$ with the step δ; $\bar{y}(t)$ is a polygonal trajectory corresponding to the control $\bar{v}(t)$ which constructs the imaginary motion of Player E as described in the definition of Π–strategy. According to Lemma 1, we have

$$\bar{D}(x,y) \supset \bar{D}\left(x\left(\frac{t_1}{2}\right), \bar{y}\left(\frac{t_1}{2}\right)\right) \supset \bar{D}(x(t_1),\bar{y}(t_1)) =$$

$$= \bar{D}(x(t_1),y(t_1)) \supset \bar{D}\left(x\left(\frac{t_2-t_1}{2}\right), \bar{y}\left(\frac{t_2-t_1}{2}\right)\right) \supset$$

$$\supset \bar{D}(x(t_2),\bar{y}(t_2)) = \bar{D}(x(t_2),y(t_2)) \supset \ldots$$

$$\ldots \supset \bar{D}(x(t_{k+1}),\bar{y}(t_{k+1})) = \bar{D}(x(t_{k+1}),y(t_{k+1})) \supset \ldots$$

From these inclusions it follows that the "capture point" of the fictitious Player $\bar{E}$ moving along $\bar{y}(t)$ belongs to the oval $\bar{D}(x,y)$. By the definition of Π-strategy, however, after the ℓ–capture of Player $\bar{E}$ the ℓ–capture of Player E takes place in the time not exceeding δ, i.e. in the ϵ–neighborhood of the set $\bar{D}(x,y)$.

Theorem 4 *For any $\epsilon > 0$ there is such $\delta > 0$ that the ϵ–neighborhood of the set $\bar{D}(x, y)$ contains the set $C_{u_\delta^\Pi(\cdot)}(x, y)$. In this case, the boundary of the set $\bar{D}(x, y)$ is the oval $D(x, y)$, representing a set of "capture points" for the straight–line motions of Player E in the situation $(x, y; u_\delta^\Pi(\cdot), v_\delta(\cdot))$, which coincides with the set $\bar{C}_{u_\delta^\Pi(\cdot)}(x, y)$. Moreover, for the convex hull of the set $\bar{C}_{u_\delta^\Pi(\cdot)}(x, y)$ the following equality holds:*

$$\bar{\bar{C}}_{u_\delta^\Pi(\cdot)}(x, y) = \bigcap_{\epsilon > 0} C_{u_\delta^\Pi(\cdot)}(x, y).$$

As a corollary of the Theorem 4 and Theorems 1–3 in this chapter, we obtain the following theorem that is basic for this chapter.

Theorem 5 *The strategy $u_\delta^\Pi(\cdot)$ is optimal in the game $\Gamma_\delta(x, y)$ for Player P in that if $\bar{D}(x, y)$ has a nonempty intersection with the complement of the set S the ℓ–capture of E in S is unposible. If one ϵ– neighborhood of $\bar{D}(x, y)$ is contained in S, then Player E is ℓ–captured in S under any strategies $v_\delta(\cdot) \in E$.*

5.5 Rejection of discrimination

Rejection of discrimination against Player E is possible in one of the following cases (in the case involved $\ell > 0$).

1. There exists such $\epsilon > 0$ that the ϵ–neighborhood of the set $\bar{D}(x, y, \ell)$ (in the section, the notation $\bar{D}(x, y, \ell)$ is used instead of $\bar{D}(x, y)$ with the emphasis on the dependence of the set $\bar{D}$ upon the ℓ–capture radius) is contained in S (by the ϵ–neighborhood of the set is meant a union of the ϵ–neighborhood of all points appearing in this set).

2. The set S does not contain the sets $\bar{D}(x, y, \ell)$

Theorem 6 *Let the point x, $y \in S$ and condition 1 be satisfied. Then the Player P can ensure the ℓ–capture of Player E independent of latter's actions, i.e. the value of the game $V(x, y) = +1$.*

Let the point x, $y \in S$ and condition 2 be satisfied. Then Player E can ensure the avoidance of the ℓ–capture by Player P independent of the latter's actions, i.e. the value of the game $V(x, y) = -1$.

Proof: We prove the first part of the theorem. Let $\bar{D}_\epsilon(x, y, \ell)$ be the ϵ–neighborhood of the set $\bar{D}(x, y, \ell)$. Note that the set $\bar{D}_\epsilon(x, y, \ell)$ varies as x, y, ℓ vary. This, in particular, means that for any $\epsilon' > 0$ there is such $\delta > 0$ that for all x', y', ℓ', for which $\rho(x', y', \ell'; x, y, \ell) < \delta$,

$$\bar{D}_\epsilon(x', y', \ell') \subset \bar{D}_{\epsilon+\epsilon'}(x, y, \ell).$$

Take ϵ', $\epsilon > 0$ such that $\bar{D}_{\epsilon+\epsilon'}(x,y,\ell) \subset S$. This can be done by Condition 1. Assume that $\bar{\epsilon} = \epsilon' + \epsilon$ and $\delta(\bar{\epsilon})$ corresponds to this $\bar{\epsilon} > 0$, and $\delta'(\epsilon)$ corresponds to $\epsilon > 0$ so that there is $C_{u^E_{\delta'}(\cdot)}(x,y) \subset \bar{D}_\epsilon(x,y,\ell)$ (Theorem 5, Ch. 5). Then for all x', y', ℓ' such that $\rho(x',y',\ell';x,y,\ell) < \delta(\bar{\epsilon})$, we have

$$\bar{D}_\epsilon(x',y',\ell') \subset \bar{D}_{\bar{\epsilon}}(x,y,\ell) \subset S.$$

Take $\ell' < \ell$, $|\ell' - \ell| < \frac{\delta(\bar{\epsilon})}{3}$ and set $\delta_1 < \min[\frac{\delta(\bar{\epsilon})}{3\beta}, \delta'(\epsilon)]$. Consider the following strategy for Pursuer P.

During the time δ_1 he stays at the point x. Next, he applies the strategy $u^\Pi_{\delta_1}(\cdot)$ (Π-strategy) to the state of Player E, i.e. at the point $y(t-\delta_1)$. Such a piecewise open–loop strategy $u^\Pi_{\delta_1}(\cdot)$ is actually realizable, since at each instant t_k Player P has information on $y(t_k)$, and hence on the line segment $[y(t_k - \delta_1), y(t_k)]$. Therefore, we may say that Player P follows Player E's trace $y(t-\delta_1)$, with discrimination being for the time $\delta_1 > 0$. Then, by Theorem 5 from the previous section, adopting the strategy $u^\Pi_{\delta_1}(\cdot)$, Player P ensures the ℓ'–capture of Player E's trace $y(t-\delta_1)$ in the set $\bar{D}_\epsilon(x,y,\ell')$. However

$$\rho(x(t),y(t)) \le \rho(x(t),y(t-\delta_1)) + \rho(y(t-\delta_1),y(t)).$$

Consider the inequality for the instant $\bar{t}$ when Player P realizes the ℓ–capture of Player E's trace $y(\bar{t}-\delta_1)$. From the definition of δ_1, we have that in the time δ_1,the point y cannot go away from the point $y(\bar{t}-\delta_1)$ to a distance exceeding $\delta_1\beta < \frac{\delta(\bar{\epsilon})}{3}$. This gives

$$\rho(x(\bar{t}),y(\bar{t})) \le \rho(x(\bar{t}),y(\bar{t}-\delta_1)) +$$

$$+\rho(y(\bar{t}-\delta_1),y(\bar{t})) \le \ell' + \frac{\delta(\bar{\epsilon})}{3} \le \ell,$$

i.e. that Player P realizes the ℓ'–capture of Player E's trace $y(t-\delta_1)$ implies the ℓ–capture of Player E by Player P. Hence Player P ensures the ℓ–capture of Player E in the set

$$\bar{D}_\epsilon(x,y,\ell') \subset \bar{D}_{\bar{\epsilon}}(x,y,\ell) \subset S.$$

This exactly means that, when Condition 1 is satisfied, $Val\Gamma(x,y) = +1$.

Assume that Condition 2 of the theorem has been satisfied. In this case, any strategy $v_\delta(\cdot)$, which in the situation $(u^\Pi_\delta(\cdot), v_\delta(\cdot))$ guaranteed Player E the ℓ–capture outside S (at the points of the set $\bar{D}_\epsilon(x,y,\ell)$ which do not belong to S), guarantees him the ℓ–capture outside S and without discrimination, since Player P's position is not improving here. This completes the proof of the theorem. ■

Corollary. The whole set of points x, $y \in S$ is broken into three sets.

1. The winning set of Player P

$$W_P = \{x, y : \exists \epsilon > 0,\ \bar{D}_\epsilon(x, y, \ell) \subset S\}.$$

2. The winning set of Player E

$$W_E = \{x, y : \bar{D}(x, y, \ell) \not\subset S\}$$

3. Neutral outcome set

$$W_0 = \{x, y : \exists \epsilon > 0,\ \bar{D}_\epsilon(x, y, \ell) \not\subset S\},$$

but there is no such $\epsilon > 0$ that $\bar{D}_\epsilon(x, y, \ell) \subset S$.

For all $x,\ y \in W_P$ the value of the game $Val\Gamma(x, y) = V(x, y) = +1$, for all $x,\ y \in W_E$ the value of the game $Val\Gamma(x, y) = V(x, y) = -1$. For any points of the set $x,\ y \in W_0$, the sets $D(x, y, \ell)$ and S have common boundary points. For all $x,\ y \in W_0$ the E-capture of Player E is not possible without discriminating him. Evidently, any trajectory $x(t)$, $y(t)$ passing from W_P to W_E must intersect W_0. In this sense, the set W_0 separates the sets W_P and W_E.

Definition 2. The set W_0 is called a barrier in the game $\Gamma(x, y)$.

In the game of kind, it is essential to define not only the optimal strategies for the players P and E but also the sets W_P and W_E. To do this, it is sufficient to construct the separating set W_0.

The barrier has the following property: for any point $x,\ y \in W_P$ there is such a strategy of Player P which guarantees him that the trajectory $x(t)$, $y(t)$ from the initial states x, y does not intersect the barrier during the game, and for any point $x, y \in W_E$ there is such a strategy of Player E which guarantees him that the trajectory $x(t)$, $y(t)$ from the initial states x, y does not intersect the barrier. These strategies are optimal for the players P and E in the game $\Gamma(x, y)$

By employing this property, R.Isaacs in [1] derived a barrier equation for the "life–line" game in the half–plane.

However, a similar definition may be provided for the barrier W_0 in the case of an arbitrary convex set S.

With the knowledge of the boundary of the set S, and having an explicit expression for the boundary of the set $\bar{D}(x, y, \ell)$ we may provide a geometrically simple construction of the barrier W_0 as a set of points $x, y \in S$, where the boundary of the set $\bar{D}(x, y, \ell)$ (that is a Cartesian oval in the case $\ell > 0$ and an Apollonius circle in the case $\ell = 0$) is tangent to the boundary of the set S.

Lemma 2 *Let $S_1 \subset S_2 \subset \ldots \subset S_n \subset \ldots$ and let $W_P^{(1)}$, $W_P^{(2)}, \ldots, W_P^{(n)}, \ldots$ and $W_E^{(1)}, \ldots, W_E^{(n)}, \ldots$ be the corresponding sequence of winning sets for the players P and E. Then*

$$W_P^{(1)} \supset W_P^{(2)} \supset \ldots \supset W_P^{(n)} \supset \ldots,$$

$$S_{k+1} \bigcap W_E^{(k)} \supset W_E^{(k+1)}.$$

Proof: The definition of $W_P^{(k)}$ suggests that for any x, $y \in W_P^{(k)}$ there is such $\epsilon > 0$ for which $\bar{D}_\epsilon(x, y, \ell) \subset S^{(k)}$. Since $S^{(k)} \supset S^{(k-1)}$, then $\bar{D}_\epsilon(x, y, \ell) \subset S^{(k-1)}$, and hence x, $y \in W_P^{(k-1)}$. If now x, $y \in W_E^{(k)} \bigcap S^{(k+1)}$, then $\bar{D}(x, y, \ell) \not\subset S^{(k)}$, and since

$$S^{(k+1)} \supset S^{(k)}, \text{ then } \bar{D}(x, y, \ell) \not\supset S^{(k+1)}.$$

■

Lemma 3 *The sets W_P and W_E are open sets if S is closed.*

Proof: Let x, $y \in W_p$. Then, by definition, there is such $\epsilon > 0$ that $\bar{D}_\epsilon(x, y, \ell) \subset S$. But the continuous dependence of the set $\bar{D}(x, y, \ell)$ on x, y, ℓ imply the existence of such $\delta(\epsilon)$ that for all x', y', for which $\rho(x', y'; x, y) < \delta(\epsilon)$,

$$\bar{D}(x', y', \ell) \subset \bar{D}_{\frac{\epsilon}{2}}(x, y, \ell).$$

Since

$$\bar{D}(x, y, \ell) \subset \bar{D}_{\frac{\epsilon}{2}}(x, y, \ell) \subset \bar{D}_\epsilon(x, y, \ell) \subset S,$$

then x', $y' \in W_P$.

Now let x, $y \in W_E$. Then $\bar{D}(x, y, \ell) \not\subset S$. Take the sequence x_n, $x_n \to x$, y and consider a corresponding sequence of sets $\bar{D}(x_n, y_n, \ell)$. From the continuous dependence of $\bar{D}(x, y, \ell)$ on x, y, ℓ we have that for any $\epsilon > 0$ there is such $N(\epsilon)$ that for all $n \geq N(\epsilon)$

$$\bar{D}(x_n, y_n, \ell) \supset \bar{D}_{-\epsilon}(x, y, \ell),$$

where $\bar{D}_{-\epsilon}(x, y, \ell) \subset \bar{D}(x, y, \ell)$ is a set of points for which the set $\bar{D}(x, y, \ell)$ is the ϵ– neighborhood. Since S and $\bar{D}(x, y, \ell)$ are closed and $\bar{D}(x, y, \ell) \not\subset S$, then there is a point $\bar{z} \in \bar{D}(x, y, \ell)$, $\bar{z} \notin S$ which is interior for $\bar{D}(x, y, \ell)$.

Denote by ϵ_0 the diameter of the maximum sphere with its center at z_0, contained in $\bar{D}(x, y, \ell)$. Evidently,

$$\bar{D}(x_n, y_n, \ell) \supset \bar{D}_{\frac{-\epsilon_0}{2}}(x, y, \ell),$$

and since $\bar{z} \in \bar{D}_{\frac{-\epsilon_0}{2}}(x, y, \ell)$, $\bar{z} \notin S$, then

$$\bar{D}_{\frac{-\epsilon_0}{2}}(x, y, \ell) \not\subset S,$$

and hence for all $n \geq N(\epsilon)$

$$\bar{D}(x_n, y_n, \ell) \not\subset S$$

This proves the lemma. ■

Lemma 4 *Let S_1, $S_2, \ldots$, $S_n, \ldots$ be a sequence of closed convex sets, and $W_P^{(1)}, W_P^{(2)}, \ldots, W_P^{(n)}$; $W_E^{(1)}, \ldots, W_E^{(n)}, \ldots$; $W_0^{(1)}, \ldots, W_0^{(n)}, \ldots$ be a corresponding sequence of winning sets for the Players P and E, and barriers. Then, if $S = \bigcap S_n$, the sets W_P, W_E where $W_P \bigcup W_0 = \bigcap(W_P^{(n)} \bigcup W_0^{(n)})$, $W_E = \bigcup W_E^{(n)}$ are winning sets for the players P and E in the game defined in the set S.*

Proof: We prove only the first part of the lemma, since the second part is proved in much the same way.

Let x, $y \in \bigcap(W_P^{(n)} \bigcup W_0^{(n)})$. This means that for all n $\bar{D}(x, y, \ell) \subset S^{(n)}$. Evidently, $\bar{D}(x, y, \ell) \subset S = \bigcap S^{(n)}$, and hence x, $y \in W_P \bigcup W_0$.

Now let x, $y \in W_P \bigcup W_0$. In this case,there exist such $\epsilon > 0$ that $\bar{D}_\epsilon(x, y, \ell) \subset S$, or the boundary of $\bar{D}(x, y, \ell)$ and S have some common points (here $\bar{D}(x, y, \ell) \subset S$). Thus, in both cases $\bar{D}(x, y, \ell) \subset S$, and hence $\bar{D}_\epsilon(x, y, \ell) \subset S^{(n)}$, $n = 1, 2, \ldots$, since $S = \bigcap S^{(n)}$. Therefore, x, $y \in W_P^{(n)} \bigcup W_0^{(n)}$.

This completes the proof of the lemma. ■

Generalization to the case of incomplete information is given in [49].

5.6 Multiplayer "life line" games

The games under study are the models of pursuit in a closed convex set S on the plane, where several players – a team of pursuers $P = \{P_1, \ldots, P_m\}$ and evaders $E_1, \ldots, E_n$– move within the set S with modulus–constrained velocities and have, at each instant of time, an opportunity to change the direction of motion. The evader E_j is said to be captured when such $j \in \{1, \ldots, m\}$ is found that the positions of the players P_i and E_j at some instant are spaced on a distance ℓ ($\ell \geq 0$) apart. Each evader seeks to reach the boundary of the set S before he is captured by one of the pursuers [24].

The pursuit is carried out on the Euclidean plane and the players' positions are determined by the point $(x^1, \ldots, x^m; y^1, \ldots, y^m)$, where $x^k = (x_1^k, x_2^k)$ is a position of pursuer P_k,a member of the team P and $y^j = (y_1^j, y_2^j)$ is a position of the evader E_j. At each instant of the time the players may choose the direction of motion (the direction of the velocity vector), and the velocity within prescribed limits (here the maximum velocity is constant and equals α_j for $P_j \in P$ and β_j for E_j).

Motion equations are of the form

$$\begin{aligned} \dot{x}_1^i &= u_1^i, & i &= 1, \dots, m \\ \dot{x}_2^i &= u_2^i, \\ \dot{y}_1^j &= v_1^j, & j &= 1, \dots, n \\ \dot{y}_2^j &= v_2^j, \\ (u_1^i)^2 + (u_2^i)^2 &\leq \alpha_i^2, & i &= 1, \dots, m \\ (v_2^j)^2 + (v_2^j)^2 &\leq \beta_j^2, & j &= 1, \dots, n \end{aligned} \tag{5.6.1}$$

Here $\alpha_i = const$, $i = 1, \dots, m$, $\beta_j = const$, $j = 1, \dots, n$, and $\min_i \alpha_i > \max_j \beta_j$

Let $\{x^1(t), \dots, x^m(t), y^1(t), \dots, y^n(t)\}$ be a solution of the system of differential Eq.(5.6.1) with the initial conditions $\xi^1, \dots, \xi^m; \eta^1, \dots, \eta^n \in S$ and $t_{S_{E_j}} = \inf\{t : \ y^j(t) \notin S\}$, $j = 1, \dots, n$. Then with $t > t_{S_{E_j}}$ $y^j(t) \notin S$, $j = 1, \dots, n$.

For any $\xi = (\xi^1, \dots, \xi^m)$, where $\xi^i \in S$, $i = 1, \dots, m$ and $\eta = (\eta^1, \dots, \eta^n)$, where $\eta^j \in S$, $j = 1, \dots, n$, we define the differential game of pursuit and denote it by $\Gamma(m, n; \xi, \eta)$. As noted before, the game $\Gamma(m, n; \xi, \eta)$ is the game with $n+1$ players – the pursuer team $P = \{P_1, \dots, P_m\}$ and the evaders E_j, $j = 1, \dots, n$. The Cartesian product of the sets $P = \prod_i^m P^i$, $P^i = \{u^i(\cdot)\}$ and the sets E^j, $j = 1, \dots, n$, $E^j = \{v^j(\cdot)\}$ are the strategy sets for the pursuer team $P = \{P_i, \dots, P_m\}$ and the evaders $E_1, \dots, E_n$, respectively.

Let $u(\cdot) \in P$ and $v^j(\cdot) \in E^j$, $j = 1, \dots, n$. Each situation $(u(\cdot), v^1(\cdot), \dots, v^n(\cdot))$ uniquely determines $n+m$ trajectories called the trajectories of pursuers $P_1, \dots, P_m$ and the evaders $E_1, \dots, E_n$.

Let

$$t_P^{ij} = \min\{t : \ \rho(x^i(t), y^j(t)) = \ell\}, \ i = 1, \dots, m, \ j = 1, \dots, n.$$

In any situation, the matrix $\{t_P^{ij}\}$ is uniquely defined, though it is evident that some of its elements may be equal to infinity. The payoff functions are given as follows:

$$K_{E_j}(\xi, \eta; u(\cdot), v^1(\cdot), \dots, v^n(\cdot)) =$$

$$= \begin{cases} +1, & \text{if } t_{S_{E_j}} < t_P^{ij},\ t_P^{ij} \neq \infty,\ i = 1, \dots, m, \\ -1, & \text{if there is such } i_0 \text{, that } t_{S_{E_j}} > t_P^{i_0 j},\ t_{S_{E_j}} \neq \infty,\ j = 1, \dots, n, \\ 0, & \text{if } t_{S_{E_j}} = \infty \text{ and } t_P^{ij} = \infty \text{ for all } i. \end{cases}$$

$$K_P(\xi, \eta; u(\cdot), v^1(\cdot), \dots, v^n(\cdot)) = -\sum_{j=1}^n K_{E_j}(\xi, \eta, u(\cdot), v^1(\cdot), \dots, v^n(\cdot)).$$

When the strategy spaces are constructed for players P and $E^j, j = 1, \ldots, n$, with the value of payoff function determined in each situation, we determine a family of games in normal form $\Gamma(m, n; \xi, \eta)$, where $\xi = (\xi^1, \ldots, \xi^m)$, $\eta = (\eta^1, \ldots, \eta^n)$, $\xi^i \in S$, $i = 1, \ldots, m$, $\eta^j \in S$, $j = 1, \ldots, n$. The game $\Gamma(m, n; \xi, \eta)$ is a generalization of the "life–line" games discussed in 1–5, Ch. 4, to the case where several pursuers and evaders take part in the process of pursuit–evasion

Consider the game $\Gamma(m, 1; \xi, \eta)$ where one evader is pursued by a team of pursuers (Player P is allowed to leave the set S prior to the capture of Player E).

Definition 3. The strategy $u^{\Pi}(\cdot) \in P = \bigcap_i^m P^i$ is called Π–strategy if

$$u^{\Pi} = (u^{1\Pi}(\cdot), \ldots, u^{m\Pi}(\cdot)),$$

where $u^{i\Pi}(\cdot)$, $i = 1, \ldots, m$ is the Π–strategy in the game $\Gamma(\xi^i, \eta)$ considered on the strategy sets P_i and E (for the definition of Π–strategy see 2 in this chapter).

We will prove some auxiliary statements.

Lemma 5 *For any $\epsilon > 0$ there is such $\delta > 0$ that in the discrimination game $\Gamma_\delta(x, y)$ (with one pursuer P and one evader E) in this situation $(u_\delta^{\Pi}(\cdot), v_\delta(\cdot))$ the points $y(t)$ with $0 \leq t \leq t_P$ belong to $\bar{D}_\epsilon(x, y, \ell)$.*

Proof: Let $y(\bar{t})$ be such a point. Consider the strategy $\bar{v}(\cdot)$ prescribing that Player E should move from the point y to the point $y(\bar{t})$ along the trajectory $y(t)$ and remain at the point $y(\bar{t})$ beginning from the instant $\bar{t}$. Since for any $\epsilon > 0$ it is possible to choose $\delta > 0$ in such a way that in any situation $(u_\delta^{\Pi}(\cdot), v(\cdot))$ the point of ℓ–capture will belong to $\bar{D}_\epsilon(x, y, \ell)$ this is valid also for the situation $(u_\delta^{\Pi}(\cdot), \bar{v}(\cdot))$ However, the "capture point" in the situation $(u_\delta^{\Pi}(\cdot), \bar{v}(\cdot))$ is exactly the point $y(\bar{t})$ and we have $y(\bar{t}) \in \bar{D}_\epsilon(x, y, \ell)$. This completes the proof of the lemma. ■

We now consider the game $\Gamma(m, 1; x, y)$. Denote by $\Gamma_\delta(m, 1; x, y)$ the game corresponding to the game $\Gamma(m, 1; x, y)$ where Player E is discriminated for the time $\delta > 0$. Let $x^1, \ldots, x^m$ be the initial positions of the players $P_1, \ldots, P_m$, respectively. Introduce the notation

$$\bar{D}(\bar{x}, y, \ell) = \bigcap_{i=1}^{m} \bar{D}(x^i, y, \ell)$$

Theorem 7 *For any $\epsilon > 0$ there is such $\delta > 0$ that in the game $\Gamma_\delta(m, 1; x, y)$ the set $C_{u_\delta^{\Pi}(\cdot)}(x, y) \subset \bar{D}_\epsilon(x, y, \ell)$.*

Proof: Let $y(t)$ be a trajectory in some situation $(u_\delta^{\Pi}(\cdot), v_\delta(\cdot))$. Let $t_P^1, \ldots, t_P^m$ be the instants, when the players $P_1, \ldots, P_m$ capture Player E, ordered so that $t_P^1 \geq t_P^2 \geq \ldots \geq t_P^m$

Consider the trajectory $y(t)$ on the interval $[0, t_P^1]$. By Lemma 5,

$$y(t) \in \bar{D}_\epsilon(x^1, y, \ell).$$

Similarly, for $t \in [0, t_P^i]$

$$y(t) \in \bar{D}_\epsilon(x^i, y, \ell)$$

Evidently, $y(t)$ for $t \in [0, t_P^m]$ belongs to all $\bar{D}_\epsilon(x^i, y, \ell)$, $i = 1, \ldots, m$, i.e. for $t \in [0, t_P^m]$

$$y(t) \in \bar{D}_\epsilon(\bar{x}, y, \ell) = \bigcap_{i=1}^{m} \bar{D}_\epsilon(x^i, y, \ell).$$

Hence, in particular, we have that the point $y(t_P^m)$ which is the "capture point" in the situation $(u_\delta^\Pi(\cdot), v_\delta(\cdot))$, belongs to $\bar{D}_\epsilon(\bar{x}, y, \ell)$. Since the strategy $v_\delta(\cdot)$ is arbitrary, then

$$\bar{D}_\epsilon(\bar{x}, y, \ell) \supset C_{u_\delta^\Pi(\cdot)}(x, y).$$

Note that the game $\Gamma(m, 1, x, y)$ is zero–sum and the payoff of the team P equals $+1$ as soon as one of the players P entering the team P approaches Player E on a distance ℓ ($\ell \geq 0$). ■

As a corollary of the previous theorem, we then have the following theorem.

Theorem 8 *In the game $\Gamma_\delta(m, 1, \xi, \eta)$ the Π–strategy is optimal for the team of pursuers $P = \{P_1, \ldots, P_m\}$ in the sense that if $\bar{D}(\bar{x}, y, \ell) \not\subset S$, then Player E has the strategy $\tilde{v}(\cdot)$ under which the ℓ–capture in S is not possible for any strategy $u_\delta(\cdot)$.*

In fact, since $\bar{D}(\bar{x}, y, \ell) \not\subset S$, then there exist $\tilde{y} \in D$ and $\tilde{y} \notin S$.

Let $\tilde{v}(\cdot)$ be a strategy prescribing Player E the following behavior move with maximum velocity in a straight line from the point y to the point $\tilde{y}$. Then the ℓ–capture of Player E on the interval $[y, \tilde{y})$ is not possible under any strategy $u_\delta(\cdot)$

Proof: Suppose the opposite is true. Let $u_\delta(\cdot)$ be a strategy under which the "capture point" in the situation $(\tilde{u}_\delta(\cdot), \tilde{v}_\delta(\cdot))$ belongs to the interval $[y, \tilde{y})$, and the capture may be performed by Player P_i. In this case, because the Π–strategy against any straight–line motion of Player E presupposes the minimum capture time in the situation $(u_\delta^\Pi(\cdot), \tilde{v}(\cdot))$, then the ℓ–capture may be performed by Player P_i on the interval $[y, \tilde{y})$ using Π–strategy. This means that there are the points of interval $[y, \tilde{y})$ belonging to the boundary of the set $\bar{D}(x^i, y, \ell)$. This, however, is not possible, since the interval $[y, \tilde{y})$ is not contained in the intersection of convex sets $\bar{D}(\bar{x}, y, \ell) = \bigcap_i^m \bar{D}(x^i, y, \ell)$ (whose boundary contains no interval). This completes the proof of the theorem. ■

As in 5, we can give up the discrimination requirement. Here we only formulate the theorem without repeating the earlier reasonings.

Theorem 9 *Suppose that in the game $\Gamma(m,1;x,y)$ there is such $\epsilon > 0$ that the set $\bar{D}_\epsilon(\bar{x},y,\ell)$ is contained in E. Then the team $P = \{P_1,\ldots,P_m\}$ can ensure the ℓ-capture of Player E (the capture of Player E by one the players P_i) independent of the latter's actions in the set S $(\ell > 0)$.*

Assume that in the game $\Gamma(m,1;x,y)$ the set S does not contain the set $\bar{D}(\bar{x},y,\ell)$ Then Player E can ensure the avoidance the ℓ-capture by the team P (the avoidance of the ℓ-capture by all players P_i independent of the actions of the team P).

In the first case the value of the game is $+1$, and in the second -1.

As in 5, we can define the notion of the winning set for the team P, the winning set for Player E, and the barrier.

For the solution of the game $\Gamma(m,n;\xi,\eta)$, we did not manage to find optimal strategies for the players and prove the existence of Nash equilibrium in pure strategies [2]. However, by employing the integer programming methods, we succeed in estimating the greatest guaranteed payoff for the team of pursuers $Val\Gamma(m,n;\xi,\eta)$.

The value $Val\Gamma(m,n;\xi,\eta)$ is estimated as follows. Assume that a set of pursuers (a team) $P = \{P_i,\ldots,P_m\}$ is divided into n groups $M_1,\ldots,M_n$ so that, prior to the game termination, each of these groups can pursue only one player $E_1,\ldots,E_n$ under the Π-strategy $(m \geq n)$.

With each fixed partitioning of the set of pursuers into the groups $M_1,\ldots,$ M_n the maximum payoff $f(M_1,\ldots,M_n)$ for the team P is obtained as a solution of the following assignment problem [4]:

$$\max \sum_{i=1}^{m} \sum_{j=1}^{n} \xi_{ij}\alpha_{ij} \tag{5.6.2}$$

provided that $\sum_{i=1}^{n} \xi_{ij} \leq 1$, $j = 1,\ldots,n$; $\sum_{j=1}^{n} \alpha_{ij} = 1$, $i = 1,\ldots,n$ (here α_{ij} is the payoff of the i-th group M_i when Player E_j is pursued in accordance with the Π-strategy. It can be found with the help of Theorem 9).

Evidently,

$$Val\Gamma(m,n;\xi,\eta) \geq \max_{M_1,\ldots,M_n} f(M_1,\ldots,M_n) \tag{5.6.3}$$

where $\{M_1,\ldots,M_n\}$ is a set of all possible partitions of the team $P_1,\ldots,P_m$ into n pursuing groups.

Note that since $n \geq Val\Gamma(m,n;\xi,\eta)$ always, then as soon as n is obtained on the right-hand side of the inequality (5.6.3) it converts to equality, and the optimal strategy for the pursuer team is the division, at the initial instant of the game, into the groups $M_1,\ldots,M_n$, giving the payoff $\max_f(M_1,\ldots,M_n)$, and the pursuit of the players E_j, for which ξ_{ij} is obtained as the solution of problem (5.6.2), carried out by the group M_i under the Π-strategy.

Remark 1. Let S be a closed convex set. Consider the game with the "life line" in the complement of the set S. Here the set $C_{u^E(\cdot)}(x,y)$ is the same as in Theorem 7. So all the reasonings holds if in the situation $(u^\Pi(\cdot), v(\cdot))$ the trajectory of Player P does not intersect the boundary of the set S before the game terminates. Otherwise the Π–strategy is inadmissible, since it does not satisfy the conditions (*) (see 2 in this chapter). This, however, is not the case if the set $M(x,y)$, which is a convex hull stretched over $C_{u^\Pi(\cdot)}(x,y)$ and y, does not intersect the set S, i.e. the following theorem holds.

Theorem 10 *Assume that the set $M(x,y)$ is contained in the complement of the set S. Then the optimal strategy for the Player P is the Π–strategy, and the ℓ–capture of the player E is always possible in the complement of the set S.*

Remark 2. Consider the game $\Gamma(x,y)$ from the initial positions $x,y \in S$ for which $C_{u^\Pi(\cdot)}(x,y) \subset S$. By theorem 5, the ℓ–capture of Player E is always possible in S. In this case the following statement of the problem has a sense Player E seeks to minimize a distance between himself and the boundary of the set S at the "capture" instant (he seeks to be captured nearer to his "own bank" – the boundary of the set S). In this case, the equilibrium is of the form $(u^\Pi(\cdot), v^*(\cdot))$, where $v^*(\cdot)$ is a strategy for Player E, prescribing that he should move along the ray joining the point y with the point of Cartesian oval D which is nearest to the boundary of the set S.

Chapter 6

Differential games with incomplete information

6.1 Pursuit games with delayed information for player P

The games of pursuit with incomplete information are an immediate generalization of the complete information games of pursuit. The simplest is the case with Player P acquires information on the phase state of Player E with the delay $\ell > 0$, and Player E has complete information. When dealing with the game with prescribed duration, a satisfactory theory is developed for such games and a structure of optimal strategies is determined.

This section described the pursuit games with prescribed duration and delayed information for Player P that are continuous counterparts of Example 3 from Ch. 1 (see [25],[26],[29]).

Let a number $\ell > 0$ be given and called the information delay. For $0 \leq t \leq \ell$, at each time instant t Player P knows his state $x(t)$, the time t and Player E's state at the initial instant y_0. For $\ell \leq t \leq T$, at each time instant t Player P knows his state $x(t)$, the time t and Player E's state at time instant $t - \ell$, $y(t - l)$, and Player E's knows his state $y(t)$, the opponents state $x(t)$ and the tme t. His payoff is equal to $\rho(x(T), y(T))$. The game is zero–sum. Denote it by $\Gamma(x_0, y_0, T)$

Pure piesewise open–loop strategies. By the pure piesewise open–loop strategy $v(\cdot)$ of Player E is meant a pair $\{\tau, \beta\}$, where τ is a partitioning of the time interval $[0, T]$ by a finite number of points $0 = t_1 < t_2 < \ldots < t_s = T$, and b is a mapping which places each state $x(t_k)$, $y(t_k)$, t_k in correspondence with the measurable open–loop control $v(\cdot)$ of Player E defined for $t \in [t_k, t_{k+1})$. By the pure piecewise open–loop strategy $u(\cdot)$ for Player P is meant a pair $\{\sigma, \alpha\}$, wher σ is an arbitrary partitioning of the time inteval $[0,T]$ by a finite number of points $0 = t_1' < t_2' < \ldots t_k' = T$ and α is a mapping whith places each state $x(t_k')$, $y(t_k' - \ell)$, t_k' for $t_k' > \ell$ and $x(t_k')$, y_0, t_k' for $t_k' \leq \ell$ in correspondens with the measurable control $u(t)$ for Player P when $t \in [t_k', t_{k+1}')$.

The set of all pure piesewise open–loop startegies for the players P and E are denoted respectively P and E.

The game developes according to the system of controlled differential equations

$$\dot{x} = f(x,u),$$
$$\dot{y} = g(y,v), \tag{6.1.1}$$

assuming that all the conditions which ensure the existense and uniqueness of a solution of the system (6.1.1) on time interval $[0,T]$ for any pair of measurable open–loop controls $u(t)$, $v(t)$ are satisfied. This ensures the existence of a unique solution of (6.1.1) when the players P and E adopt the piecewise open–loop strategies $u(\cdot) \in P$, $v(\cdot) \in E$ correspondingly from the given initial conditions x_0, y_0.

Thus, in any situation $(u(\cdot), v(\cdot))$, the payoff function is uniquely difened:

$$K(x_0, y_0; u(\cdot), v(\cdot)) = \rho(x(T), y(T)), \tag{6.1.2}$$

where $x(t)$, $y(t)$ is a solution to (6.1.1) with the initial condition x_0, y_0 in the situation $(u(\cdot), v(\cdot))$, and ρ is the Euclidean distance.

It is well known, and this may be illustrated by referring to some simple examples, that generally

$$\sup_{v(\cdot)\in E} \inf_{u(\cdot)\in P} K(x_0, y_0; u(\cdot), v(\cdot)) \neq \inf_{u(\cdot)\in P} \sup_{v(\cdot)\in E} K(x_0, y_0; u(\cdot), v(\cdot)), \tag{6.1.3}$$

since the game under study $\Gamma(x_0, y_0, T)$ is not a complete information game. From (6.1.3) it follows that, in general, the ϵ–equilibrium point exists not for all $\epsilon > 0$. Therefore, to obtain the equilibrium point, we will follow the way proposed by J.V. Neumann and O. Morgenstern for finite positional games with incomplete information [2],[8]. We extend the strategy spaces of the players P and E to the so–called mixed piecewise open–loop behavior strategies (MPOLBS) which presuppose the possibility of random choice of controls at each step. We will show that for such classes of strategies the equality (6.1.3) is satisfied.

Mixed piesewise open–loop behavior strategies. By (MPOLBS) of Player P is meant a pair $\mu(\cdot) = \{\tau, \alpha\}$, where τ is an arbitrary partitioning of the time interval $[0,T]$ by a finite number of points $0 = t_1 < t_2 < \ldots < t_k = T$, and α is a map ping which places the state $x(t_k)$, $y(t_k - \ell)$, t_k with $t_k > \ell$ and the state $x(t_k)$, y_0, t_k for $t_k \leq \ell$ in correspondence with probability distribution

$\alpha_k(\cdot)$ concentrated on a finite number of measurable open–loop controls $u(t)$ for $t \in [t_k, t_{k+1})$. Similarly, by MPOLBS of PLayer E is meant a pair $\nu(\cdot) = \{\sigma, \beta\}$, where σ is an arbitrary partitioning of the time interval $[0, T]$ by a finite number of points $0 = t'_1 < t'_2 < \ldots < t'_s = T$, and β is a mapping which places the state $x(t'_k)$, $y(t'_k)$, t'_k in correspondence with the probability distribution $\nu_k(\cdot)$ concentrated on a finite number of measurable open–loop controls $v(t)$ for $t \in [t'_k, t'_{k+1})$.

MPOLBS sets for the players P and E are denoted respectively by $\bar{P}$ and $\bar{E}$ (compare these with the "behavior strategies" from [13]).

Each pair of MPOLBS $\mu(\cdot)$, $\nu(\cdot)$ with the fixed initial conditions x_0, y_0 induces the probability distribution on the space of trajectories $x(t)$, $y(t)$, $x(0) = x_0$, $y(0) = y_0$, therefore by the payoff $M(x_0, y_0; \mu(\cdot), \nu(\cdot))$ in MPOLBS is meant the expectation of the payoff $K(x_0, y_0; u(\cdot), v(\cdot))$ averaged over the distributions over the trajectory spaces induced by MPOLBS $\mu(\cdot)$, $\nu(\cdot)$.

By defining the spaces of strategies $\bar{P}$ and $\bar{E}$ and the payoff M, we define a mixed extension $\bar{\Gamma}(x_0, y_0, T)$ of the game $\Gamma(x_0, y_0, T)$ with the initial position x_0, y_0, T.

Introduce the following auxiliary definition. Let $C_E^T(y)$ be a reachability set for Player E. Denote by $\bar{C}_E^T(y)$ a convex hull of the set $C_E^T(y)$

$$\gamma(y, T) = \min_{\eta' \in \bar{C}_E^T(y)} \max_{\eta'' \in C_E^T(y)} \rho(\eta', \eta''). \tag{6.1.4}$$

Set $\gamma(y, T)$ be achieved at the points $(\bar{\bar{y}}, \bar{y})$, so that

$$\min_{\eta' \in \bar{C}_E^T(y)} \max_{\eta'' \in C_E^T(y)} \rho(\eta', \eta'') = \rho(\bar{\bar{y}}, \bar{y}). \tag{6.1.5}$$

It follows from the definition of $\bar{\bar{y}}$ that this is a center of the minimal sphere containing the set $C_E^T(y)$. Hence the point $\bar{\bar{y}}$ is unique. At the same time, there are at least two points of tangency of this set and the minimal sphere containing it, which coincide with the points $\bar{y}$.

Jet $y(t)$ be a trajectory $y(0) = y_0$ of Player E for $0 \le t \le T$. When Player E moves along $y(t)$, the value of the quantity $\gamma(y(t), T - t)$ varies. Let $\bar{\bar{y}}(t)$ be a trajectory of the point $\bar{y}$ from (6.1.5) corresponding to the trajectory $y(t)$.

In what follows, we will be dealing with the case where for all trajectories $y(t)$ $\bar{\bar{y}}(T) \in C_P^T(x)$. The point $\bar{M}$ will be referred to as the center of pursuit if

$$\gamma(\bar{M}, \ell) = \max_{y' \in C_E^{T-\ell}(y)} \gamma(y', \ell). \tag{6.1.6}$$

is achieved at this point. Thus,

$$\gamma(\bar{M}, \ell) = \max_{y' \in C_E^{T-\ell}(y)} \left[\min_{\eta' \in \bar{C}_E^{\ell}(y')} \max_{\eta'' \in C_E^{\ell}(y')} \rho(\eta', \eta'') \right].$$

Consider an auxiliary simultaneous zero–sum game of pursuit over a convex hull of the set $C_E^T(y)$. Player P chooses a point $\eta' \in \bar{C}_E^T(y)$, and Player E a point $\eta'' \in C_E^T(y)$. The choices are made simultaneously, and Player P with the choice of η' does not know the choice of η'' by Player E, and conversely. Player E has the payoff $\rho(\eta', \eta'')$. Denote the value of this game by $V(y, T)$ to emphasize the dependence of the game value on the parameters y, T determining the strategy sets $\bar{C}_E^T(y)$ and $C_E^T(y)$ for the players P and E. The game in normal form is written as follows:

$$\Gamma(y, T) = < \bar{C}_E^T(y), C_E^T(y), \rho(y', y'') > .$$

The strategy set $\bar{C}_E^T(y)$ for Player P (minimizer) is convex as the convex hull of the set $C_E^T(y)$ is. The function $\rho(y', y'')$ is also convex in its arguments and continuous. For such games, we may employ the following theorem (see Theorem 5, Ch. I).

Theorem 1 *In the game $\Gamma(y, T)$ there is the equilibrium in mixed strategies. The optimal strategy for Player P is pure, and the one for Player E prescribes a positive probability to not more than the $(n+1)$ points of the set $C_E^T(y)$, $(C_E^T(y) \subset R^n)$. The value of the game is equal to*

$$V(y, T) = \min_{\eta' \in \bar{C}_E^T(y)} \max_{\eta'' \in C_E^T(y)} \rho(\eta', \eta''),$$

or $\gamma(y, T) = V(y, T)$.

The optimal strategy for Player P in the simultaneous game $\Gamma(y, T)$ is the choice of a center of the minimal sphere $\bar{\bar{y}}$ containing the set $\bar{C}_E^T(y)$. Indeed, for any point $\eta'' \in \bar{C}_E^T(y)$ we have

$$R = \max_{\eta'' \in C_E^T(y)} \rho(\bar{\bar{y}}, \eta'') \geq$$

$$\geq \min_{\eta' \in \bar{C}_E^T(y)} \left[\max_{\eta'' \in C_E^T(y)} \rho(\eta', \eta'') \right] = V(y, T) = R$$

where R is the radius of the minimal sphere containing the set $\bar{C}_E^T$, whence we get

$$\rho(\bar{\bar{y}}, \eta'') \leq V(y, T)$$

for all $\eta'' \in C_E^T(y)$, which exactly proves the optimality of the strategy $\bar{\bar{y}}$ for Player P.

We establish conclusively that the strategy for Player P in the game $\Gamma(y, T)$ coincides with the choice of the (unique) point $\bar{\bar{y}}$, and the optimal strategy for Player E prescribes positive probabilities to not more than $(n+1)$ points of tangency of the minimal sphere, containing the set $\bar{C}_E^T(y)$, and the set $C_E^T(y)$. The value of the game is equal to the radius of this sphere (see Example 8, Ch. 1).

Consider a simultaneous game $\Gamma(\bar{M}, \ell)$, where $\bar{M}$ is the center of pursuit defined in (6.1.6). Denote by $\bar{y}_1(\bar{M}), \ldots, \bar{y}_{k+1}(\bar{M})$ the points of the set $C_E^\ell(\bar{M})$ which enter into the mixed optimal strategy spectrum for Player E in the game $\Gamma(\bar{M}, \ell)$, and by $\bar{\bar{y}}(\bar{M})$ the center of the minimal sphere containing the set $\bar{C}_E^\ell(\bar{M})$, i.e. the optimal strategy for Player P in the game $\Gamma(\bar{M}, \ell)$.

Definition 1. The trajectory $y_k^*(t)$ is called conditionally optimal if $y^*(0) = y_0$, $y^*(T-\ell) = \bar{M}$, $y^*(T) = \bar{y}_k(\bar{M})$ for some particular k from $k = 1, \ldots, n+1$. For each k there may be several conditionally optimal trajectory for Player E.

Theorem 2 *Suppose $T \geq \ell$ and for any $\epsilon > 0$ by the time instant T Player P can ensure the ϵ–capture of the center $\bar{\bar{y}}(t)$ of the minimal sphere containing the set $C_E^\ell(y(t-\ell))$. Then the game $\Gamma(x_0, y_0, T)$ has the ϵ–equilibrium in mixed piecewise open–loop behavior strategies. The ϵ–optimal strategy for Player P is pure and coinsides with any of strategies which ensures $\frac{1}{2}$ ϵ–capture of the point $\bar{\bar{y}}(t)$ The optimal strategy for Player E is mixed: during the time $0 \leq t \leq T-\ell$ he must move to the point $\bar{M}$ along any conditionally optimal trajectory $y^*(t)$ and then, with probabilities $p_1, \ldots, p_{n+1}$ (the optimal strategy for player E in the game $\Gamma(\bar{M}, \ell)$), choose one of the conditionally optimal trajectories directed to one of the points $\bar{y}_k(M)$ from the spectrum of the optimal strategy for Player E in the game $\Gamma(\bar{M}, \ell)$. Here the value of the game is equal to $\gamma(\bar{M}, \ell)$*

Proof: Denote by $u_\epsilon^*(\cdot)$ one of the strategies for Player P whose existence is assumed in the theorem, and by $\nu^*(\cdot)$ the MPOLBS for Player E whose optimality will be proved below. We shall prove that

$$M(x_0, y_0; \mu(\cdot), \nu^*(\cdot)) + \epsilon \geq M(x_0, y_0; u_\epsilon^*(\cdot), v^*(\cdot)) \geq$$
$$\geq M(x_0, y_0; u_\epsilon^*(\cdot), \nu(\cdot)) - \epsilon \tag{6.1.7}$$

for all $\mu(\cdot)$, $\nu(\cdot)$. It is well known, however, (see Theorem 3, Ch. 1) that to prove the validity of (6.1.1) it suffices to prove (6.1.7) only for pure piecewise open–loop strategies of the players P and E, i.e. it suffices to prove that

$$M(x_0, y_0; u(\cdot), \nu^*(\cdot)) + \epsilon \geq M(x_0, y_0; u_\epsilon^*(\cdot), \nu^*(\cdot)) \geq$$
$$\geq M(x_0, y_0; u_\epsilon^*(\cdot), v(\cdot)) - \epsilon \tag{6.1.8}$$

for all $u(\cdot) \in P$, $v(\cdot) \in E$.

Denote by $x^*(t)$ Player P's trajectory in the situation $(u_\epsilon^*(\cdot), v^*(\cdot))$. Then

$$M(x_0, y_0; u_\epsilon^*(\cdot), v^*(\cdot)) = \sum_{k=1}^{n+1} p_k \rho(x^*(T), \bar{y}_k). \tag{6.1.9}$$

Let R be a radius of the minimal sphere containing the set $C_E^\ell(\bar{M})$, i.e. $R = \gamma(\bar{M},\ell)$ Then for all $k = 1,\ldots,n+1$

$$R - \frac{\epsilon}{2} \leq \rho(x^*(T),\bar{y}_k) \leq R + \frac{\epsilon}{2}.$$

From (6.1.9) we obtain $\left(\sum_{k=1}^{n+1} p_k = 1\right)$

$$R - \frac{\epsilon}{2} \leq M(x_0, y_0; u_\epsilon^*(\cdot), \nu^*(\cdot)) \leq R + \frac{\epsilon}{2}. \tag{6.1.10}$$

Let $\bar{x}(T)$, $y(T-\ell)$ be the corresponding states in the situation $(x_0, y_0; u_\epsilon^*(\cdot), \nu(\cdot))$, and $Q(\cdot)$ be a probabilitity measure induced on the set $C_E^\ell(y(T-\ell))$ From the optimality of the mixed strategy $p = (p_1,\ldots,p_{n+1})$ in the game $\Gamma(\bar{M},\ell)$ we have

$$R = \sum_{k=1}^{n+1} p_k \rho(\bar{\bar{y}}(\bar{M}),\bar{y}_k) \geq \gamma(y(T-\ell),\ell) =$$

$$= Val\Gamma(y(T-\ell),\ell) \geq \int_{C_E^\ell(y(T-\ell))} \rho\left\{\bar{\bar{y}}[y(T-\ell)],y\right\} dQ, \tag{6.1.11}$$

where $\bar{\bar{y}}[y(T-\ell)]$ is the center of the minimal sphere containing the set $\bar{C}_E^\ell(y(T-\ell))$. But $\rho(x^*(T),\bar{\bar{y}}(y(T-\ell))) \leq \frac{\epsilon}{2}$, therefore

$$\rho(\bar{x}(T),y) \leq \frac{\epsilon}{2} + \rho[\bar{\bar{y}}(y(T-\ell)),y] \leq R + \frac{\epsilon}{2} \tag{6.1.12}$$

From (6.1.10)–(6.1.12) we find

$$M(x_0,y_0;u_\epsilon^*(\cdot),\nu^*(\cdot)) \geq \int_{C_E^\ell(y(T-\ell))} \rho(\bar{x}(T),y)dQ - \epsilon, \tag{6.1.13}$$

however

$$\int_{C_E^\ell(y(T-\ell))} \rho(\bar{x}(T),y)dQ = M(x_0,y_0;u_\epsilon^*(\cdot),\nu(\cdot)). \tag{6.1.14}$$

From (6.1.13) and (6.1.14) we obtain the right–hand side of the inequality (6.1.7). The left–hand side of (6.1.7) follows from the definition of strategy u_ϵ. ■

Remark. For $T < \ell$ the solution to the game is not substantially different from the case $T \geq \ell$ and the theorem holds if instead of $C_E^\ell(y_0)$, $\bar{C}_E^\ell(y_0)$, $\gamma(\bar{M},\ell)$, $y(T-\ell)$ we understand $C_E^T(y_0)$, $\bar{C}_E^T(y_0)$, $\gamma(\bar{M},T)$, y_0 respectively.

The diameter $D[C_E^\ell(\bar{M})]$ of the set $C_E^\ell(\bar{M})$ tends to zero as $\ell \to 0$ which, in turn, makes the value of the auxiliary game $\Gamma(\bar{M},\ell)$ tend to zero. However, the value of the auxiliary simultaneous game $\Gamma(\bar{M},\ell)$ over the sets $\bar{C}_E^\ell(\bar{M})$, $C_E^\ell(\bar{M})$ equals to the value of the pursuit game with delay of information ℓ denoted by $Val\Gamma(x_0, y_0, T) = V_\ell(x_0, y_0, T)$ (here the index ℓ stands for the information delay). Hence we have

$$\lim_{\ell \to 0} V_\ell(x_0, y_0, T) = 0$$

Since $\lim_{\ell \to 0} D[C_E^\ell(\bar{M})] = 0$, then the mixed optimal strategy for Player E in the game $\Gamma(\bar{M}, \ell)$, concentrating its mass at no more than $(n+1)$ points from the set $C_E^\ell(\bar{M})$, concentrates (in the limit) its mass at one point $\bar{M}$, i.e. converts to a pure strategy. This, in turn, implies the existence of the optimal pure strategy for Player E in the game $\Gamma(x_0, y_0, T)$ as $\ell \to 0$

Example 1. Motion equations are of the form:

$$\begin{array}{lll} \text{for Player } P & \dot{x} = u, \quad |u| \le \alpha & (\alpha > \beta), \\ \text{for Player } E & \dot{y} = v, \quad |v| \le \beta & x \in R^2, \quad y \in R^2. \end{array}$$

Let x_0, y_0 be the initial state of the game,the time T satisfies the condition

$$T > \frac{\rho(x_0, y_0)}{\alpha - \beta} + \ell$$

(here ρ is the Euclidean distance between the points x_0, y_0). The reachability set $C_E^t(y_0) = \bar{C}_E^t(y_0)$ and coincides with the circle whose center is y_0 and radius βt. The value of the game $\Gamma(y, \ell)$ is equal to the radius of the circle $C_E^\ell(y)$, i.e. $Val\Gamma(y, \ell) = \beta\ell$.

It can be seen that $Val\Gamma(y, \ell)$ is independent of y, hence any point of the set $C_E^{T-\ell}(y_0)$ may serve as a point $\bar{M}$. The optimal strategy for Player P in the game $\Gamma(y, \ell)$ is the choice of the point y, and the one for Player E is mixed and is the choice of any two diametrically opposite points of the circle $C_E^\ell(y)$ with probabilities $(\frac{1}{2}, \frac{1}{2})$ (see Example 7, Ch. 1).

Accordingly, the pure optimal strategy for Player P in the game $\Gamma(x_0, y_0, T)$ is the focusing with maximal velocity on the point $y(t - \ell)$ for $\ell \le t \le T$ (on the point y_0 for $0 \le t \le \ell$) prior to this point and afterwords, remaining in the ϵ–neightbourhood of this point until the instant T. The optimal strategy for Player E the MPOLBS, is exactly the passage from the point y_0 to an arbitrary point $\bar{M} \in C_E^{T-\ell}(y_0)$ during the time $T - \ell$ and then the choice of one of two directions toward two diametrically opposite points of the circle $C_E^\ell(\bar{M})$ with probabilities $(\frac{1}{2}, \frac{1}{2})$. Moreover $Val\Gamma(x_0, y_0, \ell) = \beta\ell$.

Evidently,

$$\lim_{\ell \to 0} V_\ell(x_0, y_0, T) = \lim_{\ell \to 0} \beta\ell = 0, \tag{6.1.15}$$

and, in the limit game with complete information, for any $\epsilon > 0$ Player P ensures the ϵ–capture of Player E. At the same time, for $\ell > 0$ Player P can never ensure the ℓ–capture of his opponent.

Example 2. Motion equations are of the form:

$$\text{for Player } P \quad \begin{array}{l}\dot{x}_1 = u_1,\\ \dot{x}_2 = u_2,\end{array} \quad u_1^2 + u_2^2 \le \alpha^2, \tag{6.1.16}$$

$$\text{for Player } E \quad \begin{array}{l}\dot{y}_1 = vy_2 + y_1,\\ \dot{y}_2 = -1,\end{array} \quad -1 \le v \le +1. \tag{6.1.17}$$

The reachability set for Player P is the same as in the game from Example 1.

We find a reachability set for Player E. We have

$$y_2(t) = y_2(0) - t, \tag{6.1.18}$$

set $v = +1$ and substitute (6.1.18) into the first equation (6.1.17). Then

$$y_1(t) = t + 1 - y_2(0) + [y_1(0) + y_2(0) - 1]e^t.$$

Setting $v = -1$, we find

$$y_1(t) = -t - 1 + y_2(0) + [y_1(0) - y_2(0) + 1]e^t.$$

The reachability set $C_E^T(y_1(0), y_2(0))$ coincides with the line segment joining the points

$$A = t + 1 - y_2(0) + (y_1(0) + y_2(0) - 1)e^t,$$

$$B = -t - 1 + y_2(0) + (y_1(0) - y_2(0) + 1)e^t.$$

The radius of the minimal sphere containing $C_E^t(y_1(0), y_2(0))$, is equal to the half of the length of the interval $[A, B]$:

$$R[C_E^t(y_1(0), y_2(0))] = |1 - y_2(0) + t - (1 - y_2(0))e^t|.$$

Thus, the value of the auxiliary simultaneous game $\Gamma(y_1(0), y_2(0); \ell)$ is equal to

$$Val\Gamma(y_1(0), y_2(0); \ell) = |1 - y_2(0) + \ell - (1 - y_2(0))e^\ell|.$$

To find the value of the pursuit game, we have to determine

$$\max_{(y_1,y_2)\in C_E^{T-\ell}(y_1(0),y_2(0))} Val\Gamma(y_1, y_2; \ell) =$$

$$= \max_{(y_1,y_2)\in C_E^{T-\ell}(y_1(0),y_2(0))} |1 - y_2 + \ell - (1 - y_2)e^\ell|. \tag{6.1.19}$$

The reachability set $C_E^{T-\ell}(y_1(0), y_2(0))$, however, is composed only from the points of the form $\{y_1, y_2 = y_2(0) - T + \ell\}$, hence the maximum in (6.1.19) is equal to

$$|1 + T - y_2(0) - (1 + T - \ell - y_2(0))e^\ell|.$$

The optimal strategy of Player P is the pursuit of the center of the reachability set $C_E^\ell(y_1, y_2)$ (the middle of the interval $[A, B]$).

Let us assume that in the time T Player P can ensure the capture of that center.

The optimal strategy of Player E is arbitrary for $t \in [0, T-\ell)$. At the instant of time $T-\ell$, with probabilities $\frac{1}{2}$, $\frac{1}{2}$ Player E chooses one of the two possible values of the control $v = +1, v = -1$, and on the remaining time interval $(T-\ell, T]$ he chooses a control $v(t)$ that is identically equal to the control chosen by him at the instant $T-\ell$ as a result of application of a random mechanism.

6.2 Game with information delayed. Case of m pursuiers and one evader

This section deals with the zero–sum two–person game of pursuit between the pursuer team. $P = \{P_1, \ldots, P_m\}$ and the evader E, the duration T being prescribed. Motion equation are of the form:

$$\begin{array}{lll} \text{for the Players } P_i & \dot{x}^{(i)} = f^{(i)}(x^{(i)}, u), & i = 1, \ldots, m, \\ \text{for Player } E & \dot{y} = g(y, v), & \end{array} \qquad (6.2.1)$$

Here $x^{(i)} \in R^n$, $y \in R^n$, $u \in U^{(i)} \subset R^\ell$, $v \in V \subset R^k$.

Let a number $\ell > 0$ be given and called the information delay. For $0 \le t \le \ell$, at the instant of time t Player P knows his state $x(t)$, time t and Player E's state at the initial instant y_0. For $\ell \le t \le T$, at each instant of time t Player P knows his state $x(t)$, time t and Player E's state $y(t-\ell)$ at the instant $t-\ell$. At each instant of time t Player E knows $x(t), y(t), t$. His payoff is equal to $\frac{1}{m} \sum_{i=1}^m \rho^2(x^{(i)}(T), y(T))$. Denote this game by $\Gamma(x_0^{(i)}, \ldots, x_0^{(m)}, y_0, T)$.

Pure piecewise open–loop strategies (POLS) for Players $P = \{P_1, \ldots, P_m\}$ and Player E have the form defined in 1 of this chapter.

The game proceeds in accordance with the motion Eq. (6.2.1), assuming that all the conditions, which ensure the existence and uniqueness of a solution to the system (6.2.1) on the interval $[0, T]$ for any pair of measurable open–loop controls $u(t)$, $v(t)$ have been satisfied. This ensures the existence and uniqueness of a solution of the system (6.2.1) when the players P and E adopt the piecewise open–loop strategies $u(\cdot) \in P$, $v(\cdot) \in E$ from the initial conditions x_0, y_0.

Thus, in any situation $(u(\cdot), v(\cdot))$ with the given initial conditions $x_0 = (x_0^{(1)}, \ldots, x_0^{(m)})$, y_0 the payoff function is defined in a unique way:

$$K(x_0, y_0; u(\cdot), v(\cdot)) = \frac{1}{m} \sum_{i=1}^m \rho^2(x^{(i)}(T), y(T)), \qquad (6.2.2)$$

where $x(t) = \{x^{(i)}(t)\}$, $y(t)$ is the solution of the system (6.2.1) with the initial conditions $x_0 = (x_0^{(1)}, \ldots, x_0^{(m)})$, y_0 in the situation $(u(\cdot), v(\cdot))$, $u(\cdot) = \{u^{(i)}(\cdot)\}$ and ρ is the Euclidean distance.

In general, as in the game from 1 of this chapter,

$$\sup_{v(\cdot)\in E} \inf_{u(\cdot)\in P} K(x_0, y_0; u(\cdot), v(\cdot)) \neq \inf_{u(\cdot)\in P} \sup_{v(\cdot)\in E} K(x_0, y_0; u(\cdot), v(\cdot)) \qquad (6.2.3)$$

since $\Gamma(x_0^{(i)}, \ldots, x_0^{(m)}, y_0, T)$ is not a perfect information game. From (6.2.3) it follows that the ϵ–equilibrium in this game exists not for all $\epsilon > 0$.

We extend the strategy space P and E to the mixed piecewise open–loop behavior strategies which imply a random choice of controls at each step.

Then we show that in such class of strategies the equality (6.2.3) holds. Denote the MPOLBS for the i–th member of the team $P = \{P_1, \ldots, P_m\}$ by $\mu^i(\cdot)$. Then the MPOLBS for the team P is of the form $\mu(\cdot) = \{\mu^i(\cdot)\}$, $i = 1, \ldots, m$.

The MPOLBS sets for the players P and E are denoted respectively by $\bar{P}$ and $\bar{E}$.

With the initial conditions x_0, y_0 fixed, each MPOLBS pair induces probability distribution on the space of trajectories $x^{(i)}(t)$, $i = 1, \ldots, m$, $y(t)$, hence by the payoff $M(x_0, y_0; \{\mu^{(i)}(\cdot)\}, \nu(\cdot))$ in MPOLBS is meant the expectation of the payoff $K(x_0, y_0; \{u^{(i)}(\cdot)\}, v(\cdot))$ averaged over distribution on the trajectory spaces induced by MPOLBS $\{\mu^{(i)}(\cdot)\}, \nu(\cdot)$.

After the strategy spaces $\bar{P}$, $\bar{E}$ and the payoff M are defined, for each of the fixed initial conditions x_0, y_0, T we construct a mixed extension $\bar{\Gamma}(x_0^{(i)}, \ldots, x_0^{(m)}, y_0, T)$ of the game $\Gamma(x_0^{(1)}, \ldots, x_0^{(m)}, y_0, T)$.

Introduce the following auxiliary quantity. Let $C_E^T(y)$ be a reachability set for Player E. Denote a convex hull of the set $C_E^T(y)$ by $\bar{C}_E^T(y)$. Let

$$\gamma(y, T) = \min_{\substack{\bar{\eta} = (\eta^{(1)}, \ldots, \eta^{(m)}) \\ \eta^{(i)} \in \bar{C}_E^T(y)}} \max_{\eta \in C_E^T(y)} \frac{1}{m} \sum_{i=1}^{m} \rho^2(\eta^{(i)}, \eta). \qquad (6.2.4)$$

Consider a simultaneous auxiliary zero–sum game of pursuit on the convex hull of the set $C_E^T(y)$.

Player $P = \{P_1, \ldots, P_m\}$ choses m points $\eta^{(i)}$, and Player E a point $\eta \in C_E^T(y)$. The choices are made simultaneously and, with the choice of $\eta^{(i)} \in \bar{C}_E^T(y)$ Player $P = \{P_1, \ldots, P_m\}$ has no knowledge of the choice made by Player E, and vice versa. Player P has the payoff $-\frac{1}{m} \sum \rho^2(\eta^{(i)}, \eta)$.

Denote the value of the game by $V(y, T)$ (the game is two–person zero–sum) in order to emphasize the dependence of the game value on the parameters y,

T which determine the strategy sets $\{\bar{C}_E^T(y)\}^m$, $C_E^T(y)$ for the players P and E. The game in normal form is written as follows:

$$\Gamma(y,T) =< \{\bar{C}_E^T(y)\}^m, C_E^T(y), \frac{1}{m}\sum \rho^2(\eta^{(i)},\eta) > .$$

Here $\{\bar{C}_E^T(y)\}^m$ is the m–tuple (Cartesian product) of the sets $\bar{C}_E^T(y)$.

The strategy set $\{\bar{C}_E^T(y)\}^m$ of Player P (minimizer) is convex as the Cartesian product of convex hulls of the sets $\bar{C}_E^T(y)$. The function $\frac{1}{m}\sum_m^{i=1} \rho^2(\eta^{(i)},\eta)$ is also convex in its arguments and is continuous. For such games we can apply Theorem 5, Ch. 1. In this case, it is stated as follows.

Theorem 3 *In the game $\Gamma(y,T)$ there is an equilibrium in mixed strategies. The optimal strategy for Player P is pure, and the optimal strategy for Player E is mixed and prescribes positive probabilities to not more than $(n+1)$ points of the set $C_E^T(y)$ $(C_E^T(y) \subset R^n)$. The value of the game is equal to*

$$V(y,t) = \min_{\bar{\eta}=(\eta^{(1)},\dots,\eta^{(m)})\in\bar{C}_E^T(y)} \max_{\eta\in C_E^T(y)} \frac{1}{m}\sum_{i=1}^m \rho^2(\eta^{(i)},\eta),$$

or $\gamma(y,T) = V(y,T)$.

The solution to the game $\Gamma(y,T)$ is given in the second part of Example 8, Ch. 1.

It turns out that the optimal strategy for Player P in the game $\Gamma(y,T)$ coincides with the choice of m identical points $\bar{y}$ that are the centers of the minimal sphere S containing the set $C_E^T(y)$ (here the point $\bar{y}$ is unique), and the optimal strategy for Player E prescribed positive probabilities to not more than $(n+1)$ points from the points $\bar{\bar{y}}$ at which the minimal sphere containing the set $C_E^T(y)$ is tangent to this set. The value of the game is equal to the square of the radius of this sphere.

The point $\bar{M}$ is called the center of pursuit if

$$\gamma(\bar{M}, T-\ell) = \max_{y'\in C_E^{T-\ell}(y)} \gamma(y', T-\ell). \tag{6.2.5}$$

is achieved therein. Thus

$$\gamma(\bar{M}, T-\ell) =$$

$$= \max_{y'\in C_E^{T-\ell}(y)} \left[\min_{\substack{\bar{\eta}=(\eta^{(1)},\dots,\eta^{(m)}) \\ \eta^{(i)}\in\bar{C}_E^{\ell}(y')}} \max_{\eta\in C_E^{\ell}(y')} \frac{1}{m}\sum_{i=1}^m \rho^2(\eta^{(i)},\eta) \right]. \tag{6.2.6}$$

Consider a simultaneous game $\Gamma(\bar{M},\ell)$, where $\bar{M}$ is the center of pursuit defined in (6.2.5). Denote by $\bar{\bar{y}}_1(\bar{M}),\ldots,\bar{\bar{y}}_{n+1}(\bar{M})$ the points of the set $C_E^\ell(\bar{M})$ entering in the spectrum of mixed optimal strategies for Player E in the game $\Gamma(\bar{M},T-\ell)$, and by $\bar{y}(\bar{M})$ the center of the minimal sphere containing the set $\bar{C}_E^\ell(\bar{M})$ i.e. the optimal strategy for Player P in the game $\Gamma(\bar{M},\ell)$.

Definition 2. The trajectory $y_k^*(t)$ is referred to as conditionally optimal if $y^*(0) = y_0$, $y^*(T-\ell) = \bar{M}$, $y^*(T) = \bar{\bar{y}}_k(\bar{M})$ for some particular k from $k = 1,\ldots,n+1$. For each k there may be several conditionally optimal trajectories of Player E.

Denote by $y'(t)$ the trace of the trajectory $y(t)$ which becomes known to Player P by the instant t. By the definition of the state of information in the game $y'(t) = y_0$ for $0 \le t \le \ell$, $y'(t) = y(t-\ell)$ for $\ell \le t \le T$. Consider the quantity

$$\gamma(y'(t),\ell) = \min_{\substack{\bar{\eta} = (\eta^{(1)},\ldots,\eta^{(m)}) \\ \eta^{(i)} \in C_E^\ell(y'(t))}} \max_{\eta \in C_E^\ell(y'(t))} \left[\frac{1}{m}\sum_{i=1}^{m} \rho^2(\eta^{(i)},\eta)\right]$$

for $0 \le t \le T$. Let min max be achieved at the points

$$\gamma(y'(t),\ell) = \frac{1}{m}\sum_{i=1}^{m} \rho^2(\bar{y}^{(i)}(t),\bar{y}(t)), \quad 0 \le t \le T.$$

Each open–loop control $v(t)$ for Player E corresponds to a unique trajectory $y(t)$ the trajectory $y'(t)$, in turn, corresponds to a unique trajectory $\bar{y}^{(i)}(t)$ (evidently, $\bar{y}^i(t) = \bar{y}(t)$, $i = 1,\ldots,m$).

Let us introduce a fictitious player E' capable of moving along the trajectory $\bar{y}(t)$.

Theorem 4 *Let $T \ge \ell$, and for any $1 > \epsilon > 0$ each member of the team $P = \{P_1,\ldots,Pm\}$ can ensure the ϵ–capture of Player E' by the instant of time T. Then the game $\Gamma(x_0^{(i)},\ldots,x_0^{(m)},y_0,T)$ has the equilibrium in mixed piecewise open–loop behavior strategies. The ϵ-optimal strategy of the team $P = \{P_1,\ldots,P_m\}$ is pure and coincides with any strategy $u_\epsilon^*(\cdot) = \{u^{(1)}(\cdot),\ldots, u^{(m)}(\cdot)\}$ which ensures the $\epsilon_1 = \frac{\epsilon^*}{1+2R}$[1]–capture of the point $\bar{y}(t)$ by the instant T. The optimal strategy for Player E is mixed: during the time $0 \le t \le T-\ell$ he must move to the point $\bar{M}$ along any conditionally optimal trajectory $y^*(t)$ and then choose with probabilities $p_1,\ldots,p_{n+1}$ (the optimal strategies, of E in the game $\Gamma(\bar{M},\ell)$) one of the conditionally optimal trajectories, which carry over the point $y^*(t-\ell)$ to the points $\bar{\bar{y}}_k(\bar{M})$, $k = 1,\ldots,n+1$, entering into the spectrum of mixed optimal strategies for Player E in the game $\Gamma(\bar{M},\ell)$. The value of the game is equal to $\gamma(\bar{M},\ell)$.*

---

[1] ϵ_1 is stated from the condition $(R+\epsilon_1)^2 - R^2 < \epsilon$, $R^2 - (R-\epsilon_1)^2 < \epsilon$.

Proof: Denote by $y'(t)$ one of the strategies for Player P whose existence is assumed in the theorem, and by $\nu^*(\cdot)$ the MPOLBS for Player E whose optimality we are going to prove. We shall prove that

$$M(x_0, y_0; \mu(\cdot), \nu^*(\cdot)) + \epsilon \geq M(x_0, y_0; u^*_\epsilon(\cdot), \nu^*(\cdot)) \geq$$

$$\geq M(x_0, y_0; u^*_\epsilon(\cdot), \nu(\cdot)) - \epsilon; \;\; \mu(\cdot) = \{\mu^{(i)}(\cdot)\} \tag{6.2.7}$$

for all $\mu(\cdot)$, $\nu(\cdot)$. It is well known, however (see Theorem 3, Ch. 1), that to prove the validity of (6.2.7) it suffices to prove it only for pure POLS of the players P and E, i.e. that

$$M(x_0, y_0; u(\cdot), \nu^*(\cdot)) + \epsilon \geq M(x_0, y_0; u^*_\epsilon(\cdot), \nu^*(\cdot)) \geq$$

$$\geq M(x_0, y_0; u^*_\epsilon(\cdot), v(\cdot)) - \epsilon \tag{6.2.8}$$

for all $u(\cdot) \in P$, $v(\cdot) \in E$

Denote by $x^*(t)$ the trajectoryof the team $P = \{P_1, \dots, P_m\}$ in the situation $(u^*_\epsilon(\cdot), \nu^*(\cdot))$. Then

$$M(x_0, y_0; u^*_\epsilon(\cdot), \nu^*(\cdot)) = \sum_{k=1}^{m} P_k \left[\frac{1}{m} \sum_{i=1}^{m} \rho^2(x^{(i)^*}(T), \bar{\bar{y}}_k)\right]. \tag{6.2.9}$$

Let R be a radius of the minimal sphere containing the set $C^\ell_E(\bar{M})$, i.e. $R^2 = \gamma(\bar{M}, \ell)$. Then

$$R - \epsilon_1 \leq \rho(x^{(i)^*}(T), \bar{\bar{y}}_k) \leq R + \epsilon_1, \;\; \epsilon_1 = \frac{\epsilon}{1 + 2R} \tag{6.2.10}$$

for all $k = 1, \dots, n+1$ and all $i = 1, \dots, m$.

From (6.2.9) we get $\left(\sum_{k=1}^{n+1} p_k = 1\right)$.

$$(R - \epsilon_1)^2 \leq M(x_0, y_0; u^*_\epsilon(\cdot), \nu^*(\cdot)) \leq (R + \epsilon_1)^2. \tag{6.2.11}$$

Now let $Q(\cdot)$ be a probability measure induced by the MPOLBS $\nu(\cdot)$ on the set $C^\ell_E(M)$. From the optimality of the mixed strategy $p = (p_1, \dots, p_{n+1})$ in the game $\Gamma(\bar{M}, \ell)$ we have

$$R^2 = \sum_{k=1}^{n+1} p_k \rho^2(\bar{M}, \bar{\bar{y}}_k) \geq \int_{C^\ell_E(\bar{M})} \rho^2(\bar{M}, y) dQ. \tag{6.2.12}$$

However,

$$\rho(x^{(i)^*}(T), \bar{M}) \leq \epsilon_1 = \frac{\epsilon}{1 + 2R} \tag{6.2.13}$$

for all $i = 1, \ldots, m$. From (6.2.12), (6.2.13) we find

$$M(x_0, y_0; u_\epsilon^*(\cdot), \nu^*(\cdot)) \geq \int_{C_E^\ell(\bar{M})} \frac{1}{m} \sum_{i=1}^{m} \rho^2(x^{(i)^*}(T), y) dQ - \epsilon. \qquad (6.2.14)$$

However,

$$\int_{C_E^\ell(\bar{M})} \frac{1}{m} \sum_{i=1}^{m} \rho^2(x^{(i)^*}(T), y) dQ = M(x_0, y_0; u_\epsilon^*(\cdot), \nu(\cdot)). \qquad (6.2.15)$$

From (6.2.14), (6.2.15) we obtain the right–hand side of the inequality (6.2.9). The left–hand side of (6.2.9) follows immediately from the definition of the strategy $u_\epsilon^*(\cdot)$. ■

6.3 Existence of equilibria in mixed strategies in "princess and monster" game of pursuit

Consider the class of differential games of pursuit providing a generalization of the "princess and monster" game proposed by R. Isaacs as reseach problem in [1]. We state the problem correctly and show that such type of games have the equilibrium in mixed strategies.

The case of simple motion on the plane is discussed in some detail, although all results may be extended to a general case of arbitrary motion equations for convex vectograms of the Players P and E. First recall the problem proposed by R. Isaacs: Monster P tries to find Princess E. The time required to implement his project is the payoff. Both of them are in the absolutely dark room R (of any shape) whose boundaries are known to them. The capture takes place if the distance PE becomess less then the quantity ℓ which is small as compared to the size of the room R. Monster is assumed to be hightly intellectual and makes a simple motion with the known velocity α. Princess is free to make any move.

We give a formal description of the game.

Let a convex polyhedron S be given on the plane. Denote by S_0 the interior of S; by $S_1, \ldots, S_m$ the sides of S (without vertices) and by $S_{m+1}, \ldots, S_{2m}$ the corresponding vertexes. At the initial instant of time the "chance" choises a position $x_0 \in S$ for Player P and a position $y_0 \in S$ for Player E in accordance with the uniform distribution in S. If x_0 (y_0) belongs to S_k, $k = 0, \ldots, n$ due to a random play, then Player P (E) knows only that he is in S_k, but does not know at what exactly point of this set he is. Next, the Players P and E move into S in accordance with a simple motion

$$\dot{x} = \alpha u, \quad |u| = 1,$$
$$\dot{y} = \beta v, \quad |v| = 1$$

from the initial state $x_0 \in S$, $y_0 \in S$.

Assume that at the instant $0 \leq t \leq T$ the point $x(t) \in S_k$ and the point $y(t) \in S_\ell$. Then at this instant of time the Players P and E knows only that they are respectively in S_k and S_ℓ, but they do not know at what exactly points of this sets they are. The duration of the game is fixed and equals to T. A continuous functional $F(x_0, y_0; x(t), y(t))$ in given on set of trajectories $x(t)$, $y(t)$ for $t \in [0, T]$, and Player E's payoff is equal to F (player P's payoff is equal to $-F$).

Information sets. According with the condition of the game the players distinquish only the set S_k, $k = 0, \ldots, n$. However, being in S_k, they do not distinquish positions in this sets. Moreover, the players also know the set S. Therefore, when player P (E) is, say, on the side S_k, he knows which side is this one and hence on which side of S_k is the convex polyhedron S. If Player P (E) is at the vertex of S_k, $k = m+1, \ldots, 2m$, then he knows the position of the polyhedron S and the incident sides S_{k_i}, S_{k_ℓ} that are contingut with the vertex of S_k. If however, $x \in S_0$, then Player P (E) knows only that he is S_0 Therefore the information sets $S^{(i)}$ for Player P (E) are defined as follows:

$$\begin{aligned} S^{(0)} &= S_0, & \\ S^{(k)} &= S_k \bigcup S_0, & k = 1, \ldots, m, \\ S^{(k)} &= S_k \bigcup S_{k_i} \bigcup S_{k_\ell} \bigcup S_0, & k = m+1, \ldots, 2m \end{aligned}$$

(Fig.28). Here S_{k_i}, S_{k_ℓ} are the sides of the polyhedron S that are incident to S_k.

Define admissible controls in each of $S^{(k)}$, $k = 0, \ldots, n$. For $x \in S^{(0)}$ Player P must choose an arbitrary direction of motion (both players have a simple motion) (Fig.29). For $x \in S^{(k)}$, $k = 1, \ldots, m$, we should remember that, when on the side of S_k, Player P must choose the direction inward S, and since for all $x \in S^{(k)}$ the admissable control sets must coincide (otherwise Player P could distinquish different positions in $S^{(k)}$) (Fig.30).

If $x \in S^{(k)}$, $k = m+1, \ldots, 2m$, then for all $x \in S^{(k)}$ an admissible control set is the same as for $x \in S_k$, i.e. it represents a velosity direction inside the angle formed by the incident sides S_{k_i}, S_{k_ℓ},(Fig.31)

Although in this case the definition of information sets of Players P and E is similar to the one introduced the finitary in games (see [7]), it is different from Kuhn's definition [13] which requires that each information set chould

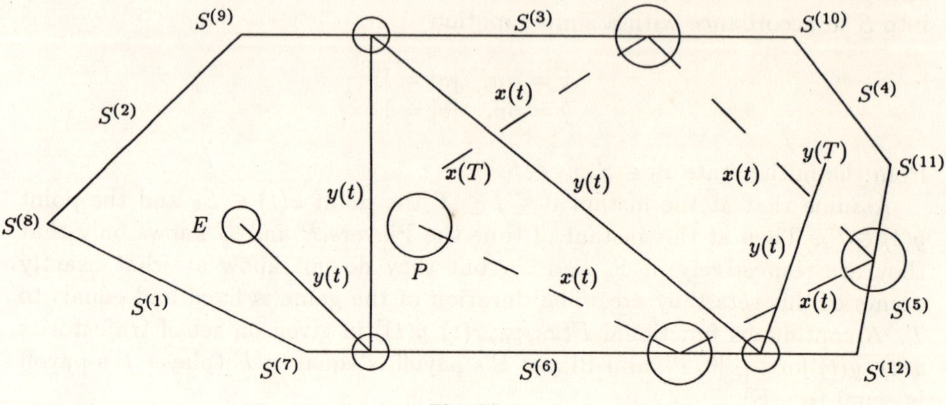

Fig. 28.

intersect the trajectory (the party) only once. In the case under study, the trajectories of the Players pass through the information sets.

Pure strategies. As in 1. Ch. 1., under the players strategy is meant the rule which associates each information state of the player with the action admissible in this information state. This nonformal description will be used to provide a formal definition of the strategy.

Since a control is chosen in accordance with the information, then it must be constant for all points of the information set $S^{(k)}$, $k = 0, \ldots, 2m$, becouse the information in all points of one information set is the same and the player knows only that he is in $S^{(k)}$.

Since, in each information set, the player must choose a control admissible in this set, for all points $x \in S^{(0)}$ Player P can choose as a control an arbitrary direction of motion. At the points $x \in S^{(k)}$, $k = 1, \ldots, m$, he must choose a control in such a way that the point remains in S. This mean that he may choose only the controls which generate the motion directed inward S. The same is with the choice of controls in $S^{(k)}$, $k = m+1, \ldots, 2m$

Denote the admissible control set in $S^{(k)}$ by U_k (as mentioned before, the sets U_k are closed in R^2). The same statement also holds for Player E.

Now we can give a formal definition of the strategy.

Definition 3. By the strategy $u(S^{(k)})$, $(v(S^{(k)}))$, $k = 0, \ldots, 2m$ is meant a mapping which places each information $S^{(k)}$, $k = 0, \ldots, 2m$ with a control $u(v) \in U_k(V_k)$ that is admissible in this information set.

Let a situation $(u(S^{(k)}), v(S^{(k)}))$ be given. The game proceeds as follows. The "chance" choses the initial points x_0, y_0 in the set S_0. Player P then choses the control $u(S^{(0)})$ prescribed by the strategy $u(S^{(k)})$, and Player E the control $v(S^{(0)})$ prescribed by the strategy $v(S^{(k)})$. Here the motion complies

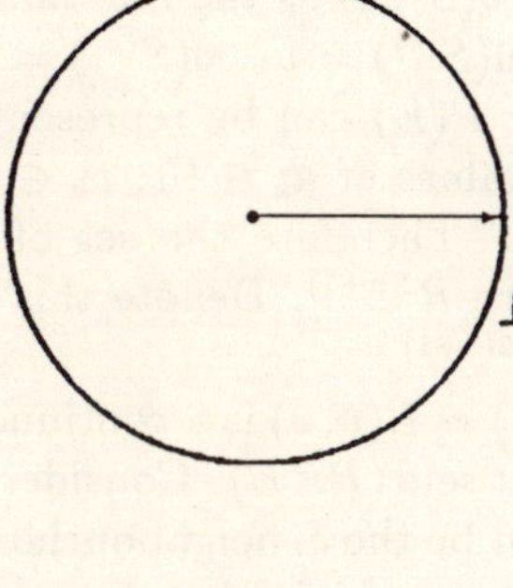

Fig. 29

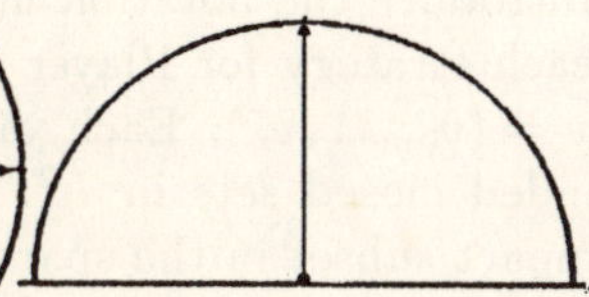

Fig. 30

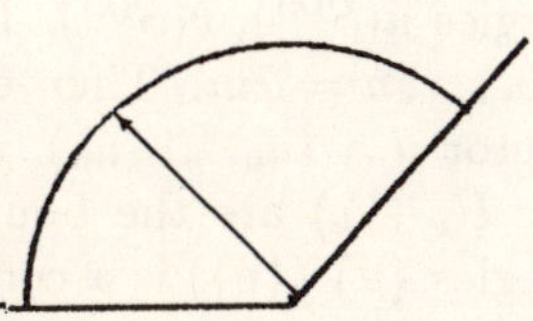

Fig. 31

with the equations

$$\dot{x} = \alpha u(S^{(0)}),$$
$$\dot{y} = \beta v(S^{(0)})$$

with the initial condition $x(0) = x_0$, $y(0) = y_0$.

Let t_1^1 (t_1^2) be the instant of time when the trajectory $x(t)$ $(y(t))$ initially passes from $S^{(0)}$ to $S^{(k_1)}$ $(S^{(k_2)})$. Then at the instant t_1^1 (t_1^2) Player P (E) choses the control $u(S^{(k_1)})$ $(v(S^{(k_2)}))$ perscribed by the strategy $u(S^{(k)})$ $(v(S^{(k)}))$ and adress to it until the game terminates or the state variable x (y) leaves the information set $S^{(k_1)}$ $(S^{(k_2)})$. The motion in $S^{(k_1)}$ $(S^{(k_2)})$ complies with the equations

$$\dot{x} = \alpha u(S^{(k_1)}), \dot{y} = \beta v(S^{(k_2)})$$

with the initial condition $x(t_1^1)$, $y(t_1^2)$. This process continues until the instant T. The game terminates at the instant T, and Player E receives the payoff

$$K(x_0, y_0; u(S^{(k)}), v(S^{(k)})) = F(x(t), y(t)),$$

where $x(t)$, $y(t)$ are the correspong trajectories of the players from the initial states x_0, y_0 in the situation $(u(S^{(k)}), v(S^{(k)}))$, and F a continuous functional given on trajectories of the Players P and E. Since the initial states x_0, y_0 are chosen in S_0 in a random way, then bevor the start of the game the player may be sure in receiving the average payoff

$$E(u(S^{(k)}), v(S^{(k)})) = \frac{1}{\mu^2(S_0)} \int_{S_0 \times S_0} K(x_0, y_0; u(S^{(k)}), v(S^{(k)}))dxdy,$$

where $\mu(S_0)$ is the Lebesque measure of the set S_0

Under the payoff function in this game is meant the average value of the payoff K, i.e. the function $E(u(S^{(k)}), v(S^{(k)}))$. We define the pursuit game with incomplete information in normal form.

Show the continuity of the functional $E(u(S^{(k)}), v(S^{(k)}))$ as the function of strategies $u(S^{(k)})$, $v(S^{(k)})$. Introduce the notation $u(S^{(k)}) = u_k$, $v(S^{(k)}) = v_k$, $k = 0, \ldots, n = 2m$. Then each strategy for Player $P(E)$ can be represented as vector $u = (u_0, \ldots, u_n)$, $v = (v_0, \ldots, v_n)$. Each values of $u_k \in U_k$, $v_k \in V_k$, where U_k (V_k) are the bounded closed sets in R^2. Therefore the set of all strategies $\{u\}$ ($\{v\}$) is a compact subset in the space $R^{2(n+1)}$. Denote this set by $P(E)$. Thus the set $P(E)$ is a compact set in $R^{2(n+1)}$.

We will show that the function $E(u(S^{(k)}), v(S^{(k)})) = E(u, v)$ is a continuous function defined on the Cartesian product of compact sets $(P \times E)$. Consider an arbitrary situation (u, v). Let D_k, $k = m+1, \ldots, 2m$ be the δ–neighbourhoods of the vertices of S_k. Denote by $S_\delta \times S_\delta$ a rectangle set of initial states x_0, y_0 for which the trajectories $x(t)$, $y(t)$ for $0 \leq t \leq T$ from the initial states x_0, y_0 in the situation (u, v) do not intersect D_k. For any $\epsilon > 0$ it is possible to find such $\delta > 0$ that

$$|\mu^2(S_\delta) - \mu^2(S)| < \epsilon$$

(here μ is the Lebesque measure of the set). Let x_0, $y_0 \in S_\delta \times S_\delta$ and $t_1, \ldots, t_r$ $(\tau_1, \ldots, \tau_\ell)$ be the instants at which the trajectory $x(t)$ $(y(t))$ falls on the sides of S_k. Since $x(t)$, $(y(t))$ do not intersect the sets D_k, then there is such $\epsilon_1 > 0$ $(\epsilon_1 < \epsilon)$ that the ϵ_1–neighbourhood of the points $x(t_1), \ldots, x(t_n)$ $(y(\tau_1), \ldots, y(\tau_\ell))$, $\gamma_\epsilon(x(t_i))$ $(\gamma'_\epsilon(y(\tau_j)))$ contains no vertices of the sets S. For any $\epsilon > 0$ there is such $\delta_1 > 0$ that as soon as $\rho(u', u) < \delta_1$, $\rho(v'v) < \delta_1$ the points, at which the trajectories $x'(t)$, $y'(t)$ intersect the sides of the set S in the situation (u', v') from the initial states x_0, y_0, belong to $\gamma_\epsilon(t_i)$, since minor variations in strategies cannot change the passage order of information sets (as compared to the situation $(u(\cdot), v(\cdot))$). Since the motions along all trajectories occur at the constant velosities α and β, then the instants of falling on the information sets are only slightly different from the instants t_i in the situation $(u(\cdot), v(\cdot))$.

It follows from the foregoing that for any $\epsilon > 0$ it is possible to find such $\delta > 0$ that as soon as $\rho(u', u) < \delta$, $\rho(v', v) < \delta$,

$$\max_{0 \leq t \leq T} |x'(t) - x(t)| \leq \epsilon, \quad \max_{0 \leq t \leq T} |y'(t) - y(t)| \leq \epsilon.$$

The latter means that the trajectories emanatung from x_0, $y_0 \in S_\delta \times S_\delta$ are continuously dependent on the strategies at the point $u(\cdot)$, $v(\cdot)$, i.e. if $u_n(\cdot)$, $v_n(\cdot)$ is a sequence of strategies converging to $u(\cdot)$, $v(\cdot)$, then for each x_0, $y_0 \in S_\delta \times S_\delta$ and $\epsilon > 0$ there is such $N(x_0, y_0, \epsilon)$ that for all $m, n > N(x_0, y_0, \epsilon)$

$$|K(x_0, y_0; u_{n+m}(\cdot), v_{n+m}(\cdot)) - K(x_0, y_0; u_n(\cdot), v_m(\cdot))| < \epsilon, \qquad (6.3.1)$$

i.e the sequence $K(x_0, y_0; u_n(\cdot), v_n(\cdot))$ is fundamental converging in itself at all points of the set $S_\delta \times S_\delta$

We now show that the sequence of integrals $\int_S\int_S K(x_0, y_0; u_n(\cdot), v_n(\cdot))d\mu d\mu$ also is fundamental (convergsing in itself) For simplicity,denote $S \times S = \bar{S}$ and $S_\delta \times S_\delta = \bar{S}_\delta$. Indeed,

$$|E(u_{n+m}(\cdot), v_{n+m}(\cdot)) - E(u_m(\cdot), v_m(\cdot))| =$$

$$= \left| \int_{\bar{S}} K(x_0, y_0; u_{n+m}(\cdot), v_{n+m}(\cdot))d\mu^2 - \int_{\bar{S}} K(x_0, y_0; u_n(\cdot), v_n(\cdot))d\mu^2 \right| =$$

$$= \left| \int_{\bar{S}\setminus\bar{S}_\delta} K(x_0, y_0; u_{n+m}(\cdot), v_{n+m}(\cdot))d\mu^2 - \int_{\bar{S}\setminus\bar{S}_\delta} K(x_0, y_0; u_n(\cdot), v_n(\cdot))d\mu^2 + \right.$$

$$\left. + \int_{\bar{S}_\delta} K(x_0, y_0; u_{n+m}(\cdot), v_{n+m}(\cdot))d\mu^2 - \int_{\bar{S}_\delta} K(x_0, y_0; u_n(\cdot), v_n(\cdot))d\mu^2 \right| \leq$$

$$\leq \left| \int_{\bar{S}\setminus\bar{S}_\delta} K(x_0, y_0; u_{n+m}(\cdot), v_{n+m}(\cdot))d\mu^2 - \int_{\bar{S}\setminus\bar{S}_\delta} K(x_0, y_0; u_n(\cdot), v_n(\cdot))d\mu^2 \right| +$$

$$+ \left| \int_{\bar{S}_\delta} K(x_0, y_0; u_{n+m}(\cdot), v_{n+m}(\cdot))d\mu^2 - \int_{\bar{S}_\delta} K(x_0, y_0; u_n(\cdot), v_n(\cdot))d\mu^2 \right|.$$

It follows from (6.3.1) that the sequence $K(x_0, y_0; u_n(\cdot), v_n(\cdot))$ converges at each point x_0, $y_0 \in \bar{S}_\delta$, hence the sequence of integrals $\int_{\bar{S}_\delta} K(x_0, y_0; u_n(\cdot), v_n(\cdot))d\mu^2$ also converges. This means that there is such $N(\epsilon)$ that with $n > N(\epsilon)$ for all m

$$\left| \int_{\bar{S}_\delta} K(x_0, y_0; u_{n+m}(\cdot), v_{n+m}(\cdot))d\mu^2 - \int_{\bar{S}_\delta} K(x_0, y_0; u_n(\cdot), v_n(\cdot))d\mu^2 \right| < \epsilon.$$

By evaluating further the first difference ($\mu(S \setminus S_\delta) < \epsilon$, $K(x_0, y_0; u(\cdot), v(\cdot)) < M$) we obtain

$$\left| \int_{\bar{S}} K(x_0, y_0; u_{n+m}(\cdot), v_{n+m}(\cdot))d\mu^2 - \int_{\bar{S}} K(x_0, y_0; u_m(\cdot), v_m(\cdot))d\mu^2 \right| \leq \epsilon + 2\epsilon M'$$

for $n > N(\epsilon)$ and all m. This means that the sequence $\int_S K(x_0, y_0; u_n(\cdot), v_n(\cdot))d\mu$ is fundamental, i.e. there is the limit

$$\lim_{n\to\infty} \int_S K(x_0, y_0; u_n(\cdot), v_n(\cdot))d\mu^2. \qquad (6.3.2)$$

Since the measure μ is continuous, then the integral $\int_S K(x_0, y_0; u(\cdot), v(\cdot))d\mu$ is a continuous function of the set. From the fact that for any $\delta > 0$

$$\lim_{n\to\infty} \int_S K(x_0, y_0; u_n(\cdot), v_n(\cdot))d\mu^2 = \int_{\bar{S}_\delta} K(x_0, y_0; u(\cdot), v(\cdot))d\mu^2,$$

it follows that the limit (6.3.2) is equal to $\int_{\bar{S}} K(x_0, y_0; u(\cdot), v(\cdot))d\mu^*$,i.e. the function $E(u, v)$ is continuous at the point $u(\cdot)$, $v(\cdot)$. Since the situation $(u(\cdot), v(\cdot))$ is arbitrary the function $E(u, v)$ is a continuous function of the strategies $u(\cdot)$, $v(\cdot)$.

Theorem 5 *The payoff function $E(u,v)$ is a continuous function on the product of compact sets $P \times E$*

It follows from Theorem 4 that any game in normal form defined on the compact sets of strategies with a continuous payoff function has the equilibrium in mixed strategies. In view of Theorem 5, this gives a theorem that is the basic for this section.

Theorem 6 *The pursuit game with incomplete information defined in 3 of this chapter, has the equilibrium in mixed strategies.*

A slight modification in the proof allows the generalization of Theorem 6 to the case of arbitrary motions

$$\dot{x} = f(x,u),$$

$$\dot{y} = g(y,v).$$

Example 3. Consider a special case where the polyhedron S is a square. Motion equations are of the form

$$\begin{array}{lll} \text{for Player } P & \dot{x}_1 = 0, & \dot{x}_2 = u_2 \\ \text{for Player } E & \dot{y}_1 = v_1, & \dot{y}_2 = 0. \end{array} \tag{6.3.3}$$

The initial positions of the players P and E coincide. The controls u_2, v_1 can take only one of two values: $+1$ or -1.

The duration of the game is fixed and equals $T = 1$. According to the conditions of the game the players cannot leave the square S, the payoff of Player E depends only on the final positions $x(1)$, $y(1)$ and is defined as folows:

$$F(x_0, y_0; x(t), y(t)) = F'(x_0, y_0; x(1), y(1)) = \begin{cases} +1, & x_2(1) \geq y_1(1) - \alpha, \\ -1, & x_2(1) < y_1(1) - \alpha. \end{cases}$$

Here $0 \leq \alpha \leq 1$.

Since the variations in the variables x_1, y_2 do not change the payoff function, then the game can be reduced to an equivalent game, though in a simpler form, by introducing the vector $z = (z_1, z_2) = (y_1, x_2)$ and the controls $u = (0, u_2)$, $v = (v_1, 0)$. Indeed, the motion equations (6.3.3) are rewritten in the form

$$z = u + v, \quad z(0) = z_0. \tag{6.3.4}$$

where v and u take the values $(+1,0)$, $(-1,0)$ and $(0,+1)$, $(0,-1)$, respectively. The payoff is of the form

$$F(z_0, z(1)) = \begin{cases} +1, & z_2 \geq z_1 - \alpha, \\ -1, & z_2 < z_1 - \alpha, \end{cases}$$

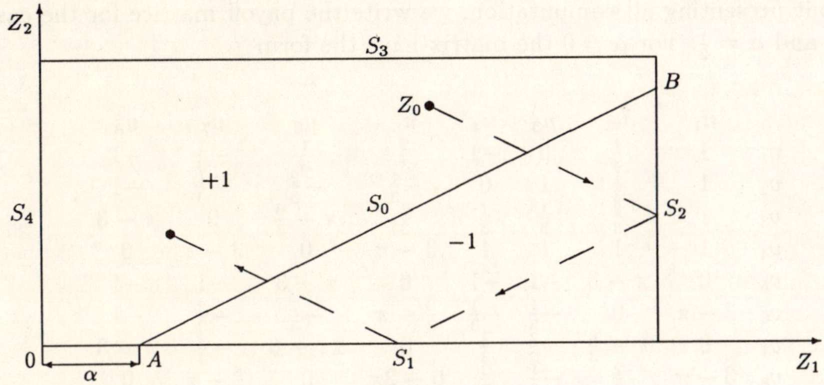

Fig. 32.

where $0 \leq \alpha \leq 1$

Assume that the initial point $z_0 \in S$ is chosen in a random way according to uniform probability distribution on S.

Let us examine Fig.32. If the game terminates at the point $z(1)$ lying above the line segment AB, then Player E's payoff is assumed to be $+1$, otherwise (-1). For simplicity, we omit the information sets which are the vertices of the square, the probability of falling on these vertex being zero. Therefore, the dimension of the strategy vector for each player is 5. Since the controls may take no more than two values, and only one value on some of the sides, then, as is readily seen, each player has exactly eight strategies.

Player E's strategy	Player P's strategy
$v_1(\cdot) = (1, 1, -1, -1, 1)$,	$u_1(\cdot) = (1, 1, 1, -1, -1)$,
$v_2(\cdot) = (-1, 1, -1, -1, 1)$,	$u_2(\cdot) = (-1, 1, 1, -1, 1)$,
$v_3(\cdot) = (1, -1, -1, -1, 1)$,	$u_3(\cdot) = (1, 1, -1, -1, 1)$,
$v_4(\cdot) = (-1, -1, -1, -1, 1)$,	$u_4(\cdot) = (-1, 1, -1, -1, 1)$,
$v_5(\cdot) = (1, 1, -1, 1, 1)$,	$u_5(\cdot) = (1, 1, -1, -1, -1)$,
$v_6(\cdot) = (-1, 1, -1, 1, 1)$,	$u_6(\cdot) = (-1, 1, 1, -1, -1)$,
$v_7(\cdot) = (1, -1, -1, 1, 1)$,	$u_7(\cdot) = (1, 1, -1, -1, -1)$,
$v_8(\cdot) = (-1, -1, -1, 1, 1)$,	$u_8(\cdot) = (-1, 1, 1, -1, -1)$.

Thus, the payoff function may be written as a payoff matrix. The elements of the matrix a_{ij} are computed by the formula

$$a_{ij} = \int_S F(z, z(1))dz,$$

where $z(t)$ is a trajectory from the initial state $z \in S$ in the situation $(u_i(\cdot), v_j(\cdot))$.

The most tedious operation is to find $z(t)$ in the situation $(u_i(\cdot), v_j(\cdot))$. Without presenting all computation, we write the payoff matrice for the case $\alpha = 0$ and $\alpha = \frac{1}{2}$. For $\alpha = 0$ the matrix have the form

	u_1	u_2	u_3	u_4	u_5	u_6	u_7	u_8
v_1	1	$\frac{1}{3}$	0	-1	$\frac{2}{3}$	$\frac{1}{3}$	$-\frac{1}{3}$	-1
v_2	1	$\frac{1}{3}$	1	0	$-\frac{1}{3}$	$-\frac{2}{3}$	$-\frac{1}{3}$	-1
v_3	1	$\frac{2}{3}$	$\frac{1}{3}$	$\frac{1}{3}$	$\frac{2}{3}$	$\pi - \frac{8}{3}$	0	$\pi - 3$
v_4	1	1	1	1	$3 - \pi$	0	$3 - \pi$	0
v_5	0	$\pi - 3$	-1	-1	0	$\pi - 3$	-1	-1
v_6	$3 - \pi$	0	$-\frac{1}{3}$	$-\frac{1}{3}$	$\frac{8}{3} - \pi$	$-\frac{2}{3}$	$-\frac{2}{3}$	-1
v_7	0	$\pi - \frac{8}{3}$	$-\frac{2}{3}$	$\frac{1}{3}$	0	$2\pi - 6$	$-\frac{2}{3}$	$\pi - 3$
v_8	$3 - \pi$	$\frac{2}{3}$	$-\frac{1}{3}$	$\frac{2}{3}$	$6 - 2\pi$	0	$\frac{8}{3} - \pi$	0

In this case, as is seen from the payoff matrix, there is the equilibrium in pure strategies $v_3(\cdot)$, $u_7(\cdot)$ and the value of the game is equal to 0.

For $\alpha = \frac{1}{2}$ the matrix is of the form:

	u_1	u_2	u_3	u_4	u_5	u_6	u_7	u_8
v_1	1	$\frac{5}{6}$	$\frac{3}{4}$	$-\frac{1}{2}$	$\frac{11}{12}$	$\frac{5}{6}$	$\frac{2}{3}$	$-\frac{1}{2}$
v_2	1	1	1	$\frac{3}{4}$	$\frac{1}{2}$	$\frac{3}{4}$	$\frac{1}{2}$	$\frac{1}{2}$
v_3	1	$\frac{11}{12}$	1	$\frac{5}{6}$	$\frac{11}{12}$	$\frac{11}{12}$	$\frac{11}{12}$	$\frac{5}{6}$
v_4	1	1	1	1	$\frac{2}{3}$	$\frac{11}{12}$	$\frac{2}{3}$	$\frac{11}{12}$
v_5	$\frac{11}{12}$	0	$\frac{1}{2}$	$-\frac{1}{2}$	$\frac{11}{12}$	$\frac{5}{6}$	$\frac{1}{2}$	$-\frac{1}{2}$
v_6	$\frac{2}{3}$	$\frac{11}{12}$	$\frac{1}{2}$	$\frac{2}{3}$	$\frac{5}{12}$	$\frac{3}{4}$	$\frac{1}{4}$	$\frac{1}{2}$
v_7	$\frac{11}{12}$	$\frac{11}{12}$	$\frac{3}{4}$	$\frac{5}{6}$	$\frac{11}{12}$	$\frac{11}{12}$	$\frac{3}{4}$	$\frac{5}{6}$
v_8	$\frac{2}{3}$	$\frac{11}{12}$	$\frac{1}{2}$	$\frac{11}{12}$	$\frac{7}{12}$	$\frac{11}{12}$	$\frac{5}{12}$	$\frac{11}{12}$

For $\alpha = \frac{1}{2}$ the equilibriumin in pure strategies does not exist and the mixed optimal strategy of Player E concentrates its measure on two pure strategies $v_3(\cdot)$, $v_4(\cdot)$, and the mixed optimal strategy of Player E on the strategies $u_7(\cdot)$, $u_8(\cdot)$. The mixed optimal strategies are of the form:

$$\begin{array}{ll} \text{for Player } E & (0, 0, \frac{3}{4}, \frac{1}{4}, 0, 0, 0, 0), \\ \text{for Player } P & (0, 0, 0, 0, 0, 0, \frac{1}{4}, \frac{3}{4}). \end{array}$$

We may show that for all $0 < \alpha < 1$ the mixed optimal strategies of the players concentrate their measures on pure strategies $v_3(\cdot)$, $v_4(\cdot)$ of Player E and on strategies $u_7(\cdot)$, $u_8(\cdot)$ of Player P.

6.4 Differential games with discrete information partition

Consider the differential game in which the players P and E have no information on their states, and know only some sets in which such states reside. Assume that a motion equation is of the form

$$\dot{x} = f(x, u, v), \quad u \in U, \quad v \in V \tag{6.4.1}$$

(f is continuous and satisfies Lipschitz condition for x uniformly with respect to u, v in a bounded set X), where $x \in R^2$, and U and V are compact sets in R^2, with the initial condition $x(0) = x_0$. For smplicity, we restrict ourselves to the games of prescribed duration T with $x \in R^2$.

Definition 4. Let $S = \{S_1, \ldots, S_m\}$ be a partition of the plane R^2 into a finite number of sets, i.e. $S_i \bigcap S_j = \Lambda$, $i \neq j$, $\bigcup_{i=1}^m S_i = R^2$, that is a diffeomorphism of sell subpartition of the same plane. (By the cell subpartition is meant any subpartition of cell partition). The point $x \in S_i$ is called a boundary point of the partition if for any $\epsilon > 0$ there is such ϵ–neightbouhood $U(x, \epsilon)$ of the point x and such $k \neq i$ that

$$U(x, \epsilon) \bigcap S_k \neq \Lambda. \tag{6.4.2}$$

The point $x \in S_i$ is called the boundary point of the first order if it is a boundary point of the partition and the set S_k from (6.4.2) is unique. The remaining boundary points are called the boundary points of the second order. The partition S is called regular if it contains only a finite number of boundary points of the second order with at least one elment of the partition being bounded. The connected set of boundary points of the first order is called a regular segment of the boundary.

Assume that, under any initial conditions and any pair of constant controls u, v, the solution $x(t)$ of the system (6.4.1) for $t > 0$ is not tangent to regular segments of the boundary.

The game proceeds as follows. A regular information partition S is given on R^2. A continuous distribution F is given on one of the bounded elements of this partition S_{k_0}. The "chance" chooses the point $x_0 \in S_{k_0}$ in accordance with the distribution F. The point x_0 serves as the initial condition for Eq.(6.4.1) and is not known to the Players P and E. The players P and E only know that the point x_0 belongs to the partition element S_{k_0}, the distibution F also being known to them. The point x moves in S_{k_0} in accordance with the motion Eq.(6.4.1) under the unfluence of controls u and v chosen respectively by the players P and E.

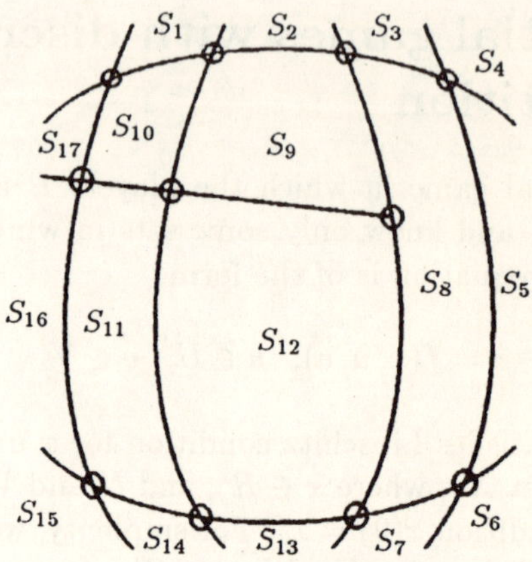

Fig. 33.

Assume that the system (6.4.1) has a unique solution, which can be continued on the interval $[0, T]$, for all initial conditions $x_0 \in S_{k_0}$ and for all measurable open-loop controls $u(t)$, $v(t)$ of the players P and E.

Let t_1 be the first instant when the point $x(t)$ falls on the boundary of the set S. At the point $x(t_1)$, the set of partition elements $\{S_{k_i}\}$, $i = 1, \ldots, \ell$, for which it is boundary point, is made known to the players. If the point x moves further along the boundary of the partition S, then the following information is furnished when $x(t)$ falls on the point of the boundary with the changed set of the neighbouring sets $\{S'_{k_i}\}$, $i = 1, \ldots, \ell$. If however, x starts moving from the point $x(t_1)$ along one of the partition S_{k_i}, the following information is furnished at the instant t_2 when $x(t_2)$ again falls on the boundary of S. In this case, the sets of partition elements, for which the point $x(t_2)$ is boundary point, are made known to the players. This process continues until the instant T. Player E gets the payoff $H(x(T))$, where the function H is a continuous function given on R^2. Player P gets the payoff $-H(x(T))$. In what follows the partition S is assumed to be regular. The regular partition S is of the form given on Fig.33. Here the boundary points of the second order are shown by circles.

As is seen from the definition of the boundary point of the first order, they serve as bounds only for two partition elements of S_k. Any regular segment of the boundary is the boundary of two elements of the partition S.

Denote a regular segment of the boundary between the sets $S_i, S_j \subset S$ by $\bar{S}_{ij}$ (if S_i and S_j have no boundary points in common, then $\bar{S}_{ij} = \Lambda$ is assumed to be equal to an empty set).

Let $\ell(S_{ij})$ be the lenght of the line segment of the curve $\bar{S}_{ij}$ (it also may be equal to infinity). If the sets S_i, S_j have a non empty regular segment of the boundary $\bar{S}_{ij}$, then between these sets there is a maximum regular segment of

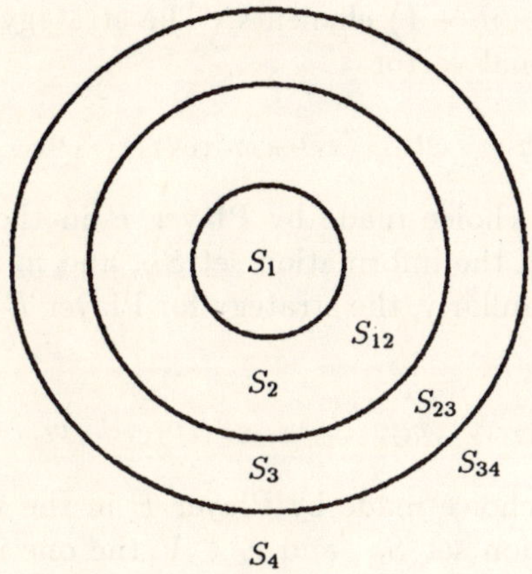

Fig. 34.

the boundary $\bar{S}_{ij}$, i.e. such that

$$\ell(S_{ij}) = \max_{\bar{S}_{ij}} \ell(\bar{S}_{ij})$$

(here $\ell(S_{ij})$ may be equal to infinity). If $\ell(S_{ij}) \neq \infty$, then the set S_{ij} is a closed curve (Fig. 34) or the ends of the segment S_{ij} are the isolated boundary points of the second order (since there is only a finite number of boundary points of the second order). If, however, $\ell(S_{ij}) = \infty$, then the segment S_{ij} has not more than one boundary point of the second order. Denote the boundary points of the second order by $\{M_\ell\}$, $\ell = 1, \ldots, S$.

The informational description of the game suggests that the regular segments of the boundary S_{ij} and the points M_ℓ play a special informational role in the game. At the points of the set S_{ij}, and at the points M_ℓ, the player acquires information on the transition of the process to one of the partition elements for which it is a boundary. Thus, after realization of a random mechanism with the distribution $F(x)$ on the set $S_{k_0} \in S$ at the initial instant of the game, the players' information states vary as the trajectory $x(t)$ passes through the set of S_{ij} and the points M_ℓ. This enables us to give the following definition.

Definition 5. The set $S_{k_0} \in S$, the regular segments of boundary S_{ij} and the isolated boundary points of the second order S_ℓ are called the information sets for the players P and E. Since the information sets are uniquely defined by the partition S, the latter is called the information partition.

Now we can define the players' strategies. Consider a collection of all information sets

$$(S_{k_0}, S_{12}, \ldots, S_{ij}, \ldots, S_{m,m-1}; M_1 \ldots, M_\ell, \ldots, M_s),$$

composed of $(s+m^2-m+1)$ elements. The strategy of Player P is $2(s+m^2-m+1)$-dimensional vector

$$u(\cdot) = (u_{k_0}, u_{12}, \ldots, u_{ij}, \ldots, u_{m,m-1}, u_1, \ldots, u_\ell, \ldots, u_s), \tag{6.4.3}$$

where $u_{k_0} \in U$ is the choice made by Player P in the information set S_{k_0}; $u_{ij} \in U$ is his choice in the information set S_{ij}, and $u_\ell \in U$ the choice in the information set M_ℓ. Similarly, the strategy for Player E is $2(s+m^2-m+1)$-dimensional vector

$$v(\cdot) = (v_{k_0}, v_{12}, \ldots, v_{ij}, \ldots, v_{m,m-1}, v_1, \ldots, v_\ell, \ldots, v_s), \tag{6.4.4}$$

where $v_{k_0} \in V$ is the choice made by Player E in the set S_{k_0}; $v_{ij} \in V$ is his choice in the information set S_{ij}, and $v_\ell \in V$ the one in the information set M_ℓ. The set of all vectors (6.4.3) for $u_{k_0} \in U$, $u_{ij} \in U$, $u_\ell \in M$ is the set of pure strategies of Player P and is denoted by $\bar{P}$. Similarly the set of vectors (6.4.4) for $v_{k_0} \in V$, $v_{ij} \in V$, $v_\ell \in V$ is the set of pure strategies for Player E and is denoted by E.

Let $(u(\cdot), v(\cdot))$ be a situation in the game. Consider the formation of $x(t)$ in the situation $(u(\cdot), v(\cdot))$ in more detail.

Since the initial choise of strategy in game belong to the information set $S_{k_0} \in S$ as a result of a random move, then at the first stage the players strategies determine the choices of controls $u(S_{k_0}) = u_{k_0}$, $v(S_{k_0}) = v_{k_0}$ which remain constant until the instant of time t_1 when the point x moving in accordance with the system

$$\dot{x} = f(x, u_{k_0}, v_{k_0}), \quad x(0) = x_0,$$

falls on the boundary of the partition S. If the point $x(t_1) \in S_{i_1,j_1}$, then its further motion occurs in accordance with the system

$$\dot{x} = f(x, u_{i_1,j_1}, v_{i_1,j_1}), \quad x(t)|_{t=t_1} = x(t_1) \tag{6.4.5}$$

where $u_{i_1,j_1} = u(S_{i_1,j_1})$ and $v_{i_1,j_1} = v(S_{i_1,j_1})$ are the players choices in the information set S_{i_1,j_1} determined by he strategies $u(\cdot)$, $v(\cdot)$, respectively. If, however, $x(t_1) = M_\ell$, then the further motion of the point orcurs in accordance with the system

$$\dot{x} = f(x, u_\ell, v_\ell), \quad x(t)|_{t=t_1} = x(t_1), \tag{6.4.6}$$

where $u_\ell = u(M_\ell)$, $v_\ell = v(M_\ell)$ are the players choices in the information set M_ℓ determined by the strategies $u(\cdot)$, $v(\cdot)$, respectively. The motion continues until the instant t_2 when the point $x(t)$ arrives at the next information set that is other than the set S_{k_0}, where the controls are changed in accordance with the strategies $u(\cdot), v(\cdot)$ and so on, until the game terminates at the time instant T. As a result, we obtain the motion trajectory $x(t)$ for $0 \le t \le T$.

Assume that to each initial condition $x_0 \in S_{k_0}$ and each situation $(u(\cdot), v(\cdot))$ where $u(\cdot) \in P$, $v(\cdot) \in E$ correspond uniquely the motion trajectory $x(t)$, $0 \leq t \leq T$, and hence the payoff function $H(x(T))$. Therefore we may note that the payoff in this game is a function of the initial condition $x_0 \in S_{k_0}$ and the strategies $u(\cdot)$, $v(\cdot)$, i.e

$$K(x_0; u(\cdot), v(\cdot)) = H(x(T)), \tag{6.4.7}$$

where $x(t)$ is the motion trajectory from the initial state x_0 in the situation $(u(\cdot), v(\cdot))$. Since the realized initial state x_0 is unknown to the players and is a random variable from the set S_{k_0} with the distribution $F(x)$, then in the situation $(u(\cdot), v(\cdot))$ Player E can reckon on the averaged payoff

$$M(u(\cdot), v(\cdot)) = \int_{S_{k_0}} K(x_0; u(\cdot), v(\cdot)) dF, \tag{6.4.8}$$

where magnitude of $K(x_0; u(\cdot), v(\cdot))$ is determined by the formula (6.4.7). Thus we have reduced the game to the normal form:

$$\Gamma = < P, E, M(u(\cdot), v(\cdot)) >, \tag{6.4.9}$$

where P, E are some compact sets in $2(S + m^2 + 1 - m)$–dimensional Euclidean space that are the strategy sets for the players P and E and $M(u(\cdot), v(\cdot))$ is a real function given on the Cartesian product $P \times E$ which is a payoff function (fot Player E).

It may be illustrated by referring to some simple examples that in the game Γ there is no equilibrium in pure strategies even in the simplest cases (for the simplest partitions $S = (S_1, \ldots, S_m)$). Therefore it seems necessary to introduce mixed strategies.

Before defining the latter, we prove the continuity of the payoff function $M(u(\cdot), v(\cdot))$ on the set $P \times E$. The proof of this assertion in similar to the one of the corresponding Theorem 3 of this chapter. But this calls for some additional assumptions.

Consider an arbitrary situation $(u(\cdot), v(\cdot))$. Denote by D_ℓ, $\ell = 1, \ldots, s$, the δ–neighbourhoods of the boundary points M_ℓ of the second order of the partition S.

Let $S(\delta) \subset S_{k_0}$ be a set of initial states x_0 for which the trajectory $x(t)$ for $0 \leq t \leq T$ does not intersect the set D_ℓ. Assume that for any $\epsilon > 0$ we can find such $\delta > 0$ that $|\mu_F(S(\delta)) - \mu_F(S_{k_0})| < \epsilon$, where μ_F is the measure generated by the distribution F on the set S_{k_0}.

Let $x_0 \in S(\delta)$ and $t_1, \ldots, t_r$ be the instants when the trajectory $x(t)$ arrives at the information sets S_{ij} (i.e. the first instants when the trajectory $x(t)$ intersects the regular boundary segments of the partition S).

Since $x(t)$ does not pass through the set D_ℓ then there is such $\epsilon_1 > 0$ ($\epsilon_1 < \delta$) that the ϵ_1–neighbourhoods of the points $x(t_1), \ldots, x(t_r)$, $\gamma_{\epsilon_1}(x(t_i))$, $i = 1, \ldots, r$, not contain the boundary points of the second order M_ℓ.

For any $\epsilon > 0$ there is such $\delta > 0$ that, when $\rho(u', u) < \delta$, $\rho(v', v) < \delta$, the intersection points of the trajectory $x'(t)$ in the situation $(u'(\cdot), v'(\cdot))$ from the initial state $x_0 \in S(\delta)$ and the boundary of the partition S, belong to $\gamma_\epsilon(t_i)$ since minor variations in strategies cannot change the passage order through the information sets (as against the situation $(u(\cdot), v(\cdot))$). Then we may show that the instants when the trajectories $x'(t)$ arrive at the information sets are slightly different from the instants t_i of the trajectory $x(t)$ arrival in the situation $(u(\cdot), v(\cdot))$.

The foregoing implies that for any $\epsilon > 0$ it is possible to find $\delta > 0$ that when $\rho(u'(\cdot), u(\cdot)) < \delta$, $\rho(v'(\cdot), v(\cdot)) < \delta$,

$$\max_{0 \le t \le T} |x'(t) - x(t)| < \epsilon.$$

The latter means that the trajectories emanatung from the points $x_0 \in S(\delta)$ are continuously dependent on the strategies at the point $(u(\cdot), v(\cdot))$. Since $H(x(T))$ is assumed to be a continuous function of x, then for any sequence $u_n(\cdot)$, $v_n(\cdot)$ converging to $u(\cdot)$, $v(\cdot)$, and for any point $x_0 \in S(\delta)$, there is such $N(x_0, \epsilon)$ that for all m, $n > N(x_0, \epsilon)$

$$|K(x_0; u_{n+m}(\cdot), v_{n+m}(\cdot)) - K(x_0; u_n(\cdot), v_n(\cdot))| < \epsilon, \tag{6.4.10}$$

i.e the sequence $K(x_0; u_n(\cdot), v_n(\cdot))$ is converging in itself at all points of the set $S(\delta)$.

We now show that the sequence of integrals $\int_{S_{k_0}} K(x_0; u_n(\cdot), v_n(\cdot))dF$ also is converging in itself. Indeed,

$$|M(u_{n+m}(\cdot), v_{n+m}(\cdot)) - M(u_m(\cdot), v_m(\cdot))| =$$

$$= \left| \int_{S_{k_0}} K(x_0; u_{n+m}(\cdot), v_{n+m}(\cdot))dF - \int_{S_{k_0}} K(x_0; u_n(\cdot), v_n(\cdot))dF \right| =$$

$$= \left| \int_{S_{k_0} \setminus S(\delta)} K(x_0; u_{n+m}(\cdot), v_{n+m}(\cdot))dF - \int_{S_{k_0} \setminus S(\delta)} K(x_0; u_n(\cdot), v_n(\cdot))dF + \right.$$

$$\left. + \int_{S(\delta)} K(x_0; u_{n+m}(\cdot), v_{n+m}(\cdot))dF - \int_{S(\delta)} K(x_0; u_n(\cdot), v_n(\cdot))dF \right| \le$$

$$\le \left| \int_{S_{k_0} \setminus S(\delta)} K(x_0; u_{n+m}(\cdot), v_{n+m}(\cdot))dF - \int_{S_{k_0} \setminus S(\delta)} K(x_0; u_n(\cdot), v_n(\cdot))dF \right| +$$

$$+ \left| \int_{S(\delta)} K(x_0; u_{n+m}(\cdot), v_{n+m}(\cdot))dF - \int_{S(\delta)} K(x_0; u_n(\cdot), v_n(\cdot))dF \right|.$$

from (6.3.1) it follows that the sequence $K(x_0; u_n(\cdot), v_n(\cdot))$ converges at each point $x_0 \in S(\delta)$, therefore the sequence of integrals $\int_{S(\delta)} K(x_0; u_n(\cdot), v_n(\cdot))dF$ also converges. This means that there is such $N(\epsilon)$ that with $n > N(\epsilon)$ for all m

$$\left| \int_{S(\delta)} K(x_0; u_{n+m}(\cdot), v_{n+m}(\cdot))dF - \int_{S(\delta)} K(x_0; u_n(\cdot), v_n(\cdot))dF \right| < \epsilon.$$

Estimating further the first difference ($\mu_F(S_{k_0} \setminus S(\delta)) < \epsilon$, $K(x_0; u(\cdot), v(\cdot)) < M$) we obtain

$$\left| \int_{S_{k_0}} K(x_0; u_{n+m}(\cdot), v_{n+m}(\cdot))dF - \int_{S_{k_0}} K(x_0; u_m(\cdot), v_m(\cdot))dF \right| < \epsilon + 2\epsilon M$$

for $n > N(\epsilon)$ and all m. This means that the sequence $\int_{S_{k_0}} K(x_0; u_n(\cdot), v_n(\cdot))d\mu_F$ is fundamental, i.e. there exist the limit

$$\lim_{n\to\infty} \int_{S_{k_0}} K(x_0; u_n(\cdot), v_n(\cdot))dF. \tag{6.4.11}$$

Since the measure F is continuous, then the integral $\int_{S_{k_0}} K(x_0; u(\cdot), v(\cdot))d\mu_F$ is a continuous function of the set. From the condition that for any $\delta > 0$

$$\lim_{n\to\infty} \int_{S(\delta)} K(x_0; u_n(\cdot), v_n(\cdot))d\mu_F = \int_{S(\delta)} K(x_0; u(\cdot), v(\cdot))d\mu_F,$$

it follows that the limit (6.4.11) is equal to $\int_{S_{k_0}} K(x_0; u(\cdot), v(\cdot))d\mu_F$, i.e. the function $M(u(\cdot), v(\cdot))$ is continuous at the point $(u(\cdot), v(\cdot))$. Since the situation $(u(\cdot), v(\cdot))$ is arbitrary, then $M(u(\cdot), v(\cdot))$ is a continuous function of strategies $u(\cdot)$, $v(\cdot)$.

Now we have the following theorem.

Theorem 7 *In the game with the regular information partition S, the payoff function $E(u(\cdot), v(\cdot))$ is a continuous function definded on the product of sets $P \times E$.*

As in the previous section, definde a mixed extention of the game Γ. Let μ, ν be a mixed strategies of the players and

$$E(\mu, \nu) = \int_P \int_E M(u(\cdot), v(\cdot))d\mu d\nu$$

be the expectation of the payoff in mixed strategies μ, ν. Denote by $\{\mu\}$ and $\{\nu\}$ the sets of all mixed strategies of the players P and E, respectively.

Employing Theorem 4 in Chapter I and Theorem 5 in this chapter, obtain the following theorem.

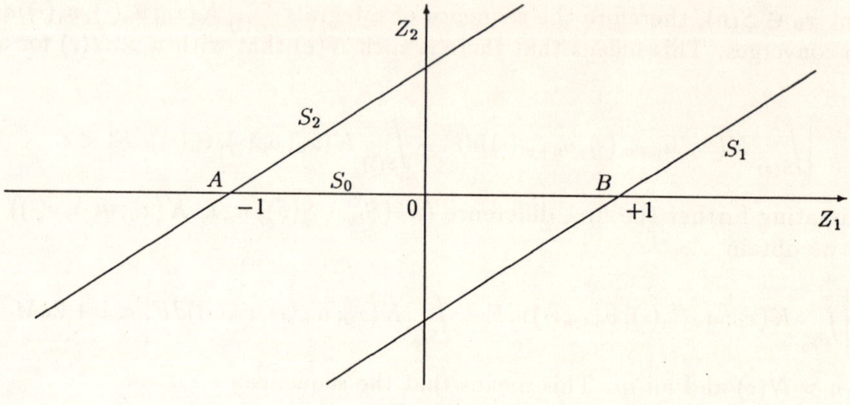

Fig. 35.

Theorem 8 *The game Γ with a regular information partition S has the equlibrium in mixed strategies.*

Example 4. The game proseeds on the plane. Motion equation are the same as in Example 3 of this chapter. The duration of the game $T = 1$. The payoff the Player E is terminal and equal to $|z_1(1)| + |z_2(1)|$. The information partition is of the form shown in Fig.35. The game starts in the information set AB (the interval $[-1, 1]$ on axis z_1) with a random choice(in accordance with a uniform distribution on AB) of the initial point z_0.

Since one of the two control ($+1$ or -1) can be chosen in each of the three information sets, then the pure strategy sets for both players coincide and are of the form

$$\begin{aligned}
&u_1(\cdot),\;\; v_1(\cdot): \;\; (1,1,1),\\
&u_2(\cdot),\;\; v_2(\cdot): \;\; (1,1,-1),\\
&u_3(\cdot),\;\; v_3(\cdot): \;\; (1,-1,1),\\
&u_4(\cdot),\;\; v_4(\cdot): \;\; (1,-1,-1),\\
&u_5(\cdot),\;\; v_5(\cdot): \;\; (-1,1,1),\\
&u_6(\cdot),\;\; v_6(\cdot): \;\; (-1,-1,1),\\
&u_7(\cdot),\;\; v_7(\cdot): \;\; (-1,1,-1),\\
&u_8(\cdot),\;\; v_8(\cdot): \;\; (-1,-1,-1).
\end{aligned}$$

To define the payoff matrix, we have to compute the points z_1, z_2 in each situation and to average the payoff over the initial data from the interval $[-1, 1]$. For the simplicity, here we omit minor computations and propose the final form of the payoff matrix:

	u_1	u_2	u_3	u_4	u_5	u_6	u_7	u_8
v_1	2	2	2	2	1,5	1,5	2	2
v_2	2	2	2	2	0,5	0,5	1	1
v_3	2	2	2	2	1,5	1,5	2	2
v_4	2	2	2	2	0,5	0,5	1	1
v_5	1	1	0,5	0,5	2	2	2	2
v_6	2	2	1,5	1,5	2	2	2	2
v_7	1	1	0,5	0,5	2	2	2	2
v_8	2	2	1,5	1,5	2	2	2	2

As is seen from the payoff matrix, the nondominated strategies (see [2,8]) are only the strategies $v_1(\cdot)$, $v_6(\cdot)$ for Player E and the strategies $u_3(\cdot)$, $u_5(\cdot)$ for Player P. The consideration of the corresponding 2×2 matrix

$$\begin{pmatrix} 2 & 1,5 \\ 1,5 & 2 \end{pmatrix}$$

shows that the above-mentioned strategies are to be chosen with equal probabilities. Thus the optimal mixed strategies for the Player P and E are of the form $(0,0,1/2,0,1/2,0,0,0)$ and $(1/2,0,0,0,0,1/2,0,0)$ respectively.

6.5 Games with mixed information state

Here, as in the previous section, we assume that a motion equation is of the form

$$\dot{x} = f(x,u,v), \tag{6.5.1}$$

where $x \in R^2$, $u \in U \subset R^2$, $v \in V \subset R^2$ (U and V are compact set) with the initial condition $x(0) = x_0$. As usual, we assume that the condition, which ensure the existence, uniqueness and $[0,T]$ continuability of the solution for any measurable open-loop controls $u(t)$, $v(t)$, $0 \leq t \leq T$, are imposed on the system (6.5.1). The game has a prescribed duration T. Player E's payoff is determined by the function $H(x(T))$, where $H(x)$ is a given continuous function (Player P's payoff is equal too $-H(x(T))$.

The convex set M wherein the players have a discrete information state, is given in R^2. When the point x falls on the boundary of M (from the inside), the game continues with complete information (i.e. the players are informed about the phase state of the system $x(t)$ at each current instant of time). Under the conditions of the game, once fallen on the boundary of the M the game preseeds with phase constraints in the set $R^2 \setminus M$.

For simplicity, the set M is assumed to be open, i.e $\overline{R^2 \setminus M} = R^2 \setminus M$. Let S' be a regular partition of the plane R^2. Denote by $S = S' \bigcap M$ a subpartition of the partition S', where S is the partitions of the set M by the sets $S'_k \bigcap M$ ($S'_k \subset S'$) that are a subpartition of the partition S'. Similary, we define regular interval of the boundary S_{ij} as an intersection $S_{ij} = S'_{ij} \bigcap M$ of regular interval of the partition boundary S' with the set M. Next let M_ℓ, $\ell = 1, 2, \ldots, s$ be the second order boundary points of the partition S' belonging to the set M.

The strategy $u(\cdot)$ of Player P consists of a pair of strategies $u(\cdot) = \{u_1(\cdot), u_2(\cdot)\}$, where $u_1(\cdot)$ is a strategy of the form (6.4.3) defined on the information set S of the partition S' (see 4 in this chapter), and $u_2(\cdot)$ is the piecewise open–loop strategy in the perfect information game with discrimination and phase constraints in the set $R^2 \setminus M$. The strategy $v(\cdot) = \{v_1(\cdot), v_2(\cdot)\}$ of Player E is defined in much the same way.

Denote by P and E the sets of all possible strategies $u(\cdot)$, $v(\cdot)$ of the players P and E, respectively.

The definition of strategy given above reflects the fact that at the first stage of te game (in M) the players actions are based on the discrete information determined by the partition S (the functions $u_1(\cdot)$, $v_1(\cdot)$ being used to form the controls), and then, after leaving M they play the perfect information game (the second components of strategies $u_2(\cdot)$, $v_2(\cdot)$ being used here).

The game proceeds as follows. Let S'_{k_0} be an element of the partition S' having a nonempty intersection with the set M. At the initial instant of time the "chance" chooses a point $x_0 \in S_k = S'_{k_0} = S'_{k_0} \bigcap M$ in accordance with the given continuous distribution F. Suppose the players have chosen the strategies $u(\cdot) = \{u_1(\cdot), u_2(\cdot)\} \in P$ and $v(\cdot) = \{v_1(\cdot), v_2(\cdot)\} \in E$. In the situation $(u(\cdot), v(\cdot))$ the motion of the point is as follows. In the set M the motion of the point x from the state $x_0 \in S_{k_0}$ realized by a random mechanism is affected only of the components $u_1(\cdot)$, $v_1(\cdot)$ of the strategies $u(\cdot)$, $v(\cdot)$, and the trajectory $x(t)$ is constructed at this stage as in 4 of this chapter.

Let $\bar{t}$ be the first instant when $x(t)$ falls on the set $\overline{R^2 \setminus M}$. From the state $x(\bar{t})$ (for $\bar{t} < T$) the motion of the point is determined by the components $u_2(\cdot)$, $v_2(\cdot)$ of the strategies $u(\cdot)$, $v(\cdot)$ and is constructed as in 2, Ch. 2. If $\bar{t} \geq T$, then the game terminates in the set M, and the trajectory $x(t)$ is constructed as in 4 of this present chapter. Thus we establish that to each randomly realized initial position and situation $(u(\cdot), v(\cdot))$ correspond a unique trajectory $x(t)$ and therefore a unique payoff

$$K(x_0; u(\cdot), v(\cdot)) = H(x(T)), \tag{6.5.2}$$

where $x(t)$ is the trajectory from the initial state x_0 in the situation $(u(\cdot), v(\cdot))$,

Since the realized initial state x_0 is unknown to the players and is a random variable with the distribution $F(x)$ on the set $S_{k_0} = S'_{k_0} \bigcap M$, then Player E

in the situation $(u(\cdot), v(\cdot))$ can only reckon on the averaged payoff.

$$M(u(\cdot), v(\cdot)) = \int_{S_{k_0}} K(x_0; u(\cdot), v(\cdot))dF, \tag{6.5.3}$$

where the magnitude of $K(x_0; u(\cdot), v(\cdot))$ is determined from the formula (6.5.2). We have thus reduced the game to normal form:

$$\Gamma = < \bar{P}, \bar{E}, M(u(\cdot), v(\cdot)) > \tag{6.5.4}$$

where $\bar{P}$, $\bar{E}$ are the strategy sets of the players P and E, and $M(u(\cdot), v(\cdot))$ is the real function (see (6.5.3)) defined on the Cartesian product $P \times E$ and representing the payoff function (Player E's payoff).

As in 6, Ch. 2, we may show that the differential game with discrimination and phase constraints in the set $\overline{R^2 \setminus M}$ for any $\epsilon > 0$ has the ϵ–equilibrium in piecewice open–loop strategies.

Denote by $V(x_0, T)$ the value function for the differential game of duration T in the set $\overline{R^2 \setminus M}$ from the initial state $x_0 \in \overline{R^2 \setminus M}$. To obtain a solution in the game Γ, we define an auxiliary game Γ'. The game Γ' differs from the game Γ by the definition of the payoff function. Let $(u(\cdot), v(\cdot))$, be the situation in Γ, $x_0 \in S_{k_0}$, be the result of choosing a random move, and $x(t)$ be a trajectory in the situation $(u(\cdot), v(\cdot))$ from the initial state $x_0 \in \underline{S_{k_0}}$. Denote by $\bar{t}$ the first instant when the trajectory $x(t)$ penetrates the set $\overline{R^2 \setminus M}$, i.e.

$$\bar{t} = \inf\{t \ : \ x(t) \notin M\}. \tag{6.5.5}$$

If there is no infimum in (6.5.5), then we set $\bar{t} = \infty$. The payoff function is defined in the game Γ' as follow:

$$K'(x_0; u_1(\cdot), v_1(\cdot)) = \begin{cases} V(x(\bar{t}), T - \bar{t}), & \text{if } 0 \le \bar{t} \le T, \\ H(x(T)) & \text{otherwise.} \end{cases}$$

Note that by the definition of the game $V(x, T)$ value $V(x, 0) = H(x)$. Next, let

$$M'(u_1(\cdot), v_1(\cdot)) = \int_{S_{k_0}} K'(x_0; u_1(\cdot), v_1(\cdot))dF.$$

The strategies in the game Γ' are the first components of the strategies $u(\cdot) = (u_1(\cdot), u_2(\cdot))$, $v(\cdot) = (v_1(\cdot), v_2(\cdot))$ determined on the elements of the partition S, i.e. $u_1(\cdot)$ and $v_1(\cdot)$ are not employed (the game terminates when the point x arrives at $\overline{R^2 \setminus M}$).

Denote the strategy sets $u_1(\cdot)$ and $v_1(\cdot)$ in the game Γ' by $\bar{P}_1$ and $\bar{E}_1$ respectively. Evidently, these sets are compact sets in the Euclidean spaces of finite dimension (see 4 in this chapter).

Theorem 9 *The payoff function $M'(u_1(\cdot), v_1(\cdot))$ is a continuous function on the set $\bar{P}_1 \times \bar{E}_1$.*

Proof: This theorem is proved in much the same way as Theorem 7. All the conditions from the 4 of this chapter are assumed to be imposed on the partition S' and the system (6.5.1).

As in 4, introduce the set $S(\delta)$ of initial states from which, in the situation $(u_1(\cdot), v_1(\cdot))$, the trajectory fails to pass through the ϵ–neighbourhoods of the second order boundary points of the partition S. We show that for $x_0 \in S(\delta)$ the payoff $K'(x_0; u_1(\cdot), v_1(\cdot))$ is continuous in the point $(u_1(\cdot), v_1(\cdot))$. To this end, consider two cases.

Case 1. The initial state $x_0 \in S(\delta)$ is such that $\bar{t} \leq T$.

Case 2. The game terminates when the players reaches the set $\overline{R \setminus M}$,i.e. $\bar{t} > T$.

In case 1 the continuity of $K'(x_0; u_1(\cdot), v_1(\cdot))$ follows from that of the function $V(x(\bar{t}), T - t)$ in x and $T - \bar{t}$ (the value function of the differential perfect information game depend continuously upon the initial states). If $\bar{t} < T$, we prove the continuity by reproducing the reasonings from 4. If, however, $\bar{t} = T$, then we employ the condition $H(x((T)) = V(x(T), 0)$ and the continuity of the function H in x.

The proof of Theorem 7 fully applies to Case 2.

The continuity of the function $M'(u_1(\cdot), v_1(\cdot))$ in the point $(u_1(\cdot), v_1(\cdot))$ follows from that of the function $K'(x_0; u_1(\cdot), v_1(\cdot))$ in the point $(u_1(\cdot), v_1(\cdot))$ for $x_0 \in S(\delta)$ (see (6.4.10), (6.4.11) and the computations of the payoff of Theorem 7). ■

Since the point $(u_1(\cdot), v_1(\cdot))$ has been chosen arbitrarily, then the payoff function $M'(u_1(\cdot), v_1(\cdot))$ of the game Γ' is continuous over the strategy set $P_1 \times E_1$. Since the strategy sets in the auxiliary game Γ' are compact sets in the Euclidean space, then employing Theorem 4, Ch. 1, we obtain the following theorem.

Theorem 10 *In the auxiliary game Γ' there exist the equilibrium in mixed strategies.*

We now focus on the original game Γ. In the game Γ the strategy sets are not compact, since each strategy contains a piecewise open–loop component representing a strategy in the perfect information game on the set $\overline{R^2 \setminus M}$. The set of such strategies is obviously non compact, since it contains the set of all measurable open–loop controls $u(t)$, $v(t)$ for $t \leq t \leq T$. At the same time, for $\epsilon > 0$ we sucseed in proving the existence of the ϵ–equilibrium in the game Γ in mixed strategies. The proof employs the condition that in the perfect information game on the set $\overline{R^2 \setminus M}$ for any $\epsilon > 0$ there exist the ϵ–equilibrium point.

Let V' be the value of the game Γ', and (μ^*, ν^*) the equilibrium in mixed strategies in this game (see Theorem 10). For a particular $\epsilon > 0$, each situation $(u_1(\cdot), v_1(\cdot))$ for $u_1(\cdot) \in \bar{P}_1$, $v_1(\cdot) \in \bar{E}_1$ in the game Γ' is accosiated with the situation $(u^\epsilon(\cdot), v^\epsilon(\cdot))$ for $u^\epsilon(\cdot) \in \bar{P}$, $v^\epsilon(\cdot) \in \bar{E}$ in the game Γ, where $u_\epsilon(\cdot) = (u_1(\cdot), u_2^\epsilon(\cdot))$, $v_\epsilon(\cdot) = (v_1(\cdot), v_2^\epsilon(\cdot))$ and $u_2^\epsilon(\cdot)$, $v_2^\epsilon(\cdot)$ are the ϵ–optimal strategies in the perfect information game of duration $T - \bar{t}$ on the set $\overline{R^2 \setminus M}$ from the initial state $x(\bar{t})$. Here $\bar{t}$ is the first instant when the trajectory $x(t)$ in the situation $(u_1(\cdot), v_1(\cdot))$ penetrates the set $\overline{R^2 \setminus M}$ (if $\bar{t} > T$, then the strategies $u_2^\epsilon(\cdot)$, $v_2^\epsilon(\cdot)$ are defined in an arbitrary way).

Theorem 11 *Let (μ^*, ν^*) be the equilibrium in mixed strategies in the game Γ'. Then the same situation (μ^*, ν^*) is the $\epsilon_1 = 3\epsilon$–equilibrium in the game Γ if by μ^* is meant the probability measure on the strategy set $\bar{P}_1$ of the form $u_\epsilon(\cdot) = (u_1(\cdot), u_2^\epsilon(\cdot))$, and by ν^* the probability measure on the strategy set $\bar{E}_1$ of the form $v_\epsilon(\cdot) = (v_1(\cdot), v_2^\epsilon(\cdot))$.*

Proof: By the definition of the game Γ'

$$K'(x_0; u_1(\cdot), v_1(\cdot)) = \begin{cases} V(x(\bar{t}), T - \bar{t}), & \text{for } \bar{t} \le T, \\ H(x(T)) & \text{fot } \bar{t} > T. \end{cases}$$

but $|V(x(\bar{t}), T - \bar{t}) - K(x_0; u_\epsilon(\cdot), v_\epsilon(\cdot))| < \epsilon$, since the pair $u_2^\epsilon(\cdot)$, $v_2^\epsilon(\cdot)$ forms the ϵ-equilibrium in the perfect information game with the duration $T - \bar{t}$ on the set $\overline{R \setminus M}$ from the initial state $x(\bar{t})$. Therefore

$$|K'(x_0; u_1(\cdot), v_1(\cdot)) - K(x_0; u_\epsilon(\cdot), v_\epsilon(\cdot))| < \epsilon, \tag{6.5.6}$$

and, consecuently,

$$|M'(u_1(\cdot), v_1(\cdot)) - M(u_\epsilon(\cdot), v_\epsilon(\cdot))| < \epsilon$$

for all $u_1(\cdot)$, $v_1(\cdot)$ and $u_\epsilon(\cdot)$, $v_\epsilon(\cdot)$. Then

$$|E'(\mu, \nu) - E(\mu, \nu)| < \epsilon, \tag{6.5.7}$$

where

$$E'(\mu, \nu) = \int_{P_1}\int_{E_1} M'(u(\cdot), v(\cdot)) d\mu d\nu,$$

$$E(\mu, \nu) = \int_{P}\int_{E} M(u_\epsilon(\cdot), v_\epsilon(\cdot)) d\mu d\nu,$$

We show that

$$E(u(\cdot),\nu^*)+3\epsilon \geq E(\mu^*,\nu^*) \geq E(\mu^*,v(\cdot))-3\epsilon \qquad (6.5.8)$$

for all $u(\cdot)\in P$, $v(\cdot)\in E$.

By Theorem 3,Ch. 1, this implies that (μ^*,ν^*) is the ϵ_1–equilibrium point $\epsilon_1=3\epsilon$ in the game Γ. Prove the left hand side of (6.5.8):

$$E(u(\cdot),\nu^*)=\int_E M(u(\cdot),v_\epsilon(\cdot))d\nu^*=\int_E\int_{S_{k_0}} K(x_0;u(\cdot),v_\epsilon(\cdot))dFd\nu^*=$$

$$=\int_E\int_{S_{k_0}} K\{x_0;[u_1(\cdot),u_2(\cdot)],[v_1(\cdot),v_2^\epsilon(\cdot)]\}dFd\nu^*\geq$$

$$\geq\int_E\int_{S_{k_0}} K\{x_0;[u_1(\cdot),u_2^\epsilon(\cdot)],[v_1(\cdot),v_2^\epsilon(\cdot)]\}dFd\nu^*-\epsilon=$$

$$=\int_E\int_{S_{k_0}} K(x_0;u_\epsilon(\cdot),v_\epsilon(\cdot))dFd\nu^*-\epsilon.$$

Here $u_\epsilon(\cdot)\in P_\epsilon$. From (6.5.6),(6.5.7) we have

$$\int_E\int_{S_{k_0}} K(x_0;u_\epsilon(\cdot),v_\epsilon(\cdot))dFd\nu^*-\epsilon\geq$$

$$\geq\int_E\int_{S_{k_0}} K'(x_0;u_1(\cdot),v_1(\cdot)dFd\nu^*-2\epsilon=$$

$$=E'(u_\epsilon(\cdot),\nu^*)-2\epsilon\geq E'(\mu^*,\nu^*)-2\epsilon\geq E(\mu^*,\nu^*)-3\epsilon.$$

We have thus proved the left–hand side of inequality (6.5.8). The proof of the right–hand side of (6.5.8) is omitted here, since it is carried out in much the same way. ■

Conclusively we obtain the following theorem.

Theorem 12 *In the game Γ for any $\epsilon>0$ there exist the ϵ–equilibrium in mixed strategies.*

Example 5. The set M is an open unit circle with its center lying in the origin of coordinates. The regular partition S' of the plane R^2 is composed of two sets S'_1 and S'_2, where $S'_1\supset M$, and $S'_2=R^2\setminus S'_1$. Thus, the subpartition S of the partition S' is composed of a unique set $S_1=S'\bigcap M=M$. The motion equations in the case involved are the same as in the example 3, 4 of this chapter,i.e.

$$\dot z=u+v,\quad z\in R^2$$

The sets of control variables U, V, however are richer than in the previous examples, viz:

$$V = U = \{(1,0), (-1,0), (0,1), (0,-1)\}.$$

The initial state z_0 is chosen in accordance with uniform distribution on M. As in 5, the players have perfect information out off M, i.e. they know the state z at each instant of time. The duration of the game is fixed and equal $T = 2$. The payoff is terminal and equals

$$F(z_0, z(2)) = |z_1(2)| + |z_2(2)|.$$

As shown before, to solve the game, we have to find first a solution of the prfect information subgame on the set $R^2 \setminus M$.

Suppose the initial state z belongs to the boundary of the set M. Consider the case $z_1 \geq 0$, $z_2 \geq 0$. The perfect information game from other initial states in solved in much the same way. Assume that Player E adopts the constant strategy $v^*(\cdot) = (1,0)$ irrespective of time and realized states z. The best responce strategy by Player P then is to minimize the grouth rate of the component z_1, i.e. to adopt the strategy $u^*(\cdot) = (-1,0)$ (any other strategy involves the grouth $|z_1| + |z_2|$). As a result the point stays where it is until the game terminates. At the same time, we may see that the best responce of Player E to the strategy $u^*(\cdot) = (-1,0)$ is the choise of the strategy $v^*(\cdot) = (1,0)$. Thus, the situation $(u^*(\cdot), v^*(\cdot))$ in the region of $z_1 \geq 0$, $z_2 \geq 0$ forms the equilibrium in the perfect information game, and the value of this game is equal to $|R\cos\varphi| + |R\sin\varphi|$, where (R, φ) are polar coordinates of the initial point $z = (z_1, z_2)$ in the perfect information subgame from the initial states $z = (z_1, z_2)$ belonging to the set M proves to be equal to

$$|z_1| + |z_2| = |R\cos\varphi| + |R\sin\varphi|$$

and is independent of the duration of this subgame.

Optimal strategies in the perfect information game are of the form

$$v^*(\cdot) = \begin{cases} (1,0) & \text{for } z_1 \geq 0, \\ (-1,0) & \text{for } z_1 < 0, \end{cases}$$

$$u^*(\cdot) = \begin{cases} (-1,0) & \text{for } z_1 \geq 0, \\ (1,0) & \text{for } z_1 < 0. \end{cases}$$

For the complete solution of the game it remains to define optimal strategies in the information set M. Since the partition S contains only one information set, then the strategy sets for both players coincide, each containing four strategies:

$$\begin{aligned} u_1(\cdot),\ v_1(\cdot) &: \ (1,0), \\ u_2(\cdot),\ v_2(\cdot) &: \ (0,1), \\ u_3(\cdot),\ v_3(\cdot) &: \ (-1,0), \\ u_4(\cdot),\ v_4(\cdot) &: \ (0,-1). \end{aligned}$$

Note that in the computation of the payoff matrix in all situations $(u(\cdot), v(\cdot))$ except $(u_1(\cdot), v_3(\cdot))$, $(u_2(\cdot), v_4(\cdot))$, $(u_4(\cdot), v_2(\cdot))$, $(u_3(\cdot), v_1(\cdot))$, for all initial states $z_0 \in M$ the trajectory falls on the boundary of the set M. From the solution of the perfect information game it follows that the trajectory remains on the boundary of M until the game termination. Therefore, the payoff in any of the indicated situations is the mathematical expectation of the sum of absolute values of coordinates taken over the initial positions z_0 from the set M if the trajectory intersects the boundary of M in a suitable situation.

We shall not provide all computation, though it should be noted that the payoff M' concide in the following situations:

$$M'(u_1(\cdot), v_1(\cdot)) = M'(u_2(\cdot), v_2(\cdot)) =$$

$$= M'(u_3(\cdot), v_3(\cdot)) = M'(u_4(\cdot), v_4(\cdot)) = a,$$

$$M'(u_1(\cdot), v_2(\cdot)) = M'(u_1(\cdot), v_4(\cdot)) =$$

$$= M'(u_2(\cdot), v_1(\cdot)) = M'(u_2(\cdot), v_4(\cdot)) =$$

$$= M'(u_4(\cdot), v_1(\cdot)) = M'(u_4(\cdot), v_3(\cdot)) =$$

$$= M'(u_3(\cdot), v_2(\cdot)) = M'(u_3(\cdot), v_4(\cdot)) = b,$$

$$M'(u_1(\cdot), v_3(\cdot)) = M'(u_2(\cdot), v_4(\cdot)) =$$

$$= M'(u_4(\cdot), v_2(\cdot)) = M'(u_3(\cdot), v_1(\cdot)) = c.$$

Now for the solution of the game Γ' we have to solve the following auxiliary matrix game:

	$u_1(\cdot)$	$u_2(\cdot)$	$u_3(\cdot)$	$u_4(\cdot)$
$v_1(\cdot)$	a	b	c	b
$v_2(\cdot)$	b	a	b	c
$v_3(\cdot)$	c	b	a	b
$v_4(\cdot)$	b	c	b	a

The mixed optimal strategies in the matrix game Γ' under consideration prescribe equal probabilities $(1/4, 1/4, 1/4, 1/4)$ to each pure strategy of the players P and E. The mixed optimal strategies in the original game are constructed with the help of the solution of the game Γ' and the perfect information subgame on the set $R^2 \setminus M$ just as it was in Theorem 11.

6.6 Pursuit game with prescribed duration and delayed information for both players

Consider a differential zero–sum pursuit game of prescribed duration T between the pursuer team $P = \{P_1, \ldots, P_m\}$, acting as one player, and the evader E. Motion equations are of the form:

$$\begin{array}{lll} \text{for players } P_i & \dot{x}^{(i)} = f^{(i)}(x,u), & u \in U^{(i)} \subset Comp\, R^k, i = 1, \ldots, m, \\ \text{for Player } E & \dot{y} = g(y,v), & v \in V \supset Comp\, R^\ell, \end{array} \tag{6.6.1}$$

where $x^{(i)} \in R^n$, $y \in R^n$, $x^{(i)}(0) = x_0^{(i)}$, $y(0) = y_0$. All the conditions, which ensure the existence, uniqueness and $[0,T]$ continuability of the solution from the initial states $x_0^{(i)}$, y_0 for any pair of measurable open–loop controls $u(t) = \{u^{(i)}(t)\}$, $v(t)$, are imposed on the right–hand sides of equations (6.6.1).

The games under consideration differ from the imperfect information games of pursuit presented in 1 of this chapter in the non–convexity of the payoff function and delayed information about the variable of Player $P = \{P_i\}$ (aside from delayed information on about the Player E for Player P). In 2 convexity of the payoff implies the existence of the pure optimal strategy for Player P and the mixed one for Player E. In the present case both players have only mixed optimal strategy. Let us define more presicely the state of information in game.

Assume that the numbers $\ell_1 > 0$, $\ell > 0$ ($\ell_1 > \ell$) are given and represent a delay in acquiring information on Player P by Player E, and vice versa. For $0 \le t \le T$, at each time instant t Player P knows his state $x(t) = \{x^{(i)}(t)\}$, the time t and Player E's state at the initial time instant y_0. For $\ell \le t \le T$, at each instant t Player P knows his state $x(t) = \{x^{(i)}(t)\}$, the time t and Player E's state at the instant $t - \ell$. For $0 \le t \le \ell_1$, at each instant t Player E knows the time t, his state $y(t)$ and Player P's state $x_0 = \{x_0^{(i)}\}$ at the initial instant of time. For $\ell_1 \le t \le T$, at each instant t Player E knows his state $y(t)$, the time t and the state $x(t-\ell_1) = \{x^{(i)}(t-\ell_1)\}$ of Player $P = \{P_i\}$ at the instant $t - \ell_1$.

Player E's payoff is defined as

$$\min_i \rho(x^{(i)}(T), y(T)), \tag{6.6.2}$$

where $\rho(x^{(i)}, y)$ is the Euclidean distance between the points $x^{(i)}$, y.

Denote the defined game by $\Gamma(x_0, y_0, T)$.

As in 1, 2, introduce mixed and pure piecewise open–loop strategies.

Pure piecewise open–loop strategies. By the pure piecewise open–loop strategy $v(\cdot)$ of Player E is meant a pair $\{\tau, b\}$, where τ is a partition of the time interval $[0,T]$ by a finite number of points $0 = t_1 < t_2 < \ldots < t_s = T$, and b is mapping which places each state t_k, x_0, $y(t_k)$ for $0 \le t_k < \ell_1$ in correspondence with the measurable open–loop control $v(t)$ for $t \in [t_k, t_{k+1})$ and places each state t_k, $x(t_k - \ell_1)$, $y(t_k)$ for $\ell_1 \le t_k \le T$ in correspondence with the measurable open–loop control $v(t)$ for $t \in [t_k, t_{k+1})$. By the pure piecewise open–loop strategy $u(\cdot)$ of Player P is meant a pair $\{\sigma, a\}$, where σ is a partition of the time interval $[0,T]$ by a finite numberof points $0 = t'_1 < t'_2 < \ldots < t'_q = T$, and a is a mapping which places each state t'_k, y_0, $x(t'_k)$ for $0 \le t'_k < \ell$ in correspondence with a measurable open–loop control $u(t) = \{u^{(i)}(t)\}$ for $t \in [t'_k, t'_{k+1})$, and places each state t'_k, $y(t'_k - \ell)$, $x(t'_k)$ for $\ell \le t'_k \le T$ in correspondence with the measurable open–loop control $u(t) = \{u^{(i)}(t)\}$ for $t \in [t'_k, t'_{k+1})$.

Mixed piecewise open–loop behavior strategy. By the mixed piecewise open–loop behavior strategy $\nu(\cdot)$ of Player E is meant a pair $\{\tau, c\}$, where τ is a partition of the time interval $[0,T]$ by a finite number of points $0 = t_1 < t_2 < \ldots < t_s = T$, and c is a mapping which places each state t_k, x_0, $y(t_k)$ for $0 \le t_k < \ell_1$ in correspondence with the probability measure ν (that is dependent on t_k, x_0, $y(t_k)$) on Player E's reachability set $C_E^{t_{k+1}-t_k}(y(t_k))$ from the state $y(t_k)$ in the time $t_{k+1} - t_k$, and places each state t_k, $y(t_k)$, $x(t_k - \ell_1)$ for $\ell_1 \le t_k \le T$ in correspondence with the probability mesure ν on Player E's reachability set $C_6^{t_{k+1}-t_k}(y(t_k))$. By the mixed piecewise open–loop behavior strategy $\mu(\cdot)$ of team $P = \{P_1, \ldots, P_m\}$ is a meant a pair $\{\sigma, d\}$, where σ is a partition of the time interval $[0,T]$ by a finite number of points $0 = t'_1 < t'_2 < \ldots < t'_q = T$ and d is a mapping which places each state t'_k, y_0, $x(t'_k)$ for $0 \le t'_k \le \ell$ in correspondence with the probability measure μ (that is dependent on t'_k, y_0, $x(t'_k)$) on Player P's reachability set

$$C_P^{t'_{k+1}-t'_k}(x(t'_k)) = \prod_i C_{P_i}^{t'_{k+1}-t'_k}(x^{(i)}(t'_k))$$

in the time $t'_{k+1} - t'_k$, and places each state t'_k, $y(t'_k - \ell)$, $x(t'_k)$ for $\ell \le t'_k \le T$ in correspondence with the probability measure μ (that is dependent on t'_k, $y(t'_k - \ell)$, $x(t'_k)$) on Player P's reachability set

$$C_P^{t'_{k+1}-t'_k}(x(t'_k)) = \prod_i C_{P_i}^{t'_{k+1}-t'_k}(x^{(i)}(t'_k))$$

in the time $t'_{k+1} - t'_k$.

For each POLS pair $(u(\cdot), v(\cdot))$ we may define the payoff function

$$K(x_0, y_0; u(\cdot), v(\cdot)) = \min_i \rho(x^{(i)}(T), y(T)), \tag{6.6.3}$$

where $x^{(i)}(t)$, $y(t)$ are the players' trajectories realized when employing the strategies $u(\cdot)$, $v(\cdot)$ in the game $\Gamma(x_0, y_0, T)$. Since the game is an imperfect information game, the equilibrium in pure POLS's may not exist therein, hence the introduction of MPOLBS is absolutely necessary. Each pair of strategies $\mu(\cdot)$, $\nu(\cdot)$ with the fixed initial conditions x_0, y_0 induces the probability distribution on the final position of the game

$$C_P^T(x_0) = \prod_{i=1}^{m} C_{P_i}^T(x_0^{(i)}), \;\; C_E^T(y_0),$$

hence by the payoff is meant the mathematical expectation (6.6.3) which is denoted by $M(x_0, y_0; \mu(\cdot), \nu(\cdot))$. Under the solution of the game $\Gamma(x_0, y_0, T)$ we mean the finding of the equilibrium in the MPOLBS class.

The motion under MPOLBS $\mu(\cdot)$, $\nu(\cdot)$ from the initial states x_0, y_0 is constructed as follows. Let $\mu(\cdot) = \{\sigma, d\}$, $v(\cdot) = \{\tau, c\}$, where $\sigma = \{t'_k\}$, $\tau = \{t''_k\}$. At the instant $t'_1 = t''_1 = 0$ the players P and E realize the measures μ, ν dictated by the maps c and d in the state x_0, y_0, 0 and make transitions to the points $x^{(i)}(t'_2) \in C_P^{t'_2-t'_1}(x_0^{(i)})$ and $y(t''_2) \in C_E^{t''_2-t''_1}(y_0)$ using any open-loop controls which carry over the point $x^{(i)}(t'_1)$ to $x^{(i)}(t'_2)$ and $y(t''_1)$ to $y(t''_2)$ (here the points $x^{(i)}(t'_2)$ and $y(t''_2)$ are chosen randomly in accordance with the probability measures μ, ν entering into the strategies $\mu(\cdot)$, $\nu(\cdot)$ as behaviors). The probability measures μ, ν dictated by the maps c, d are again realized in the states $x(t'_2) = (x^{(i)}(t'_2))$ and $y(t''_2)$. In these information state the players make transitions to the randomly chosen points $x(t'_3) = (x^{(i)}(t'_3))$, where $x^{(i)}(t'_3) \in C_P^{t'_3-t'_2}(x^{(i)}(t'_2))$ and $y(t''_3) \in C_E^{t''_3-t''_2}(y(t''_2))$ etc. Such sequential choice of behavior implies the realization of random trajectories $x(t)$, $y(t)$ from the initial states x_0, y_0 correspoding to the situation in MPOLBS $\mu(\cdot)$, $\nu(\cdot)$.

Let $C_P^T(x_0^{(i)})$, $C_T^E(y_0)$ be the reachability sets of the team member P_i and Player E from the initial states $x_0^{(i)}$, y_0 at the time T. For simplicity, assume that (though it is not important for what follows) the set $C_E^T(y)$ for any y, T is convex and compact.

Consider a simultaneous auxiliary zero–sum game of pursuit Γ_y on the set $C_E^\ell(y)$. The game proceeds as follows. Player P chooses m points $\eta_i \in C_E^\ell(y)$, and Player E the point $\xi \in C_E^\ell(y)$. The choices are made simultaneously and independently of one another, and Player E receives the payoff defined by the formula (6.6.2). Since the function (6.6.2) is continuous, and the pure strategy sets in the auxiliary game Γ_y are compact, then the game has the equilibrium in the class of mixed strategies. Denote the value of the game by V_y in order to emphasize its dependence on the strategy sets $[C_E^\ell(y)]^m$, $C_E^\ell(y)$ of the players P and E. Let (μ_y^*, ν_y^*) be the equilibrium in the game Γ_y. For $y \in C_E^{T-\ell}(y_0)$, the following requirements are imposed on the family of games Γ_y.

1. For any $\epsilon > 0$ there is such N that in the game Γ_y Player P has the mixed ϵ–optimal strategy $\mu_y^{*\epsilon}$ prescribing equal probabilities $(\frac{1}{N})$ to some

N points (m collections of points) $\xi_1(y), \ldots, \xi_N(y)$ of the set $[C_E^\ell(y)]^m$. Moreover, the number N chosen by a given ϵ is independent of y for all $y \in C_E^{T-\ell}(y_0)$.

2. Let $y(t)$ be an arbitrary admissable trajectory of Player E for $0 \leq t \leq T$. Then there are such continuous nonintersecting trajectories $\xi_j[y(t-\ell)]$, $j = 1, \ldots, N$, that $\xi_j[y(t-\ell)] \in [C_E^\ell(y(t-\ell))]^m$, and each point $\xi_j[y(t-\ell)]$ is a point of the strategy spectrum $\mu_{y(t-\ell)}^{*\epsilon}$ in the game $\Gamma_{y(t-\ell)}$.

Theorem 13 *Assume that the set $C_E^\ell(y)$ is convex and compact for any $y \in C_E^{T-\ell}(y_0)$, the above-mentioned conditions 1,2 are satisfied and each team partner P_i by the time instant $T - \ell_1$ can ensure fulfilment of the inclusion*

$$C_{P_i}^{\ell_1}(x^{(i)}(T-\ell_1)) \supset C_E^\ell(y(T-\ell)) \tag{6.6.4}$$

for all $y(T-\ell) \in C_E^{\ell_1}(y(T-\ell-\ell_1))$ irrespective of Player E's actions, proceeding from the information acquired in the game $\Gamma(x_0, y_0, T)$ only. And the team P acting further on the time interval $[T-\ell_1, T]$ can ensure the ϵ-capture (is meant the arrival of the point $x(x^{(1)}, \ldots, x^{(m)})$ in the ϵ-neighbourhood of the point $\xi_j(\xi_j^{(1)}, \ldots, \xi_j^{(m)})$) of any point $\xi_j[y(T-\ell)]$ for any $\epsilon > 0$ by the instant T (here ξ_j, m are the collections of points defined by the conditions 1, 2.)

Player E then has the optimal MPOLBS which prescribes for $0 \leq t \leq T-\ell$ an open-loop transition to the point $\bar{y}$, for which the value of the auxiliary game $V(\bar{y})$ is maximal. Further, at the instant $T-\ell$, the optimal MPOLBS prescribes the choice of any point $y(T)$ in accordance with the mixed strategy ν_y^* which is optimal on the game $\Gamma_{\bar{y}}$ (a random choice is to be made), and an open-loop transition to the point $y(T)$ realized as a result of a random choice. For any $\epsilon > 0$ Player P has the ϵ-optimal MPOLBS which, for $0 \leq t \leq T$, prescribes that he should act so (in a determined way) as to ensure the fulfilment of the inclusion (6.6.4), and then to choose one of the points $\xi_j[y(T-\ell-\ell_1)]$ with the probability $\frac{1}{N}$ ensuring the ϵ-capture of the point ξ_j moving along the trajectory $\xi_j[y(t-\ell)]$ by the instant T (by the assumptions of the theorem this is possible).

Proof: Let $\mu^{*\epsilon}(\cdot)$, $\nu^{*\epsilon}(\cdot)$ be the strategies whose optimality will be proved. It suffices to prove that the inequalities

$$M(x_0, y_0; u(\cdot), \nu^{*\epsilon}(\cdot)) - \epsilon \geq M(x_0, y_0; \mu^{*\epsilon}(\cdot), \nu^{*\epsilon}(\cdot)) \geq$$

$$\geq M(x_0, y_0; \mu^{*\epsilon}, v(\cdot)) + \epsilon \tag{6.6.5}$$

are satisfied for all pure piecewise open-loop strategies $u(\cdot)$, $v(\cdot)$. From the definition of the strategies $\mu^{*\epsilon}(\cdot)$ $\nu^{*\epsilon}(\cdot)$ is follows immediately that

$$|M(x_0, y_0; \mu^{*\epsilon}(\cdot), \nu^{*\epsilon}(\cdot)) - V(\bar{y})| < \epsilon \tag{6.6.6}$$

Therefore it suffices to prove that the inequalities

$$M(x_0 y_0; u(\cdot), \nu^{*\epsilon}(\cdot)) \geq V(\bar{y}) + \epsilon, \tag{6.6.7}$$

$$M(x_0, y_0; \mu^{*\epsilon}(\cdot), v(\cdot)) \leq V(\bar{y}) - \epsilon. \tag{6.6.8}$$

are satisfied.

To prove (6.6.7), consider two cases.

1. Assume that the strategy $u(\cdot)$ fails to ensure fulfilment of the inclusion (6.6.4) by the instant $T - \ell_1$. Then the inequality (6.6.7) follows from the monotonic dependence of the payoff (6.6.2) on ρ and the convexity of the set $C_E^\ell(y)$.

2. If, however, the strategy $u(\cdot)$ ensures the fulfilment of the inclusion (6.6.4) and then directs the pursuit not towards the points ξ_j entering into the ϵ–optimal strategy spectrum of Player P in the game Γ_y, then the inequality (6.6.7) follows from the optimality of the strategy $\nu_{\bar{y}}$ of Player E in the game $\Gamma_{\bar{y}}$. The inequality (6.6.8), however, follows immediately from the fact that the optimal strategy of Player E contains the optimal strategy $\nu_{\bar{y}}^*$ in the game $\Gamma_{\bar{y}}$ as a behavior strategy, and the maximum value of the auxiliary game is achieved at the point $\bar{y}$.

To complete the proof, note that the conditions 1, 2 on the class of games Γ_y are not rigid and are satisfied in most of the cases involved when the payoff in each game Γ is continuously dependent upon the parameter y. The other conditions of the theorem are the conditions concerning the perfect information games. ■

6.7 Delayed information for both players when the evader team takes part in game

Consider a generalization of the problem given in the previos section, to the case when several evaders, acting as one player, take part in the game. We a have a zero–sum differential pursuit–evasion game of prescribed duration between a team of pursuers $P = \{P_1, \ldots, P_m\}$, acting as one player, and a team of evaders $E = \{E_1, \ldots, E_n\}$, also acting as one player.

Motion equations are of the form

$$\begin{aligned} &\text{for } P_i\text{:} \quad \dot{x}^{(i)} = f^{(i)}(x^{(i)}, u^{(i)}), \quad u^{(i)} \in U^{(i)} \subset R^p, \quad i = i, \ldots, m; \\ &\text{for } E_j\text{:} \quad \dot{y}^{(j)} = g^{(j)}(y^{(j)}, v^{(j)}), \quad v^{(j)} \in V^{(j)} \subset R^q, \quad j = 1, \ldots, n; \end{aligned} \tag{6.7.1}$$

$$x^{(i)} \in R^k, \; y^{(j)} \in R^k, \; x^{(i)}(0) = x_0^{(i)}; \; y^{(j)}(0) = y_0^{(j)}$$

All the conditions which ensure the existence, uniqueness and $[0,T]$ continuability of the solution (6.7.1) are supposed to be ssatisfied for any measurable open–loop controls $u(t) = \{u^{(i)}(t)\}$, $v(t) = \{v^j(t)\}$.

The information state in the game is as follows, given are the numbers $\ell > 0$ and $\ell_1 > 0$ representing Player P delayed information about Player E and Player E's delayed information about Player P, respectively. This means that, for $0 \leq t \leq \ell$, at each instant of time t Player P knows his state $x(t) = \{x^{(i)}(t)\}$, the time t and Player E's state at the initial instant $y_0 = \{y_0^{(i)}\}$. For $\ell \leq t \leq T$, Player P knows his state $x(t) = \{x^{(i)}(t)\}$, the time t and Player E's state $y(t-\ell) = \{y^{(j)}(t-\ell)\}$ at the time instant $t-\ell$. Similarly, for $0 \leq t \leq \ell_1$, at each instant of time t Player E knows his state $y(t) = \{y^{(j)}(t)\}$, the time t and Player P' state $x_0 = \{x_0^{(i)}\}$ at the initial instant. For $\ell_1 \leq t \leq T$, Player E knows his state $y(t) = \{y^{(j)}(t)\}$, the time t and Player P's state at the time instant $t - \ell_1$.

Player E's payoff is terminal and is defined as follows. Consider the matrix $A(x,y) = (\rho(x^{(i)}, y^{(j)}))$, where $\rho(x^{(i)}, y^{(i)})$ is the Euclidean distance between the points $x^{(i)}$, $y^{(j)} \in R^k$; $x = (x^{(1)}, \ldots, x^{(m)})$, $y = (y^{(1)}, \ldots, y^{(n)})$. Introduce an order relation on the matrix set. We say that $A > B$ if all elements b_{ij} of the matrix B do not exceed the corresponding elements a_{ij} of the matrix A and there exist such indexes (i_0, j_0) that $a_{i_0,j_0} > b_{i_0,j_0}$. The real function F defined on the set of $m \times n$ matrices, is monotonically increasing if $F(A) \geq F(B)$ for $A > B$. Let the payoff of Player E be equal to $K(x,y) = F(A(x,y))$, where F is a real, continuous, monotonically increasing function.

By the pure piecewise open–loop strategy $u(\cdot)$ of Player E is meant a pair $\{\sigma, a\}$, where σ is an arbitrary partition of the time interval $[0,T]$ by a finite number of points $0 = t_1 < t_2 < \ldots < t_\tau = T$ and a is a mapping which places each state $x(t_k) = \{x^{(i)}(t_k)\}$, $y_0 = \{y_0^{(j)}\}$, t_k for $0 \geq t_k \leq \ell$ in correspondence with the open–loop control $u(t) = \{u^{(i)}(t)\}$, $[t_k, t_{k+1})$, and for $\ell \leq t_k \leq T$ it palces each state $x(t_k)$, $y(t_k - \ell) = \{y^{(j)}(t_k - \ell)\}$, t_k in correspondence with the measurable open–loop control $u(t) = \{u^{(i)}(t)\}$ defined on $[t_k, t_{k+1})$. Similarly, by the pure piecewise open–loop strategy $v(\cdot)$ for Player E is meant a pair $\{\tau, b\}$, where τ is an arbitrary partition of the time interval $[0,T]$ by a finite number of points $0 = t'_1 < t'_2 < \ldots < t'_s = T$ and b is a mapping which places each state x_0, $y(t'_k)$, t'_k for $0 \leq t'_k \leq \ell_1$ in correspondence with a measurable open–loop control $v(t) = \{v^{(j)}(t)\}$ defined on the time–interval $[t'_k, t'_{k+1})$, and for $\ell_1 \leq t'_k \leq T$ it places each state $x(t'_k - \ell_1)$, $y(t'_k)$, t'_k in correspondence with a measurable open–loop control $v(t)$ defined in the interval $[t'_k, t'_{k+1})$. The sets of all pure piecewise open–loop strategies for the players P and E are denoted by $\mathcal{P}$ and $\mathcal{E}$, respectively. Since, for every fixed pair of initial of states x_0, y_0, there exists a unique solution of the system (6.7.1), for each pair of strategies

$u(\cdot) \in \mathcal{P}$, $v(\cdot) \in \mathcal{E}$, then the payoff function is defined uniquely

$$K(x_0, y_0; u(\cdot), v(\cdot)) = F(A(x(T), y(T))) \tag{6.7.2}$$

where $x(t)$, $y(t)$ constitute a solution of (6.7.1) from the initial states x_0, y_0 in the situation $(u(\cdot), v(\cdot))$. Denote the game in hand by $\Gamma(x_0, y_0, T)$.

By the mixed piecewise open-loop behavior strategy (MPOLBS) $\mu(\cdot)$ of Player P is meant a pair $\{\sigma, d\}$, where σ is an arbitrary partition of the time interval $[0, T]$ by a finite number of points $0 = t_1 < t_2 < \ldots < t_r = T$ and d is a mapping which for $0 \le t_k < \ell$, places each state $x(t_k)$, y_0, t_k in correspondence with the probability measure μ on Player P's reachability set in the time $t_{k+1} - t_k$ from the state $x(t_k)$, $C_P^{t_{k+1}-t_k}(x(t_k)) = \prod_{i=1}^m C_{P_i}^{t_{k+1}-t_k}(x^{(i)}(t_k))$, and for $\ell \le t_k \le T$ it places each state $x(t_k)$, $y(t_k - \ell)$, t_k in correspondence with the probability measure μ on Player P's reachability set $C_P^{t_{k+1}-t_k}(x(t_k))$. Similarly, by the MPOLBS $\nu(\cdot)$ of Player E is meant a pair $\{\tau, c\}$, where τ is an arbitrary partition of the time interval $[0, T]$ by a finite number of points $0 = t'_1 < t'_2 < \ldots < t'_s = T$ and c is a mapping which places each state x_0, $y(t'_k)$, t'_k for $0 \le t'_k < \ell_1$ in correspondence with the probability measure ν on Player E's reachability set from the state $y(t'_k)$ in the time $t'_{k+1} - t'_k$, $C_E^{t'_{k+1}-t'_k}(y(t'_k)) = \prod_{j=1}^n C_{E_j}^{t'_{k+1}-t'_k}(y^{(j)}(t'_k))$, and places each state $x(t'_k - \ell)$, $y(t'_k)$, t'_k for $\ell_1 \le t'_k \le T$ in correspondence with the probability measure on the reachability set $C_E^{t'_{k+1}-t'_k}(y(t'_k))$. The sets of all MPOLBS of the players P and E are denoted respectively by $\bar{\mathcal{P}}$ and $\bar{\mathcal{E}}$. With the initial states fixed, each MPOLBS pair $\mu(\cdot)$, $\nu(\cdot)$ induces probability distribution over the final states of the game $C_P^T(x_0)$, $C_E^T(y_0)$ therefore by the payoff of Player E we mean the mathematical expectation $M(x_0, y_0; \mu(\cdot), \nu(\cdot))$ of the payoff $K(x_0, y_0; u(\cdot), v(\cdot))$. We have thus defined a mixed extension $\bar{\Gamma}(x_0, y_0, T)$ of the game $\Gamma(x_0, y_0, T)$.

Let us describe how in the situation $(\mu(\cdot), \nu(\cdot)) \in \bar{\mathcal{P}} \times \bar{\mathcal{E}}$ the motion from the initial states x_0, y_0 is constructed. Denote the union of partitions $\sigma = \{t_k\}$ and $\tau = \{t'_k\}$ by $\omega = \{t''_k\}$. At the instant $t''_1 = 0$ the players P and E realize the measures μ and ν prescribed respectively by the mappings d and c in the state x_0, y_0, and make transitions to the points $x(t_2) = \{x^{(i)}(t_2)\}$, $x^{(i)}(t_2) \in C_{P_i}^{t_2-t_1}(x_0^{(i)})$, $i = 1, \ldots, m$, and $y(t'_2) = \{y^{(j)}(t'_2)\}$, $y^{(j)}(t'_2) \in C_{E_j}^{t_2-t_1}(y_0^{(j)})$, $j = 1, \ldots, n$ using any open-loop controls which carry over the point $x_0^{(i)}$ to $x^{(i)}(t_2)$ and $y_0^{(i)}$ to $y^{(j)}(t'_2)$. Here $x(t_2)$ and $y(t'_2)$ are chosen randomly in accordance with the probability measures μ and ν generated by the strategy $\mu(\cdot)$ and $\nu(\cdot)$. In the states $x(t_2)$ and $y(t'_2)$, the probability measures μ and ν dictated by the maps d and c are again realized, and the players make transitions to the randomly chosen points $x(t_3) \in C_P^{t_3-t_2}(x(t_2))$, $y(t'_3) \in C_E^{t'_3-t'_2}(y(t'_2))$, etc. As a result of such sequential choice of behaviors, the random trajectories $x(t)$ and $y(t)$ from the initial states x_0 and y_0 corresponding to the situation $(\mu(\cdot), \nu(\cdot))$

are realized.

Let all reachability sets $C^{T}_{E_j}(y^{(j)})$ be compact, and $D^{\ell}_{E}(y)$ be a convex hull of the set $\bigcup_{i=1}^{n} C^{\ell}_{E_j}(y^{(j)})$. Consider a simultaneous auxiliary zero-sum game $\Gamma(y) =< [D^{\ell}_{E}(y)]^m, C^{\ell}_{E}(y), K >$. The game proceeds as follows. Player $P = \{P_1, \ldots, P_m\}$ chooses m points $\xi^{(i)}$ of the set $D^{\ell}_{E}(y)$, and Player E chooses n points $\eta^{(j)} \in C^{\ell}_{E_j}(y^{(j)})$. The choices are made simultaneously and independently of one another. Player E receives the payoff $F(A(\xi^{(1)}, \ldots, \xi^{(m)}; \eta^{(1)}, \ldots, \eta^{(n)}))$. Since the payoff function of Player E is continuous, and the sets of pure strategies of the players are compact, then the game $\Gamma(y)$ has the equilibrium in the class, of mixed strategies. Let $Val\Gamma(y) = V(y)$, $(\mu^{\epsilon}_{y}, \nu^{\epsilon}_{y})$ be the ϵ-equilibrium in the game $\Gamma(y)$. Suppose the following two conditions are satisfied.

1. For any $\epsilon > 0$ there is such number N that in the game $\Gamma(y)$ Player P has the ϵ-optimal strategy μ^{ϵ}_{y} prescribing equal probabilities $\frac{1}{N}$ to the points $\xi_1, \ldots, \xi_n$. from the set $[D^{\ell}_{E}(y)]^m$, the number N being independent of y for all $y \in C^{T-\ell}_{E}(y_0)$.
2. Let $y(t) = \{y^{(j)}(t)\}$ be an arbitrary admissible motion of Player E for $0 \leq t \leq T$. Then there are such continuous nonintersecting trajectories $\xi_j[y(t-\epsilon)]$, $j = 1, \ldots, N$ that $\xi_j[y(t-\ell)] \in [D^{\ell}_{E}(y(t-\ell))]^m$ and each of $\xi_j[y(t-\ell)]$ is a point of the strategy spectrum $\mu^{\epsilon}_{y(t-\ell)}$ in the game $\Gamma(y(t-\ell))$.

Theorem 14 *Assume that using his accessible information, by the time instant $T-\ell_1$, Player P can ensure the inclusions*

$$D^{\ell}_{E}(y(T-\ell)) \subset C^{\ell_1}_{P_i}(y(T-\ell-\ell_1)) \tag{6.7.3}$$

for all points $y(T-\ell) \in C^{\ell_1}_{E}(y(T-\ell-\ell_1))$ and, acting further on the time interval $[T-\ell, T]$, he can ensure by the instant of time T for any number $\epsilon_1 > 0$ the ϵ_1-capture of any of thepoints $\xi_j[y(T-\ell_1]$ determined by conditions 1 and 2. Then Player $E = \{E_1, \ldots, E_n\}$ has the optimal MPOLBS which is formulated as follows. For $0 \leq t \leq T-\ell$ he has to make an open-loop transition to the point $\bar{y} = (y^{(1)}, \ldots, y^{(n)})$ where the auxiliary game value $V(y)$ is maximal; at the instant of time $T-\ell$ he has to choose any of the points $y' \in C^{\ell}_{E}(\bar{y})$ in accordance with the mixed strategy $\nu^{*}_{\bar{y}}$ that is optimal in the game $\Gamma(y)$ (making a random choice) and make an open-loop transition to the point $y' = y(T)$ realized as a result of this mixed strategy ν^{*} (random choice). For any number $\epsilon > 0$ there is the following 4ϵ-optimal MPOLBS for Player $P = \{P_1, \ldots, P_m\}$ For $0 \leq t \leq T-\ell_1$ he has to act (in a determinininistic way) so as to ensure inclusions (6.7.3) and then, with probability $\frac{1}{N}$, to choose one of the points $\xi_j[y(T-\ell-\ell_1]$ and ensure by the instant of time T the ϵ_1-capture ($\epsilon_1 = \epsilon_1(\epsilon)$) of the points ξ_j moving along the trajectory $\xi_j[y(t-\ell)]$.

Proof: Note that the fulfilment of inequalities $\rho(x^{(i)}, \xi_j^{(i)}) < \epsilon_1$ for all $i = 1, \ldots, m$ is called the ϵ_1–capture of the point $\xi_j = \{\xi_j^{(i)}\}$ by the point $x = \{x^{(i)}\}$. Let $\mu^\epsilon(\cdot)$, $\nu^*(\cdot)$ are the strategies described in the theorem. We prove that

$$M(x_0, y_0; \mu^\epsilon(\cdot), \nu(\cdot)) - 4\epsilon \le M(x_0, y_0; \mu^\epsilon(\cdot), \nu^*(\cdot)) \le$$
$$\le M(x_0, y_0; \mu(\cdot), \nu^*(\cdot)) + 4\epsilon \tag{6.7.4}$$

for all MPOLBS $\mu(\cdot)$, $\nu(\cdot)$. We may show that (2.7.4) follow from inequalities

$$M(x_0, y_0; \mu^\epsilon(\cdot), v(\cdot)) - 4\epsilon \le M(x_0, y_0; \mu^\epsilon(\cdot), \nu^*(\cdot)) \le$$
$$\le M(x_0, y_0; u(\cdot), \nu^*(\cdot)) + 4\epsilon \tag{6.7.5}$$

for all piecewise open–loop strategies $u(\cdot)$, $v(\cdot)$. Let $x_\epsilon(t)$ and $y_\epsilon(t)$ be the trajectories of the players P and E respectively resulting from employment of strategies $\mu^\epsilon(\cdot)$, $\nu^*(\cdot)$. Then the point $x_\epsilon(T)$ with probability $\frac{1}{N}$ falls within the ϵ_1–neighbourhood of the point $\xi_j[y(T-\ell)]$ from the strategy $\mu_{\bar{y}}^\epsilon$ that is ϵ–optimal in the game $\Gamma(\bar{y})$. The point $y_\epsilon(T)$ is a realization of the strategy $\nu_{\bar{y}}^*$ that is optimal for Player E in the game $\Gamma(\bar{y})$. Since the payoff function is continuously dependent on $\rho(x^{(i)}, y^{(i)})$, then for any number $\epsilon > 0$ there is the number $\epsilon_1 > 0$ such that as soon as the point $x_\epsilon(T)$ with probability $\frac{1}{N}$ falls within the ϵ_1–neighbourhood of the point $\xi_j[y(T-\ell)]$, the inequality

$$|M(x_0, y_0; \mu^\epsilon(\cdot), \nu^*(\cdot)) - M(\mu_{\bar{y}}^\epsilon, \nu_{\bar{y}}^*)| < \epsilon, \tag{6.7.6}$$

where $M(\mu_{\bar{y}}^\epsilon, \nu_y^*)$ is the mathematical expectation of payoff in the game $\Gamma(\bar{y})$ when the mixed strategies $\mu_{\bar{y}}^\epsilon$, $\nu_{\bar{y}}^*$ are used is satisfied. On the other hand, the situation $(\mu_{\bar{y}}^\epsilon, \nu_{\bar{y}}^*)$ is ϵ–optimal in the game $\Gamma(\bar{y})$ and hence

$$|M(\mu_{\bar{y}}^\epsilon, \nu_{\bar{y}}^*) - V(\bar{y})| < \epsilon \tag{6.7.7}$$

From the formulas (6.7.5)–(6.7.7) it follows that to prove this theorem it suffices to show that the inequalities

$$M(x_0, y_0; \mu^\epsilon(\cdot), v(\cdot)) - 2\epsilon \le V(\bar{y}) \tag{6.7.8}$$

$$M(x_0, y_0; u(\cdot), \nu^*(\cdot)) + 2\epsilon \ge V(\bar{y}) \tag{6.7.9}$$

are satisfied for all pure piecewise open–loop strategies $u(\cdot)$, $v(\cdot)$.

We first prove inequality (6.7.8). Let Player P adopt the strategy $\mu^\epsilon(\cdot)$, and Player E an arbitrary pure piecewise open–loop strategy $v(\cdot)$. Then, under the strategy $\mu^\epsilon(\cdot)$, Player P makes a random choice at the instant of time $T - \ell_1$ and, by the game termination instant T, he appears with probability $\frac{1}{N}$ in the ϵ_1–neighbourhood of the point $\xi_j[y(T-\ell)]$ entering into the spectrum of the strategy $\mu^{\epsilon}_{y(T-\ell)}$ that is ϵ–optimal in the game $\Gamma(y(T-\ell))$. Whatever the strategy $v(\cdot) \in \mathcal{E}$ of Player E and its corresponding trajectory $y(t)$ may be, the inclusion $y(T) \in C^{\ell}_{E}(y(T-\ell))$ is satisfied, i.e. $y(t)$ is a pure strategy for Player E in the game $\Gamma(y(T-\ell))$. Since the delay of Player E's information on Player P is equal to $\ell_1 > 0$, then the result of Player P' random choice remains unknown to Player E until the game terminates and Player P's behavior proves to be strategically equivalent to the equiprobable choice of points in the instant T. Next, since the payoff function is continuously dependent on $\rho(x^{(i)}, y^{(j)})$, and the set $C^{\ell}_{E}(y(T-\ell))$ is compact, then for any number $\epsilon > 0$ there is such number $\epsilon_1 > 0$ that as soon as $\rho(x^{(i)}(T), \xi^{(i)}_j[y(T-\ell)]) < \epsilon_1$, $i = 1, \ldots, m$, then the inequality

$$|M(x_0, y_0; \mu^{\epsilon}(\cdot), v(\cdot)) - M(\mu^{\epsilon}, y(T))| < \epsilon, \tag{6.7.10}$$

where $M(\mu^{\epsilon}, y(T))$ is the mathematical expectation of payoff in the auxiliary game $\Gamma(y(T-\ell))$, and $y(t)$ is the trajectory resulting from employment of the strategy $v(\cdot)$, is satisfied. On the other hand, the strategy μ^{ϵ} is ϵ-optimal in the game $\Gamma(y(T-\ell))$:

$$M(\mu^{\epsilon}, y(T)) - \epsilon \leq V(y(T-\ell)) \tag{6.7.11}$$

It follows from inequality (6.7.10),(6.7.11) that $M(x_0, y_0; \mu^{\epsilon}(\cdot), v(\cdot)) - 2\epsilon \leq V(y(T-\ell))$, here the number ϵ_1 can be chosen the same for all strategies $v(\cdot) \in \mathcal{E}$. Since $V(y(T-\ell)) \leq V(\bar{y})$, then this implies inequality (6.7.8).

We now prove inequality (6.7.9). Let Player E makes a random choice at the time instant $T-\ell$ and hence Player P does not know the result of this choice until the game terminates; therefore he may not move to the point realized as a result of this random choice. Consider two cases.

1. Let the strategy $u(\cdot)$ of Player P be such that $x(T) \in [D^{\ell}_{E}(\bar{y})]^m$, where $x(t)$ is the pursuer's trajectory corresponding to this strategy. Then $x(T)$ is a pure strategy for Player P in the game $\Gamma(\bar{y})$ and hence the inequality (6.7.9) is satisfied because of the optimality of strategy $\nu^*_{\bar{y}}$ in the game $\Gamma(\bar{y})$.

2. Let $u(\cdot)$ be such strategy of Player P that at least one of the points $x^{(i)}(T)\bar{\in} D^{\ell}_{E}(\bar{y})$. Assume that $x^{(1)}(T) \notin D^{\ell}_{E}(\bar{y})$. Consider the quantity

$$\min_{\xi \in D^{\ell}_{E}(\bar{y})} \rho(x^{(1)}, \xi) = \rho(x^{(1)}, \xi_0),$$

where ξ_0 is a boundary point of the set $D^{\ell}_{E}(\bar{y})$. Since the set $D^{\ell}_{E}(\bar{y})$ is convex, then the inequality $< x^{(1)} - \xi_0, y - \xi_0 > \leq 0$, where $< x, y >$ is the scalar product of vector $x, y \in R^k$, holds for any point y of this set. Therefore

$$\rho^1(x^{(1)}, y) - \rho^2(\xi_0, y) =< x^{(1)} - y, x^{(1)} - y > - < \xi_0 - y, \xi_0 - y >=$$

$$=< x^{(1)} - \xi_0, x^{(1)} - \xi_0 > +2 < x^{(1)} - xi_0, \xi_0 - y > > 0.$$

Consequently, $\rho(x^{(1)}, y) > \rho(\xi_0, y)$ for any point $y \in D_E^\ell(\bar{y})$ and, in particular, for the points setting up the spectrum of strategy $\nu_{\bar{y}}^*$ in the game $\Gamma(\bar{y})$. Then the matrix $A(x,y) = (\rho(x^{(i)}(T), y^{(i)}(T))$ can be compared with the matrix $B(\bar{x}, y)$ different from the former in the first row, where the point ξ_0 lies instead of the point $x^{(1)}$ and $A(x,y) > B(\bar{x}, y)$. The function F increases monotonicaly and hence, for the strategy $u(\cdot)$, Player P may select such dominating strategy $\bar{u}(\cdot)$ that $\bar{x}^{(i)}(T) \in D_E^\ell(\bar{y})$, $i = 1, \ldots, m$, where $\bar{x}(t) = \{\bar{x}^{(i)}(t)\}$ is Player P's trajectory corresponding to the strategy $\bar{u}(\cdot)$, i.e. we have made a transition to the first case.

■

We now prove two theorems which enable us to isolate a sufficiently wide class of auxiliary games satisfijing condition 1 and 2. Consider an infinite zero–sum two–person game Γ. Let X, Y be the pure strategy sets respectively for the minimizer and the maximizer, and $K(x,y)$ be the kernel of the game;

$$M(\xi, \nu) = \int_X \int_Y K(x,y)\xi(dx)\nu(dy)$$

is the expectation of the maximizaer's payoff, where ξ is the probability measure on the Borel field of subsets in X, and ν is the probability measure on the Borel field of subsets in Y.

Theorem 15 *Let X and Y be compact subsets respectively in the spaces R^n and R^m, and K be a continuous function defined on the set $X \times Y$. Then for any number $\epsilon > 0$ there are such number N and M that in the game Γ the players have the mixed ϵ–optimat strategies ξ_N^ϵ and ν_M^ϵ prescribing equal probabilities $\frac{1}{N}$ and $\frac{1}{M}$ respectively to N and M points $x_i \in X$, $i = 1, \ldots, N$, and $y_i \in Y$, $j = 1, \ldots, M$.*

Proof: Let as define the Borel field $\hat{\mathcal{X}}$ of subsets in the set X with the help of the metric

$$\rho_x(x_1, x_2) = \max_{y \in Y} |K(x_1, y) - K(x_2, y)|,$$

and the Borel field $\hat{\mathcal{Y}}$ of subsets in the set Y with the help of the metric

$$\rho_y(y_1, y_2) = \max_{x \in Y} |K(x, y_1) - K(x, y_2)|.$$

We shall say that the strategies x_1 and x_2 (y_1 and y_2), for which $\rho(x_1, x_2) = 0$ ($\rho(y_1, y_2) = 0$),are equivalent. A. Vald shows that if the strategy space of one of the player in the zero–sum game Γ is conditionally compact in the topology under considiration, then for any number $\epsilon > 0$ the game has the ϵ–equilibrium, and each of the players has the ϵ–optimal strategy which is a mixture of a finite number of pure strategies. From the conditions of the theorem it follows that this assertion is applicable to the game Γ.

Let $\xi^{\frac{\epsilon}{8}} = (p_1, \ldots, p_r)$ be a finite $\frac{\epsilon}{8}$–optimal strategy for the minimizer, p_i being the probability of choosing a pure strategy $x_i \in X$. We may alwais suppose that $p_i = \frac{t_i}{g_i}$, where t_i, q_i are integers. All p_i are reduced to a common denominator and are written as $p_i = \frac{\ell_i}{N}$, where $\sum_{i=1}^{r} \ell_i = N$. Since the function K is uniformly continuous on the compact $X \times Y$, then for any number $\epsilon > 0$ there is such number $\delta > 0$ that the inequality $\rho(x', x'') < \delta$ for any $y \in Y$ implies the inequality $|K(x', y) - K(x'', y)| < \frac{\epsilon}{4}$. The strategy ξ_N^{ϵ} is constructed as follows. Suppose the strategy prescribes equal probabilities $\frac{1}{N}$ to the nonco-incident points x_i' such that $\rho(x_i', x_k) < \delta$ with $i = \sum_{s=0}^{k-1} \ell_s + 1, \ldots, \sum_{s=0}^{k-1} \ell_s + \ell_k$, $k = 1, \ldots, r$ $(\ell_0 = 0)$. Thus, the points x_i' are grouped in the δ–neighbourhoods of the points x_k, with ℓ_k points in each neighbourhood. The number δ is chosen in such a way that their neighbourhoods do not intersect. Then for any element $y \in Y$ the following condition is satisfied:

$$|M(\xi^{\frac{\epsilon}{4}}, y) - M(\xi_N^{\epsilon}, y)| = \left| \sum_{k=1}^{r} \frac{\ell_k}{N} K(x_k, y) - \sum_{i=1}^{N} \frac{1}{N} K(x_i', y) \right| \le$$

$$\le \frac{1}{N} \sum_{k=1}^{r} \sum_{i=(\sum_{s=0}^{k-1} \ell_s + 1)}^{(\sum_{s=0}^{k} \ell_s)} |K(x_k, y) - K(x_i', y)| \le \frac{1}{N} N \frac{\epsilon}{4} = \frac{\epsilon}{4}. \tag{6.7.12}$$

For the strategy $\xi^{\frac{\epsilon}{8}}$ we have

$$M(\xi^{\frac{\epsilon}{4}}, y) \le V + \frac{\epsilon}{4}, \tag{6.7.13}$$

where V is the value of the game Γ. From the inequalities (6.7.12),(6.7.13) it follows that $M(\xi_N^{\epsilon}, y) \le V + \frac{\epsilon}{2}$ for all $y \in Y$. The maximizer's strategy η_M^{ϵ} is constructed in much the same way, satisfying the inequality $M(x, \eta_M^{\epsilon}) \ge V - \frac{\epsilon}{2}$ for all $x \in X$. Then $M(\xi_N^{\epsilon}, y) - \epsilon \le M(\xi_N^{\epsilon}, \eta_M^{\epsilon}) \le M(x, \eta_M^{\epsilon}) + \epsilon$ for all pure strategies x, y. This completes the proof of the theorem. ■

The Cartesian product of the sets $Q^{(\delta)}$ of all points $q \in R^n$, for each of which there is a point $q' \in Q_j$ such that $\rho(q, q') < \delta$, will be called the δ–neighbourhood $Q^{(\delta)} = \prod_{j=1}^{n} Q_j^{(\delta)}$ of the set $Q = \prod_{j=1}^{n} Q_j$. The sets $Q(x) = \prod_{j=1}^{n} Q_j(x^{(j)})$ are said to be continuous in the inclusion at the point $x^* = (x_*^{(1)}, \ldots, x_*^{(n)})$ if for any number $\delta > 0$ there is such number $\Delta > 0$ that $Q(x^*) \subset [Q(x)]^{(\delta)}$ and $Q(x) \subset [Q(x^*)]^{(\delta)}$ are satisfied for all points $x = (x^{(1)}, \ldots, x^{(n)})$ satisfying the inequality $\max_j \|x^{(j)} - x_*^{(j)}\| < \Delta$.

Theorem 16 *Let $C^{T-\ell}_{E_j}(y^{(j)}_0)$, $C^{\ell}_{E_j}(y^{(j)})$, $j = 1, \ldots, n$ be the compact and convex sets for $y^{(j)} \in C^{T-\ell}_{E_j}(y^{(j)}_0)$, and the sets $C^{\ell}_E(y)$ be continuous in the inclusion, as functions of y for any $y \in C^{T-\ell}_E(y_0)$. Then for every $\epsilon > 0$ the number N of spectrum points of the strategy μ^{ϵ}_y may be selected the same for all $y \in C^{T-\ell}_E(y_0)$.*

We first prove the following statement.

Lemma 1 *In the conditions of Theorem 16 the value of the game $\Gamma(y)$ is continuously dependent on the points $y \in C^{T-\ell}_E(y_0)$.*

Proof: For simplicity, assume that $m = 1$, $n = 1$. We seek to prove that for any number $\epsilon > 0$ there is such number $\delta > 0$ that if $\rho(y, \tilde{y}) < \delta$, then the inequality $|V(y) - V(\bar{y})| < \epsilon$ is satisfied. The set $C^{\ell}_E(y)$ continuously vary in the inclusion and hence for any number $\Delta > 0$ there is such number $\delta_1 > 0$ that if $\rho(y, \tilde{y}) < \delta_1$, then $C^{\ell}_E(\tilde{y}) \subset [C^{\ell}_E(y)]^{(\Delta)}$ and $C^{\ell}_E(y) \subset [C^{\ell}_E(\tilde{y})]^{(\Delta)}$. Consider a simultaneous zero–sum game $\hat{\Gamma}(y)$ with the payoff function $K(x, y) = F(A(x, y))$ on the closure S_Δ of the set $[C^{\ell}_E(y)]^{(\Delta)}$. In this game for any number $\epsilon_1 > 0$ there is a pair of ϵ_1–optimal strategies $\hat{\mu}^{\epsilon_1}$, $\hat{\nu}^{\epsilon_1}$ with finite spectrum which is constructed as follows. Let us set up finite ϵ_1–nets $\hat{\xi}_1, \ldots, \hat{\xi}_N$ and $\hat{\eta}_1, \ldots, \hat{\eta}_N$ in correspondings Helly metrics ρ_x and ρ_y on the set S_Δ. Then $\hat{\mu}^{\epsilon_1}$, $\hat{\nu}^{\epsilon_1}$ are the optimal strategies of the players in the matrix game $\hat{\Gamma}^{\epsilon_1}(y)$. On these ϵ_1–nets construct the strategy μ^{ϵ_1} as follofs. If $\hat{\xi}_i \in C^{\ell}_E(y)$, then $\xi_i = \hat{\xi}_i$ is a point from the spectrum μ^{ϵ_1}, this being chosen with the same probability p_i as $\hat{\xi}_i$ in the strategy $\hat{\mu}^{\epsilon_1}$. If $\hat{\xi}_i \in S_\delta \setminus C^{\ell}_E(y)$ then as ξ_i we choose any point $\xi_i \in C^{\ell}_E(y)$ such that $\rho_x(\xi_i, \hat{\xi}_i) < \Delta$ and it appears in the strategy μ^{ϵ_1} with the same weight p_i as point $\hat{\xi}_i$ in the strategy $\hat{\mu}^{\epsilon_1}$. Note that in view of the uniform continuity of the payoff function K on the compact set S_Δ such point $\xi_i \in C^{\ell}_E(y)$ exists for sufficiently small numbers Δ.

The strategy ν^{ϵ_1} is constructed in much the same way. Then μ^{ϵ_1} and ν^{ϵ_1} are strategies in the games $\hat{\Gamma}(y)$ and $\Gamma(y)$. For any point $\eta \in S_\Delta$ and, in particular, for $\eta \in C^{\ell}_E(y)$ we have

$$|M(\hat{\mu}^{\epsilon_1}, \eta) - M(\mu^{\epsilon_1}, \eta)| = \left| \sum_{i=1}^{N} p_i (K(\hat{\xi}_i, \eta) - K(\xi_i, \eta)) \right| \le$$

$$\le \sum_{i=1}^{N} p_i |K(\hat{\xi}_i, \eta) - K(\xi_i, \eta)| < \epsilon_1 \tag{6.7.14}$$

Similarly, for any point $\xi \in S_\Delta$, and specifically for $\xi \in C^{\ell}_E(y)$, the inequality

$$|M(\xi, \hat{\nu}^{\epsilon_1}) - M(\xi, \nu^{\epsilon_1})| < \epsilon_1 \tag{6.7.15}$$

is satisfied. Moreover

$$|M(\hat{\mu}^{\epsilon_1}, \hat{\nu}^{\epsilon_i}) - M(\mu^{\epsilon_1}, \nu^{\epsilon_i})| = \left| \sum_{j=1}^{M} \sum_{i=1}^{N} p_i q_j (K(\hat{\xi}_i, \hat{\eta}_j) - K(\xi_i, \eta_j)) \right| \leq$$

$$\leq \sum_{j=1}^{M} \sum_{i=1}^{N} (|K(\hat{\xi}_i, \hat{\eta}_j) - K(\xi_i, \hat{\eta}_j)| + |K(\xi_i, \hat{\eta}_j) - K(\xi_i, \eta_j)|) p_i q_j < 2\epsilon_1 \quad (6.7.16)$$

(here q_j are the weights of the points $\hat{\eta}_j$ in the strategy $\hat{\nu}^{\epsilon_1}$). From the inequalities (6.7.14)– (6.7.16) and the ϵ_1–optimality of the strategies $\hat{\mu}^{\epsilon_1}$, $\hat{\nu}^{\epsilon_1}$ in the game $\hat{\Gamma}(y)$ it follows that for any points $\xi, \eta \in S_\Delta$ (and hence for $\xi, \eta \in C_E^\ell(y)$)

$$M(\mu^{\epsilon_1}, \eta) - 4\epsilon_1 \leq M(\hat{\mu}^{\epsilon_1}, \eta) - 3\epsilon_1 \leq M(\hat{\mu}^{\epsilon_1}, \hat{\nu}^{\epsilon_1}) - 2\epsilon_1 \leq M(\mu^{\epsilon_1}, \nu^{\epsilon_1}) \leq$$
$$\leq M(\hat{\mu}^{\epsilon_1}, \hat{\nu}^{\epsilon_1}) + 2\epsilon_1 \leq M(\xi, \hat{\nu}^{\epsilon_1}) + 3\epsilon_1 \leq M(\xi, \nu^{\epsilon_1}) + 4\epsilon_1$$

Thus the strategy pair $(\mu^{\epsilon_1}, \nu^{\epsilon_1})$ forms the $4\epsilon_1$–equilibrium in the $\Gamma(y)$. Therefore $|V(y) - M(\mu^{\epsilon_1}, \nu^{\epsilon_1})| \leq 4\epsilon_1$ and $|\hat{V} - M(\hat{\mu}^{\epsilon_1}, \hat{\nu}^{\epsilon_1})| \leq \epsilon_1$ ($\hat{V}$ being the value of the game $\hat{\Gamma}(y)$), which, together with inequality (6.7.16) gives an estimate $|\hat{V} - V(y)| < 7\epsilon_1$. If the strategies $\tilde{\mu}_1^\epsilon$, $\tilde{\nu}_1^\epsilon$ in the game $\Gamma(\hat{y})$ are now constructed similarly to the strategies μ^{ϵ_1}, ν^{ϵ_1} (by "adjusting" the strategies $\hat{\mu}^{\epsilon_1}$, $\hat{\nu}^{\epsilon_1}$) we obtain the inequality $|\hat{V} - V(\tilde{y})| < 7\epsilon_1$ and hence $|V(y) - V(\tilde{y})| < 14\epsilon_1$. We choose ϵ_1 in such a way that $14\epsilon_1 = \epsilon$, and finding the corresponding $\delta = \delta_1$. This completes the proof of the lemma. ■

Proof: of Theorem 16. It suffices to show that the number of points $N = N(y)$ of the strategy spectrum μ_y^ϵ may be chosen in such a way that

$$\sup_{y \in C_E^{T-\ell}(y_0)} N(y) < +\infty.$$

Suppose the oposite is true, i.e. let a number $\epsilon > 0$ and a sequence of points $\{y_s\} = \{y_s^{(1)}, \ldots, y_s^{(n)}\} \in C_E^{T-\ell}(y_0)$ be such that $\lim_{s\to\infty} N_s = \infty$ for any choice of a number $N_s = N(y_s)$. Since the set $C_E^{T-\ell}(y_0)$ is compact, then there is a convergent subsequence of the sequence $\{y_s\}$. For simplicity, it is assumed to be the sequence $\{y_s\}$ itself, then $\bar{y} \in C_E^{T-\ell}(y_0)$, where $\bar{y} = \lim_{s\to\infty} y_s$.

Consider an auxiliary game $\Gamma(\bar{y})$. By Theorem 15 in this game for any number $\epsilon_1 > 0$ there is the ϵ_1–optimal strategy $\mu_{\bar{y}}^{\epsilon_1}$ for Player P prescribing equal probabilities to the points $\xi_1, \ldots, \xi_N$ of the set $[D_e^\ell(\bar{y})]^m$. Fix an arbitrary number $\delta > 0$. Because of the continuity in the inclusion of the sets $C_E^\ell(y)$ at the point $\bar{y}$ there is such number $\Delta > 0$ that the inclusions $C_E^\ell(y) \subset [C_E^\ell(\bar{y})]^{(\delta)}$ and $C_E^\ell(\bar{y}) \subset [C_E^\ell(y)]^{(\delta)}$ are satisfied under the condition $\max_j \|y^{(j)} - \bar{y}^{(j)}\| \leq \Delta$ In particular, these inclusions are satisfied for all y_s with $s > S$ for a sufficiently large number S, i.e.

$$C_E^\ell(y_s) \subset [C_E^\ell(\bar{y})]^{(\delta)}, \;\; C_E^\ell(\bar{y}) \subset [C_E^\ell(\bar{y}_s)]^{(\delta)} \quad (6.7.17)$$

for all $s > S$; here $\max_j \|y_s^{(j)} - \bar{y}^{(j)}\| < \Delta$. The strategies $\mu_{y_s}^*$ of Player P in the games $\Gamma(y_s)$ $(s > S)$ are constructed in the following way. Let $\mu_{y_s}^*$ prescribe equal probabilities to the points $\xi_1', \xi_2', \ldots, \xi_N'$ of the set $[D_E^\ell(y_s)]^m$; moreover, if $\xi_i \in [D_E^\ell(y_s)]^m$, then $\xi_i' = \xi_i$. If $\xi_i \bar{\in} [D_E^\ell(y_s)]^m$, then as ξ_i' we take any point $\xi_i' \in [D_E^\ell(y_s)]^m$ for which

$$\max_r \|\xi_i^{(r)\prime} - \xi_i^{(r)}\| < \delta. \tag{6.7.18}$$

Evidently, ξ_i' may be selected pairwise unequal to one another.

Let us fix an arbitrary point $y \in C_E^\ell(y_s)$. From the continuity of the payoff function it follows that for any nuymber $\epsilon_2 > 0$ there is such number $\delta_2 > 0$ that $|K(\xi_i', y) - K(\xi_i, y)| \le \epsilon_2$, $i = 1, \ldots, N$ for $\max_r \|\xi_i^{(r)\prime} - \xi_i^{(r)}\| \le \delta_2$, the number δ_2 being independent of $y \in C_E^\ell(y_s)$. Thus,

$$\left| M(\mu_{y_s}^*, y) - \sum_{i=1}^N \frac{1}{N} K(\xi_i, y) \right| \le \frac{1}{N} \sum_{i=1}^N |K(\xi_i', y) - K(\xi_i, y)| \le \epsilon_2,$$

where $M(\mu_{y_s}^*, y)$ is the mathematical expectation of payoff in the situation (μ_y^*, y) in the game $\Gamma(y_s)$. This implies the inequality

$$M(\mu_{y_s}^*, y) \le \sum_{i=1}^N \frac{1}{N} K(\xi_i, y) + \epsilon_2. \tag{6.7.19}$$

Estimate the sum on the right side. We first consider the case when $y \in C_E^\ell(y) \setminus C_E^\ell(\bar{y})$. Because of inclusions (6.7.17) there is such point $\bar{\bar{y}} \in C_E^\ell(\bar{y})$ that $\max_j \|\bar{\bar{y}}^{(j)} - \bar{y}^{(j)}\| \le \delta$. From the continuity of the payoff it follows that for any number $\epsilon_3 > 0$ there exists such number $\delta_3 > 0$ that as soon as $\max_j \|\bar{\bar{y}}^{(j)} - \bar{y}^{(j)}\| < \delta_3$ the inequality holds

$$\left| \sum_{i=1}^N \frac{1}{N} K(\xi_i, y) - M(\mu_{\bar{y}}^{\epsilon_1}, \bar{\bar{y}}) \right| = \left| \sum_{i=1}^N \frac{1}{N} K(\xi_i, y) - \sum_{i=1}^N \frac{1}{N} K(\xi_i, \bar{\bar{y}}) \right| \le$$

$$\le \sum_{i=1}^N \frac{1}{N} |K(\xi_i, y) - K(\xi_i, \bar{\bar{y}})| < \delta_3, \tag{6.7.20}$$

the number δ_3 being independent of the choice of a point $y \in C_E^\ell(y_3)$. The ϵ_1–optimality of the strategy $\mu_{\bar{y}}^{\epsilon_1}$ of Player P in the game $\Gamma(\bar{y})$ and (6.7.20) suggest that

$$\sum_{i=1}^N \frac{1}{N} K(\xi_i, y) \le V(\bar{y}) + \epsilon_1 + \epsilon_2. \tag{6.7.21}$$

when $y \in C_E^\ell(\bar{y}) \bigcap C_E^\ell(y_s)$ the ϵ_1–optimality of the strategy $\mu_{\bar{y}}^{\epsilon_1}$ in the game $\Gamma(\bar{y})$ implies the inequality

$$\sum_{i=1}^{N} \frac{1}{N} K(\xi_i, y) = M(\mu_{\bar{y}}^{\epsilon_1}, y) \leq V(\bar{y}) + \epsilon_1 \leq V(\bar{y}) + \epsilon_1 + \epsilon_2.$$

By Lemma 1, for any number $\epsilon_4 > 0$ there is such number $\Delta_4 > 0$ that

$$|V(y_s) - V(\bar{y})| < \epsilon_4, \tag{6.7.22}$$

if $\max_j \|y_s^{(j)} - y_s^{(j)}\| < \Delta_4$. Let the number S be chosen in such a way that for $s > S$ (6.7.17)–(6.7.22) are satisfied. The inequalities (6.7.19),(6.7.21) and (6.7.22) suggest that

$$M(\mu_{y_s}^*, y) \leq V(y_s) + \epsilon_1 + \epsilon_2 + \epsilon_3 + \epsilon_4 \tag{6.7.23}$$

for all $y \in C_E^\ell(y_s)$, $s > S$. Since the numbers ϵ_1, ϵ_2, ϵ_3, ϵ_4 are chosen arbitrarily, then from the inequality (6.7.23) it follows that for any number $\epsilon > 0$ we may find such strategy $\mu_{y_s}^*$ that $M(\mu_{y_s}^*) \leq V(y_s) + \frac{\epsilon}{2}$ for all $y \in C^\ell\delta(y_s)$, $s > S$. Thus, there exists such number S that the strategy $\mu_{y_s}^*$ with the spectrum composed of $N = N_s$ points is ϵ–optimal in the game $\Gamma(y_s)$ for $s > S$, which is inconsistent with the assumption that $\lim_{s\to\infty} N_S = \infty$. This completes the proof of the theorem. ■

We now give two examples of differential zero–sum pursuit–evasion games of prescribed duration with delayed information for both players.

Example 1. Consider the zero–sum game on the plane between the pursuer team $P = \{P_1, P_2\}$ and the Evader E. Motion equations are of the form

$$\begin{array}{ll} P_i: \ \dot{x}^{(i)} = u^{(i)}, & \|u^{(i)}\| \leq \alpha^{(i)} \\ E: \ \dot{y} = v, & \|v\| \leq \beta; \end{array} \tag{6.7.24}$$

$$x^{(i)}, y \in R^2; \quad \alpha^{(i)} > \beta; \quad i = 1, 2.$$

Player E's payoff is defined as

$$\min_i \rho^2(x^{(i)}(T), y(T)). \tag{6.7.25}$$

Consider the auxiliary simultaneous game of pursuit $\Gamma(y)$ on the set $C_E^\ell(y)$ which is a circle with center at the point y. As shown by Satimov et. al., $V(y) = \beta^2\ell^2 - 4\beta^2\ell^2/\pi^2$ and the players posses the following optimal strategies. For Player E, the uniform distribution on the circle of radius $\beta\ell$ with its center at the point y is optimal. For Player P, the uniform distribution of points $\xi^{(i)}$ $(i = 1, 2)$ on the circle of radius $r_0 = 2\beta\ell/\pi$ with its center at the same point

y, the points $\xi^{(1)}$ and $\xi^{(2)}$ being at the opposite ends of the diameter of this circle which is denoted by $C(y, r_0)$

Let $\mu^2_{N,y}$ be a mixed strategy of Player P which prescribed equal probabilities $\frac{1}{N}$ to be points $\xi_i = (\xi_j^{(1)}, \xi_j^{(2)})$ such that the points $\xi_j^{(1)}$, $\xi_j^{(2)}$ are lying at the opposite ends of the diameter of the circle $C(y, r_0)$, the point $\xi_1^{(1)}$ is chosen arbitrarily on this circle, and other points $\xi_j^{(1)}$ devide the circle into equal arcs. Such sequence of strategies $\{\mu^\epsilon_{N,y}\}$ converges to an optimal strategy $\{\mu^*_y\}$ of Player P in the game $\Gamma(y)$ and since the payoff function is continuous, $\lim_{n\to\infty} M(\mu^\epsilon_{N,y}, \nu) = M(\mu^\epsilon_y, \nu)$. Since the sets $C^\ell_E(y)$ are congruent, then the number N is independent of y and hence the game $\Gamma(y)$ for all $y \in C^{T-\ell}_E(y_0)$ satisfies condition 1.

Let $y(t)$ be an arbitrary admissible motion of Player E for $0 \le t \le T$. Denote by $\xi_j[y(t-\ell)]$ the points of the strategy spector $\mu^\epsilon_{N,y(t-\ell)}$ in the game $\Gamma(y(t-\ell))$. Suppose that the motion of the circle $C^\ell_E(y(t-\ell))$ is translational when the center of the circle moves along the trajectory $y(t-\ell)$. Then the trajectories of all points on the circle represent the curves which are congruent to the trajectory $y(t-\ell)$. In particular the trajectories $\xi_j^{(1)}[y(t-\ell)]$, $\xi_j^{(2)}[y(t-\ell)]$ are also such curves. Thus condition 2 is also satisfied. Assume that each team partner of P stays at the point $x_0^{(i)}$ until the time instant ℓ, and then employs a deterministaic strategy which is a linear pursuit of the point $y(t-\ell)$, i.e.the strategy $u(\cdot) = \{\sigma, a\}$ is such that the mapping a associates with each state $x(t_k) = (x^{(1)}(t_k), x^{(2)}(t_k))$, $y(t_k - \ell)$ the control $u(t) = (u^{(1)}(t), u^{(2)}(t))$, $t_k < t \le t_{k_1}$, where

$$u^{(i)}(t) = \alpha^{(i)} \frac{y(t_k - \ell) - x^{(i)}(t_k)}{\rho(x^{(i)}(t_k), y(t_k - \ell))}, \quad i = 1, 2. \tag{6.7.26}$$

If however the inequality

$$T \ge \frac{\rho(x_0^{(i)}, y_0)}{\alpha^{(i)} - \beta} + \ell + \ell_1, \quad \alpha^{(i)}\ell_1 \ge \beta(\ell + \ell_1), \quad i = 1, 2, \tag{6.7.27}$$

is satisfied, then by the time instant $T - \ell_1$ each pursuer P_i ensures that the inclusion

$$C^\ell_E(y(T-\ell)) \subset C^{\ell_1}_{P_i}(x^{(i)}(T - \ell_1)) \tag{6.7.28}$$

holds for all $y(T-\ell) \in C^{\ell_1}_E(y(T - \ell - \ell_1))$.

Next, consider one of the points $\xi_j = \xi_j[y(t-\ell)] = \{\xi_j^{(1)}[y(t-\ell)], \xi_j^{(2)}[y(t-\ell)]\}$, $t \in [T - \ell_1, T]$. These points move along the curves which are congruent to the trajectory $y(t-\ell)$. Moreover, at each time instant t for $T - \ell_1 \le t \le T$ Player P knows the coordinates of these points. Thus the problem of Player

P_i's approach to the point $\xi_j^{(i)}$ $(i = 1,2)$ converts to the problem of pursuit with perfect information. Motion equations for this problem are of the form

$$P_i: \ \dot{x}^{(i)} = u^{(i)}, \quad \|u^{(i)}\| \leq \alpha^{(i)};$$
$$E: \ \dot{\xi}_j^{(i)} = v, \quad \|v\| \leq \beta;$$

where $\tilde{E}_i$ is a "dummi" evader "handling" the motion of the point $\xi_j^{(i)}$, $\alpha^{(i)}$ and β are the same as in the equations (6.7.24), and the duration of the game is fixed and equals ℓ_1. Denote these dummi games by $\tilde{\Gamma}_i$. In the games $\tilde{\Gamma}_i$ the choice of the control v made by Player $\tilde{E}_i$ coinsides with the choice of control made by Player E in the game $\Gamma(x_0, y_0, T)$ at any instant of time. Recall that the inclusion (6.7.28) is satisfied, which means that in the game $\tilde{\Gamma}_i$ the reachability set of the evader $\tilde{E}_i$ in the time ℓ_1 from the initial state $\xi_j^{(i)}[y(T-\ell-\ell_1)]$ is contained in the reachability set of the pursuer P_i in the time ℓ_1 from the initial state $x^{(i)}(T-\ell)$. This means that the pursuer P_i can approach the point $\xi^{(i)}$ moving along any trajectory $\xi_j^{(i)}[y(t-\ell)]$ in the time which does not exceed ℓ_1 if the strategy of linear pursuit (6.7.26) is employed.

To be noted is that the function (6.7.25) is a special case of the payoff function (6.7.2). Thus, in this example all conditions of Theorem 1 are satisfied if the conditions (6.7.27) are satisfied, i.e. if the velocities of the team partners are sufficiently high as compared to the evader's velocity. Let us describe optimal strategies for the players. Since the value of the auxiliary game $\Gamma(y)$ is independent of the choice of the point y, then the optimal strategy for Player E is as follows. For $0 \leq t \leq T-\ell$, Player E makes an open-loop transition from the point y_0 to any point $y \in C_E^{T-\ell}(y_0)$, at the time instant $T-\ell$, in accordance with the uniform distribution he choses any point η on the circle of radius $\beta\ell$ with its center at the point y and makes an open-loop transition to the point η in the time ℓ. The following MPOLBS is 4ϵ-optimal for Player $P = \{P_1, P_2\}$. Until the time instant ℓ the pursuers P_1, P_2 move towards the point y_0 or stand idle, for $\ell \leq t \leq T-\ell_1$ the pursuer P_i pursues the point $y(t-\ell)$ directing his velocity vector on the point $y(t-\ell)$ and thereby ensures that inclusion (6.7.28) is satisfied. At the time instant $T-\ell_1$ some ϵ-optimal strategy $\mu_{N,y}^{\epsilon}$ is fixed and Player P with probability $\frac{1}{N}$ selects one of the points ξ_i (a pair of points being at the opposite ends of the diameter of the circle with $r_0 = 2\beta\ell/\pi$ as its radius), and during the remaining time $T-\ell_1 \leq t \leq T$ Player P_i pursues (directing his velocity vector on the evading point) the point $\xi_j^{(i)}$ moving along the trajectory $\xi_j^{(i)}[y(t-\ell)]$ until the ϵ_1-capture of this point; thus, when the game terminates the point $x(T) = (x^{(1)}(T), x^{(2)}(T))$ is in the ϵ_1-neighbourhood of the point $(\xi_j^{(1)}[y(t-\ell)], \xi_j^{(2)}[y(t-\ell)])$. The value of the game is equal to $\beta^2\ell^2 - 4\beta^2\ell^2/\pi^2$.

Example 2. Consider a zero-sum differential game of pursuit on the plane between the pursuer team $P = \{P_1, \ldots, P_m\}$, acting as one player, and the Evader E. Motion equations are of the form

$$\begin{aligned} P_i: \ & \dot{x}^{(i)} = u^{(i)}, && \|u^{(i)}\| \le \alpha^{(i)} && i = 1, \ldots, m; \\ E: \ & \dot{y}_1 = vy_2 + y_1, && -1 \le v \le +1, \\ & \dot{y}_2 = -1. \end{aligned}$$

Players E's payoff is defined as $\min_i \rho(x^{(i)}(T), y(T))$. The partners of the team P have a simple motion, and from the motion equations for the evader we have

$$y_2(t) = y_2(0) - t, \tag{6.7.29}$$

$$y_1(t) = t + 1 - y_2(0) + [y_1(0) + y_2(0) - 1]e^t \text{ for } v = +1, \tag{6.7.30}$$

$$y_1(t) = -t - 1 + y_2(0) + [y_1(0) - y_2(0) + 1]e^t \text{ for } v = -1. \tag{6.7.31}$$

It may be shown that the reachability set $C_E^t(y_1(0), y_2(0))$ is the line segment joining the point $A_t(y(0))$ and $B_t(y(0))$, the coordinates of the point $A_t(y(0))$ being defined by Eq.(6.7.29) and Eq.(6.7.30), and those of the point $B_t(y(0))$ by Eq.(6.7.29) and Eq.(6.7.31). Half of this line segment is equal to $d_t(y(0)) = t + 1 - y_2(0) + [y_2(0) - 1]e^t$. Thus the set $C_E^t(y(0))$ is the line segment running parallel to the abcissa axis whose length is a continuous time function.

Consider a simultaneous zero-sum auxiliary game $\Gamma(y)$ on the set $C_E^\ell(y)$ (on the interval $[B_\ell(y), A_\ell(y)]$). The game proceeds as follows. Player P chooses m points $\xi^{(1)}, \ldots, \xi^{(m)}$ from the set $C_E^\ell(y_1, y_2)$, and Player E the point $\eta \in C_E^\ell(y_1, y_2)$. The choices are made simultaneously and independently of one another. Player E receives the payoff $\min_i \rho(\xi^{(i)}, \eta)$. As is known, the value of the game here is equal to $V(y) = d_\ell(y_1, y_2)/(2m - 1)$, and the players' optimal strategy for Player P is the choice with equal probabilities $\frac{1}{2}$ such two collections of points $\xi_1 = \{\xi_1^{(1)}, \ldots, \xi_1^{(m)}\}$ and $\xi_2 = \{\xi_2^{(1)}, \ldots, \xi_2^{(m)}\}$ that $\xi_1^{(1)} = B_\ell(y)$, $\xi_2^{(m)} = A_\ell(y)$ and the points $\xi_1^{(1)}$, $\xi_2^{(1)}, \ldots, \xi_1^{(m)}$, $\xi_2^{(m)}$ devide the interval $[B_\ell(y), A_\ell(y)]$ into equal parts. For Player E there are several optimal strategies one of which is the choice of the points $\xi_1^{(1)}$, $\xi_2^{(1)}, \ldots, \xi_1^{(m)}$, $\xi_2^{(m)}$ with equal probabilities $\frac{1}{2m}$. Thus, this game satisfies condition 1.

Let $y(t)$ be an arbitrary admissible motion of Player E for $0 \le t \le T$. Then the set $C_E^\ell(y(t-\ell))$ is the line segment moving parallel to the abcissa axis, its length

$$2d_t(y(t-\ell)) = 2\{t + 1 - y_2(t-\ell) + [y_2(t-\ell) - 1]e^t\} \tag{6.7.32}$$

being independent of the coordinates $y_1(t-\ell)$. In this case, all points $\xi_1^{(i)}$, $\xi_2^{(i)}$ $(i = 1, \ldots, m)$ describe the nonintersecting trajectories, for at any instant of time t $\rho(\xi_1^{(i)}[y(t-\ell)], \xi_2^{(i)}[y(t-\ell)]) = 2d_t(y(t-\ell))/(2m-1)$ and hence condition 2 is also satisfied. From formulas (6.7.29) and (6.7.32) it follows that

$$\max_{y \in C_E^{T-\ell}(y(0))} V(y) = \frac{1}{2m-1} \max_{y \in C_E^{T-\ell}(y(0))} \{\ell + 1 - y_2(T-\ell) + [y_2(T-\ell) - 1]e^{\ell}\} =$$

$$= \frac{1}{2m-1}\{T + 1 - y_2(0) + [y_2(0) - T - 1 + \ell]e^{\ell}\}$$

and the point $(y_1, y_2) \in C_E^{T-\ell}(y(0))$ may be chosen arbitrarily.

Now, if the condition of Theorem 14 are satisfied, then the players' optimal strategies are constructed in the following way. Each partner of the team pursues the middle point of the interval $[B_\ell(y(t-\ell)), A_\ell(y(t-\ell))]$ until the time instant $T - \ell_1$ in order to ensure that the inclusion (6.7.3) is satisfied. At the instant $T - \ell_1$, with probability $\frac{1}{2}$ Player P chooses one of the m-collections of points $\xi_1[y(T-\ell-\ell_1)]$, $\xi_2[y(T-\ell-\ell_1)]$, and on the remaining time interval each partner P_i pursues the point $\xi_j^{(i)}$ moving along the trajectory $\xi_j^{(i)}[y(t-\ell)]$ to ensure by the time instant T the arrival of the point $x^{(i)}$ at the ϵ_1-neighbourhood of the point $\xi_j^{(i)}$. For $0 \le t \le T-\ell$, Player E can move in an arbitrary way, at the time instant $T-\ell$ he chooses with probability $\frac{1}{2m}$ one of the points $\xi_j^{(i)}$ dividing the interval $[B_\ell(y(T-\ell)), A_\ell(y(T-\ell))]$ into equal parts, and in the time $T-\ell \le t \le T$ he makes an open-loop transition to the point resulting from a random choice.

In order to find out the parameters for which the conditions of Theorem 1 are satisfied, we have to solve the auxiliary perfect information game of pursuit and make appropriate estimates. These estimates are rather cumbersome and are not given here.

6.8 One multistage game with delayed information

In the general case, even the games of pursuit, where at each stage the players have complete information on the opponents' actions, turn out to be very complex. Therefore, to analyze such games it is not unusual to consider relatively simple situations. Because of the considerable complexity of the incomplete information games of pursuit, it is more important to find nontrivial examples which might be relatively free from the complicating factors. In this respect, of interest is the problem formulated by R. Isaacs, where a ship carries out maneuvers so as to minimize the probability of being stricken by a bomber.

The problem is idealized since it is assumed that the ship (Player E or the evader) is a point on the plane and the bomber (Player P or the pursuer)

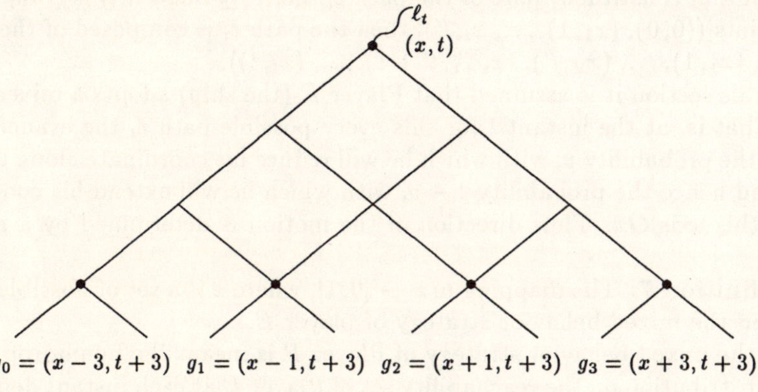

Fig. 36.

has only one bomb to strike any point on the plane (the sea surface). The survivability of the ship is not taken into consideration, i.e. hitting the ship is said to be equivalent to its destriction. It is assumed that after its release the bomb reached the water suface in k time units (k is a natural number), and in the time unit the ship must necessary move a distance unit either south–east or south–west. Let us introduce a system of coordinates xOt: the axis Ox is directed from west to east; the axis Ot is directed from north to south; and the origin of coordinates is placed at the point where the ship was at the initial instant of time. Since in a time unit Player E has to move southward, the axis Ot sewes to reckon the time. At the initial instant of time the ship can move to the point $(-1,1)$, or to the point $(1,1)$ if at the time instant t (where t is a natural number) it is at the point (x,t), then at the $(t+k)$ instant of time it must be at one of the points $g_0 = (x-k, t+k)$, $g_1 = (x-k+2, t+k), \ldots,$ $g_j = (x-k+2j, t+k), \ldots,$ $g_k = (x+k, t+k)$, (see Fig.36). In what follows this set of points is called the reachability set in k steps from the position (x,t) and is denoted by $D_k(x,t)$. At the time instant t both players know the position occupied by Player E at this instant. Next it is assumed that Player P may drop the bomb only at the instants t (where t is a natural number or o), and the bomb reaches the sea surface at the instants $t+k$. Evidently, if the bomber drops the bomb at the instant t, then it hits one of the points of the set $D_k(x,t)$ due to a bomb fall delay for k time units. Thus the bomb fall delay serves as a normal information delay for Player P

Definition 6. The path ℓ_t of the length t is represented by a collection of the points $((0,0), (x_1,1), (x_2,2), \ldots, (x_t,t)$, where $x_i - x_{i+1} = \pm 1$ for $i = 0,1,\ldots,t-1$, and $S(\ell_t)$ stands for the point(x_t,t), i.e. $S(\ell_t)$ is a finite culminating point of the path ℓ_t. The notation $\ell_{t'} \leq \ell_t$ means that the path $\ell_{t'}$

is the initial constituent part of the path ℓ_t, i.e. $t' \leq t$ and if ℓ_t is composed of the points $((0,0), (x_1, 1), \ldots, x_t, t')$, then the path ℓ_t is composed of the points $((0,0), (x_1, 1), \ldots, (x_{t'}, t'), (x_{t'+1}, t'+1), \ldots, (x_t, t))$.

In this section it is assumed that Player E (the ship) adopts a mixed strategy. That is, at the instant t for this every possible path ℓ_t the evader determines the probability p, with which he will reduce his coordinate along the axis Ox, and hence the probability $1-p$, with which he will extend his coordinate along this axis Ox. Thus direction of the motion is determined by a random test.

Definition 7. The mapping $u; z \to [0,1]$, where z is a set of possible paths, is called the mixed behavior strategy of player E.

By the mixed behavior strategy of Player P is meant the bombdrop probability distribution on the reachability set of Player E at each instant depending on the path ℓ_t. For any position (x,t) (the point) we will number the elements of the set $D_k(x,t)$ and denote them by $A_1(x,t), \ldots, A_{k+1}(x,t)$, respectively. If $(y, t+k) \in D_k(x,t)$, then $\bar{y}(x)$ denotes a number of the point $(y, t+k)$ in the set $D_k(x,t)$. Consider the set $D = \{A_1, \ldots, A_{k+1}\}$ and denote by $\bar{D} = D \bigcup \Theta$, where Θ is an abstract element substantially standing for a bomb drop omission.

Definition 8. The behavior strategy for Player P is a mapping V which corresponds any path $\ell_t \in Z$ the probability distribution on the set $\bar{D}$ equal to $(V(\ell_t, A_1), \ldots, V(\ell_t, A_{k+1}), V(\ell_t, \Theta))$

The number $V(\ell_t, A_i)$ with $i = 1, \ldots, k+1$ signifies the probability of dropping a bomb to a position $A_i(x,t)$, and $V(\ell_t, \Theta)$ the probability of bomb drop omission at that instant of time. In what follows the following notation is also admissible: $A_i(\ell_t) = V(\ell_t, A_i)$ and $\Theta(\ell_t) = V(\ell_t, \Theta)$ (evidently, $V(\ell_t, \Theta) = 1 - \sum_{i=1}^{k+1} V(\ell_t, A_i)$).

For Player E (the ship) the natural problem is to ensure maximum uncertainty about his true position (for Player P) at the bomb release moment. It can be stated as follows:

1. If the ship has travelled the path ℓ_t and arrived at the point (x,t), then under behavior strategy the conditional probability of its arrival at any point of the set $D_k(x,t)$ should be close to $\frac{1}{(k+1)}$.

This condition is the intuitive formulation of the optimality criterion for Player E's behavior strategy. Indeed, if it is possible to define such behavior strategy $u(\cdot)$ that the conditional probabilities mentioned in condition 1 are equal to $\frac{1}{(k+1)}$, then at any instant of time t there is the $(k+1)$ point for the bomber where the ship is suspected to stay at the instant $t+k$. All these points are equivalent even though the path ℓ_t travelled by the ship is known to Player P. But under the game conditions the evader must actually stay at one of these points, i.e. he cannot additionally cheat on the pursuer. Thus, this behavior strategy u proves to be the best for Player E. Unfortunately for $k > 1$ and sufficiently large values of T such behavior strategies do not exist.

Theorem 17 *For any natural number $k > 1$ there is such number $N = N(k)$ there does not exist such behavior strategy that the conditional probabilities from condition 1 are exactly equal to $\frac{1}{(k+1)}$*

Proof: Suppose the opposite is true, namely: there exist such number $k > 1$ that a behavior strategy u ensures equal conditional probabilities from condition 1(i.e. $\frac{1}{k+1}$). Let $\mu_1, \ldots, \mu_k$ be the probabilities (under a given behavior strategy) corresponding to a transition from the position $(1-i, i-1)$ to the position $(-i, i)$, and $\bar{\mu}_1, \ldots, \bar{\mu}_k$ be the probabilities corresponding to a transition from the position $(i-1, i-1)$ to the position (i, i) where $i = 1, \ldots, k$ (since probability of the path leading to these positions $(-k, k)$ and (k, k) are equal to $\frac{1}{(k+1)}$, therefore $\prod_{i=1}^{k} \mu_i = \prod_{i=1}^{k} \bar{\mu}_i = \frac{1}{(k+1)}$), but using the rules of the game we have $\mu_1 + \bar{\mu}_1 = 1$ and $\mu_i, \bar{\mu}_i \in [0, 1]$, $i = 1, \ldots, k$. Hence we obtain the inequalities $\mu_1 \geq \mu_1 \prod_{i=2}^{k} \mu_i = \frac{1}{(k+1)}$ and $\bar{\mu}_1 \geq \bar{\mu}_1 \prod_{i=2}^{k} \bar{\mu}_i = \frac{1}{(k+1)}$. Consequently, $1 = \mu_1 + \bar{\mu}_1 \geq \mu_1 + \frac{1}{(k+1)}$ or $\mu_i \leq \frac{k}{(k+1)}$ (similar inequalities hold for probabilities $\bar{\mu}_i$). Thus $\mu_1, \bar{\mu}_1 \in [\frac{1}{(k+1)}, \frac{k}{(k+1)}]$. And this conditions mast be imposed on the probabilities μ_i ($\bar{\mu}_i$) assuming that the path $((0,0), \ldots, (1-i, i-1))$ $(((0,0), \ldots, (i-1, i-1)))$ has been realized $\mu_i, \bar{\mu}_i \in [\frac{1}{(k+1)}, \frac{k}{(k+1)}]$, $i = 1, \ldots, k$. But then $\mu_1 (\frac{k}{k+1})^{(k-1)} \geq \mu_1 \prod_{i=2}^{k} \mu_i = \frac{1}{(k+1)}$, and we get the inequality $\mu_1 \geq (k+1)^{(k-2)} \cdot (k)^{(1-k)}$ (similar reasonings apply the probability $\bar{\mu}_1$), then $1 = \mu_1 + \bar{\mu}_1 \geq \mu_1 + (k+1)^{(k-2)} \cdot k^{(1-k)}$ and $\mu_1 \leq 1 - (k+1)^{(k-2)} \cdot k^{(1-k)}$, whence we get $\mu_1, \bar{\mu}_1 \in [(k+1)^{(k-2)} \cdot k^{(1-k)}, 1 - (k+1)^{(k-2)} \cdot k^{(1-k)}]$, etc. As noted above, however, we may show that the same restrictions should satisfy the probabilities $\mu_i, \bar{\mu}_i, i = 1, \ldots, k$. Assume that at particular stage of similar reasonings we have the estimate $\mu_i, \bar{\mu}_i \in [\alpha, 1-\alpha]$, where $\alpha \leq \frac{1}{2}$,then at the next stage we have the estimate $\mu_i, \bar{\mu}_i \in [(1-\alpha)^{(1-k)}/(k+1), 1-(1-\alpha)^{(1-k)}/(k+1)]$, i.e. the lower bound for the value of the probability μ_1 is raised by $\delta(\alpha, k) = [(1-\alpha)^{(1-k)} - \alpha(k+1))]/(k+1)$. Consider the guantity $\delta(\alpha, k)$ as a function α. It is minimal when $\alpha = \alpha_0 = 1 - (k-1)^{1/k} \cdot (k+1)^{-1/k}$ (tested immediately by taking a derivative) and $\delta(\alpha_0, k) = ((k-1)/(k+1))^{1/k}((k+1)^{(k-2)} \cdot (k-1)^{(1-k)} + 1) - 1 > 0.1$. In fact, a derivative of the expression $(k+1)^{1/k} \cdot (k-1)^{-1/k}$ with respect to k is positive for $k \geq 2$ and is monotone, and a derivative of $(k+1)^{(k-2)} \cdot (k-1)^{(1-k)}$ with respect to k is also positive and it maximal when $k = (3+\sqrt{13})/2$. Consequently, the value of the function $\delta(\alpha_0, k)$ is minimal at $k = 2$, which implies the inequality $\delta(\alpha_0, k) \geq \delta(\alpha_0, 2) = \sqrt{1/3 \cdot (1+1)} - 1 > 0.1$. Thus, for any k five stages of similar reasonings enable us to have a conclusion that the system of restrictions for the probability μ_1 is contradictory,i.e.

$\mu_1 > \frac{1}{2}$ and $\mu_1 < \frac{1}{2}$. Since it was obviously assumed that the number of steps in the game considered in the above reasonings, exceeds the number of steps at the earlier stages by k, then we have the upper bound for the length of the game T which enables the existance of the behavior strategy garanteeing the value equal to $\frac{1}{(k+1)}$ to the conditional probabilities from condition 1. This completes the proof of the theorem. ■

The proximity of conditional probabilities to the number $\frac{1}{(k+1)}$ may be interpreted in various ways. Let $\alpha = (\frac{1}{(k+1)}), \dots, \frac{1}{(k+1)})$ be a vector of $(k+1)$ dimension. For each behavior strategy and for each path ℓ_t $(t \leq T-k)$, the vector $\beta = (\beta_0, \beta_1, \dots, \beta_k)$ is formed from the conditional probabilities, the vector being placed in Q correspondence with this path, i.e. $Q(\ell_t) = \beta$. Denote by $G_r(\ell_t)$ the value of the expression $(\sum_{i=0}^{k}(|\beta_i - \alpha_i|)^r)^{1/2}$ for $r > 1$ (i.e. the norm $\beta - \alpha$ in the space L_r), or $(\sum_{i=0}^{k}\beta_i \ln \beta_i + \ln(k+1))$ for $r = 0$ (the amount of information on Player E's position). Thus, $G(\ell_t)$ is the expexted gain from the step t if the path ℓ_t has already been realised. If the index r is omitted, this implies any of its values from the set $(1, \infty)$ $V\{0\}$. Let

$$W_r(U^T) = \max_{\ell_t \in Z,\, t \leq T-k} \{G(\ell_t)\}.$$

The optimal behavior strategy in the sense of the criterion r is such strategy on which a minimum of the function $W_r(U^T)$ is achieved in the class of behavior strategies. Consider various techniques of solving this game for $k = 2$ by referring to various approaches to the notion of "proximity". In what follows this game is called the "bomber versus ship" game.

Denote by p the probability of arrival at the position $(x-1, t+1)$ from the point (x, t), by b and c the probabilities of arrival at the positions $(x-2, t+2)$ and $(x, t+2)$ from the points $(x-1, t+1)$ and $(x+1, t+1)$, respectively; by $q_1 = pb$, $q_2 = p(1-b) - (1-p)c$, $q_3 = (1-p)(1-c)$ the conditional probabilities from condition 1.

The optimal behavior strategy in this section is such behavior strategy on which a minimum is achieved under the behavior strategies for a maximum with respect to all possible lengths not exceeding $T-2$ of the largest of the numbers q_1, q_2, q_3 for each path. Since duration of the process is equal to T, then Player P has to drop the bomb during this time. This payoff is unity if the ship is hit, otherwise it is zero. Denote this game by Γ_1. If Player P releases the bomb at the initial instant, then the expected value of his payoff equals q_1, q_2 or q_3 depending on what exactly he is aiming at. If the bomb has not been released, then the expected payoff equals $pg_{T-1}(b) + (1-p)g_{T-1}(c)$, where $g_{T-1}(y)$ is the value of the game for the pursuer if there are $T-1$ steps until the game terminates, and the probability chosen by the evader to reduce his coordinate along the axis Ox is equal to y. Evidently, the largest expected payoff of Player P is equal to the maximum of these four quantities, and since Player E seeks to minimize this payoff, then the following equality holds

$$g_T(p) = \min_{b,c\in[0,1]} \max \begin{cases} pb, \\ p(1-b)+(1-p)c, \\ (1-p)(1-c), \\ pg_{T-1}(b)+(1-p)g_{T-1}(c) \end{cases} \tag{6.8.1}$$

(here $p \in [0,1]$, $g_0 = 0$). The analysis and solution of functional equations (6.8.1) are employed to find a solution to the infinite game. Let $val\Gamma_n = \min_{p\in[0,1]} g_n(p)$.

Lemma 2 (1) *The function g_n is simmetric about the point $p = \frac{1}{2}$ and $g_n(\frac{1}{2}) = val\Gamma_n$.*

Proof: Since g_0 is a constant, then statements of the lemma hold for it. If we assume that the function g_i is symmetric, and replace respectively p by $1-p$, b by $1-c$, c by $(1-b)$, then the equality $g_{i+1}(p) = g_{i+1}(1-p)$ follows from the formula (6.8.1). If $g_n = \min_{p\in[0,1]} g_n(p)$, then $g_n(1-p) = \min_{p\in[0,1]} g_n(p)$, and since these probabilities are components of the optimal strategy for the evader, then their mixture is also optimal and hence $g_n(\frac{1}{2}) = \min_{p\in[0,1]} g_n(p) = val\Gamma_n$. ■

Lemma 3 *The inequality $|g_n(p\pm\epsilon) - g_n(p)| \le \epsilon$ holds for any positive number ϵ and number n.*

This statement follows from the lemma proved in what follows.

Lemma 4 $g_n(p) \le g_{n+1}(p)$

Proof: In this game, using $(n+1)$ steps, the pursuer can ensure the payoff $g_n(p)$ if he adopts the behavior strategy optimal in the n–step lenth game. ■

Lemma 5 *The function g_n uniformly converges to a continuous function g satisfying the equation*

$$g_T(p) = \min_{b,c\in[0,1]} \max \begin{cases} pb, \\ p(1-b)-(1-p)c, \\ (1-p)(1-c), \\ pg(b)+(1-p)g(c) \end{cases} \tag{6.8.2}$$

Proof: It follows from Lemma 4 that for any fixed value of the argument p the sequence $\{g_n(p)\}$ does not decrease and is bounded above, say by unity. Consequently, the pointwise limit exists. From Lemma 3, and from Arzela theorem, it follows that this convergence is uniform and the limit function g is continuous. Passing to the limit, as in (6.8.1), we get (6.8.2). This completes the proof of the lemma.

Denote by U_0 the following strategy: at the first step Player E reduces his coordinate along the axis Ox with an arbitrary probability $p \in [p_0, 1-p_0]$, where $p_0 = (3-\sqrt{5})/2$, and then continues his motion in the direction along which he moved at the earlier step with probability p. Such behavior strategy is called Markovian, since the choice of control at each step is only dependent on the earlier step and is independent of the controls chosen before. The following statement is varified immediately. ■

Lemma 6 *If Player E adopts the behavior strategy U_0, then the expected maximum payoff of Player P is equal to p_0.*

Note that since time does not take part in definition of the strategy U_0, then this lemma holds for the game with any number of steps, specifically for the infinite game.

Theorem 18 *The minimum value of the function $g(p)$ equals p_0 and is achieved on the interval $[p_0, 1-p_0]$.*

Proof: By Lemma 4, the values of $val\Gamma_n = \min_{p\in[0,1]} g_n(p)$ converges to the quantity $val\Gamma$. Moreover, the function $g(y)$ is symmetric as a limit of symmetric functions. Let p_1 be the least value of p for which $g(p) = val\Gamma (p_1 \le \frac{1}{2})$ and b, c are the corresponding values on which a minimum in expression (6.8.2) is achieved, i.e.

$$val\Gamma_n = g(p_1) = \max \begin{cases} p_1 b, \\ p_1(1-b) + (1-p_1)c, \\ (1-p_1)(1-c), \\ p_1 g(b) + (1-p_1)g(c). \end{cases} \tag{6.8.3}$$

It follows from (6.8.3) that $g(p_1) \ge p_1 g(b) + (1-p_1)g(c)$ and hence the inequalities $g(p_1) \ge g(b)$ and $g(p_1) \ge g(c)$ holds. However, from the fact that p_1 is a minimum value of $p \ : \ g(p) = val\Gamma$, and from the symmetry of the function g about the point $p = \frac{1}{2}$ it follows that $p_1 \le 1-b$ and $c \ge p_1$ (otherwise there would have been $g(p_1) < g(b)$ or $g(p_1) < g(c)$). Consequently, $p_1(1-b) + (1-p_1)c \ge p_1^2 + p_1(\ell - p_1) = p_1$. At the same time, from (4.8.3) we have that $(1-b)p_1 + c(1-p_1) \le g(p_1) = val\Gamma$, thus

$p_1 \leq val\Gamma$. Assume that $p_1 < val\Gamma$, then $val\Gamma \geq (1-c)(1-p_1)$ and $val\Gamma \geq (1-b)p_1+(1-p_1)c \geq p_1^2+c(1-p_1)$. Adding together these inequalities, we get $2val\Gamma \geq p_1^2 + c(1-p_1)$. Since $p_1^2 - p_1 + 1$ is a stricly decreasing (for $p_1 \in [0, \frac{1}{2}]$) function and $p_1 < val\Gamma \leq \frac{1}{2}$, then $2val\Gamma > (val\Gamma)^2 - val\Gamma + 1$ and hence $val\Gamma > (3-\sqrt{5})/2 = p_0$. But by Lemma 5 we have the inequality $val\Gamma \leq p_0$. This yieds a contradiction which precisely shows that $p_1 = val\Gamma$. We have thus proved the theorem. ■

Since $(1-b)p_1 + (1-p_1)c \geq p_1^2 + p_1(1-p_1) = p_1 = val\Gamma$ and $(1+b)p_1 + c(1-p_1) \leq val\Gamma$ (see formula (6.8.3)), then $p_1 = 1-b$, $c = p_1$ and $val\Gamma = p_1 = p_0 = (3-\sqrt{5})/2$ Using the same formula (6.8.3), it can be readily seen that $g(p) = val\Gamma$ if and only if $p_0 \leq p_1 \leq 1 - p_0$ and $c = 1 - b = p_0$. Since any truncated (i.e. finite step) game is an ordinary matrix game, whose matrix is composed of zeros and unities, then its value is a rational number. Because of Lemma 4, $val\Gamma_n \leq val\Gamma$, and $val\Gamma$ is an irrational number, hence $val\Gamma_n < val\Gamma$ and $\lim_{n\to\infty} val\Gamma_n = val\Gamma$, i.e. for any positive number ϵ there is such number N that $val\Gamma - val\Gamma_n < \epsilon$. By employing the optimal behavior strategy in the game of length N in the infinite game, we obtain the ϵ–optimal behavior strategy for Player P. The following statement holds.

Theorem 19 *The infinite "bomber versus ship" game has the value* $val\Gamma = (3-\sqrt{5})/2$. *The optimal strategy for Player* E *coinsides with the strategy* U_0. *For any number* $\epsilon > 0$ *there exist the* ϵ*–optimal behavior strategy for Player* P.

Chapter 7

Noncooperative differential games

7.1 Game on a finite graph tree

In the majority of the conflictly controlled desission making problems, the passage to the normal form, i.e. the transformation of the problem to a single instantaneous choice of pure strategies fails to involve the efficient solution methods, alhough it enables to illustrate one or another optimality principle. In some cases, the general existence theorems for the solutions of normalized games do not allow one to find or even specify the optimal behavior in the original games, for which the normalization was provided. For instance as is seen below, in "shess" there exist a saddle point in pure strategies. At the same time, dealing with a corresponding matrix game which is a normalization of the game of chess, one may only claim the existence of a solution in the class of mixed strategies. This becomes more apparent in the studies of the differential game of pursuit for which in some cases, a solution turns out to be found explicitly, but the normal form of the differential game is general to such extend that to obtain results proves to be practically impossible.

The theory of positional games studies the mathematical models of conflict where dynamic is taken into account. The simplest class of positional games is the class of finite–stage positional games with complete information. The definition of these games involves a certain knowledge of the graph theory.

Let X be a finite set; the rule f setting up the correspondence between each element $x \in X$ and the element $f(x) \in X$ is called the single valued mapping of X into X or the function defined on X and taking values in X. The point to set mapping γ of X into X is the rule which sets up a correspondence between each element $x \in X$ and a subset $\gamma(x) \subset X$. Here $\gamma(x) = \emptyset$ is also possible.

The set $\gamma(A) \rightleftharpoons \bigcup_{x \in A} \gamma(x)$ is called the image of the set $A \subset X$. It is fairly easy to see that if $A_i \subset X$ $(i = 1, \ldots, n)$, then

$$\gamma\left(\bigcup_{i=1}^{n} A_i\right) = \bigcup_{i=1}^{n} \gamma(A_i), \quad \gamma\left(\bigcap_{i=1}^{n} A_i\right) \subset \bigcap_{i=1}^{n} \gamma(A_i)$$

Let $\gamma^2(x) \rightleftharpoons \gamma(\gamma(x)), \ldots, \gamma^{i+1}(x) \rightleftharpoons \gamma(\gamma^i(x))$ $(i = 2, 3, \ldots)$. The mapping $\hat{\gamma}$ is called a transitive closure of the mapping γ if

$$\hat{\gamma}(x) = \{x\} \bigcup \gamma(x) \bigcup \gamma^2(x) \bigcup \cdots \bigcup \gamma^n(x) \bigcup \cdots$$

The mapping γ^{-1} is the inverse of the mapping γ and is defined as follows: $\gamma^{-1}(y) \rightleftharpoons \{x \in X \mid y \in \gamma(x)\}$. Let $(\gamma^{-1})^2(y) \rightleftharpoons \gamma^{-1}(\gamma^{-1}(y)), \ldots, (\gamma^{-1})^{i+1}(y) \rightleftharpoons \gamma^{-1}((\gamma^{-1})^i(y))$ $(i = 2, 3, \ldots)$ and $\gamma^{-1}(B) = \{x \in X \mid \gamma(x) \bigcap B \neq \emptyset\}$ for all $B \subset X$.

Example 1. **Draughts.** Each position on the draught–board is determined by the arrangement of draughtsmen at a given moment, and by the indication as to whose move it is. Let X be a set of positions and $\gamma(x)$ a set of positions which can be realized immediately after the position $x \in X$; if in the position x the number of white or black draughtsmen is zero or the position is drawn, then $\gamma(x) = \emptyset$. Next, $\gamma^k(x)$ is a set of positions which can be obtained from x in k moves, $\hat{\gamma}(x)$ a set of all positions which can be obtained from x and $\gamma^{-1}(A)$ a set of positions from which it is possible to make in one move a transition to the positions from the set $A \subset X$. Depicting the positions by dots and drawing the arrow joining two position x and y, where $y \in \gamma(x)$, theoretically it is possible to construct a plane figure called the graph of game. Because of a very large number of positions, hovewer it is practically impossible to draw such a graph. The use of point to set mappings makes it possible to represent a structure of numerous games: chess, go, etc.

The pair $G =< X, \gamma >$ is a graph if X is a finite set, and γ a point to set mapping of X into X. In what follows, the elements of the set X are represented by the points on the plane, and the pairs of point x and y, for which $y \in \gamma(x)$, are joined by a continuous line with the arrow directed from x to y. Therefore, each element of the set X is called the vertex or node of a graph, and the pair of elements $z = (x, y)$, where $y \in \gamma(x)$, is the arc of the graph. We say that x and y arc the boundary nodes of the arc $z = (x, y)$, with x as the origin and y as the end point. The arcs z_1 and z_2 are called contingent if they are different and have a boundary point in common. The path in the graph $G =< X, \gamma >$ is such sequence of arcs $\mu = (z_1, \ldots, z_i, \ldots)$ such that the end point of each preceeding arc coincides with the origin of the subsequent one. The length of a finite path $\mu = (z_1, \ldots, z_k)$ is the number $\ell(\mu) = k$ of its arcs, and in the case of an infinite path μ we set $\ell(\mu) = +\infty$.

The set of arcs in the graph $G =< X, \gamma >$ is denoted as U. Representation of the set of arcs determines the mapping γ, and vice versa, the mappings

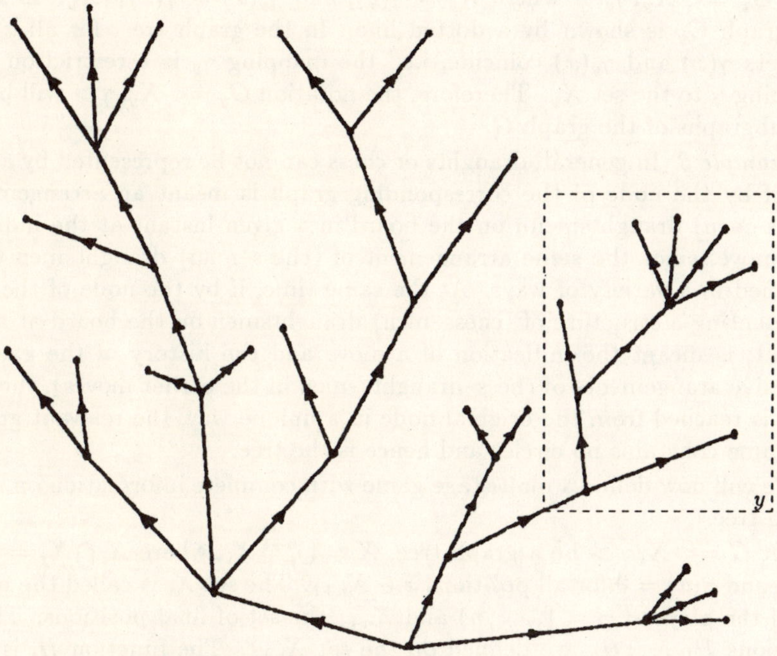

Fig. 37.

γ determines the set U; hence the graph $G =< X, \gamma >$ can be written as $G =< X, U >$ The edge of the graph $G =< X, \gamma >$ is the pair of two elements $\tilde{z} = \{x, y\}$, for which $(x, y) \in U$ or $(y, x) \in U$. As distance from the arc, the orientation plays no role in the edge. The set of edges of the graph $G =< X, U >$ is denoted by $\hat{U}$. By the chain is meant a sequence of edges $(\tilde{z}_1, \tilde{z}_2, \ldots, \tilde{z}_i, \ldots)$ where one of the boundary nodes of each edge $\tilde{z}_i$ is boundary for $\tilde{z}_{i-1}$, and the other for $\tilde{z}_{i+1}$ $(i \geq 2)$. The circle is a finite chain starting in a node x and terminating in the same node. The graph is called connected if any two of its nodes may be connected by a chain. By definition the tree, or the graph tree, is a finite connected graph without circles which has at least two nodes. In any graph tree there is a unique node x_0 such that $\hat{\gamma}(x_0) = X$. The node x_0 is called the initial node of the graph.

Example 2. Fig.37. shows the tree, or the graph tree,with the origin x_0. The nodes $x \in X$ or vertices of the graph are marked by dots. The arcs are depicted as thre arrowed segments emphasizing the origin and the end point of the arc.

Let $y \in X$, the subgraph of the graph tree $G =< X, \gamma >$ is a graph of the form $G_y =< X_y, \gamma_y >$ where $X_y = \hat{\gamma}(y)$, and $\gamma_y(x) = \gamma(x) \bigcap X_y$. In Fig.37, the graph G_y is shown by a dotted line. In the graph tree, for all $x \in X_y$ the sets $\gamma(x)$ and $\gamma_y(x)$ coincide, i.e. the mapping γ_y is a restriction of the mapping γ to the set X_y. Therefore, the notation $G_y =< X_y, \gamma >$ will be used for subgraphs of the graph G.

Example 3. In general, draughts or chess can not be represented by a graph tree if by the node of the corresponding graph is meant an arrangement of (chess–men) draughtsmenn on the board in a given instant at the indication of a move, since the same arrangement of (chess–man) draughtsmen can be obtained in a variety of ways. At the same time, if by the node of the graph representing a structure of (chess–men) draughtsmen on the board at a given instant, is meant the indication of a move and the history of the game (all succesive arangements of chess–draughts–men in the earlier moves), then each node is reached from the original node in a unique way, the relevant graph of the game constains no circles and hence is the tree.

We will now define a multistage game with complete information on a finite graph tree.

Let $G =< X, \gamma >$ be a graph tree, $X = \bigcup_{i=1}^{n+1} X_i$, where $X_i \bigcap X_j = \emptyset$ with $i \neq j$ and $\gamma(x) = \emptyset$ for all positions $x \in X_{n+1}$. The set X_i is called the priority set of the player i $(i = 1, \ldots, n)$ and X_{n+1} the set of final positions. The real functions $H_1, \ldots, H_n$ are defined on the set X_{n+1}. The function H_i is called the payoff of the player i $(i = 1, \ldots, n)$.

The game proceeds as follows. Suppose we have n players labelled with natural numbers $1, \ldots, n$. Assume that $x_0 \in X_{i_1}$, then in the node (position) x_0 the player i_1 "makes a move" and selects the node $x_1 \in \gamma(x_0)$; if $x_1 \in X_{i_2}$, then in the node x_1 the player i_2 "makes a move" and selects the next node $x_2 \in \gamma(x_1)$ etc. If the position $x_{n-1} \in X_{i_k}$ is realized at the k–th step, then in this position the player i_k "makes a move" and selects the next node (position) x_k from the set $\gamma(x_{k-1})$. The game terminates as soon as the node $x_\ell \in X_{n+1}$ is reached. Such step–by–step selection implies a unique realization of nodes $x_0, \ldots, x_\ell$ which determines the path in the graph tree G emanating from the initial position x_0 and reaching one of the final positions in the game. In what follows, such a path is called the play of the game. Because of the treelike structure of the graph G, each position $x_\ell \in X_{n+1}$ uniquely determines the play reaching this position. In the final position x_ℓ, each of the players $i \in \{1, \ldots, n\}$ receives the payoff $H_i(x_\ell)$.

When making a decision in the position $x \in X_i$, the player i is assumed to know this position and hence, because of the treelike structure of the graph

G, he can restore all previous positions. In this case, the players are said to have complete information provided by chess and draughts since chess– and draught–players may put down their moves, and hence they are said to know the history of the game in making each move in turn.

The function u_i, which places each position $x \in X_i$ in correspondence with a position $y \in \gamma(x)$, is called a strategy for the player i. The set of all possible strategies for the player i is denoted by D_i $(i = 1, \ldots, n)$. Thus the strategy for the i-th player prescribes him, in any position x from his priority set X_i, a unique choice of the next position.

The set of n–tipes of strategies $u = (u_1, \ldots, u_n)$, where $u_i \in D_i$, is called a situation. Each situation uniquely defines a play of the game and hence the players' payoffs. Indeed, let $x_0 \in X_{i_1}$, then in the situation $u = (u_1, \ldots, u_n)$ the next position x_1 is uniquely determined by the rule $x_1 = u_{i_1}(x_0)$; now let $x_1 \in X_{i_2}$, then x_1 is uniquely determined by the rule $x_2 = u_{i_2}(x_1)$; if the position $x_{k-1} \in X_{i_k}$ is realized at the k-th step, then x_k is uniquely determined by the rule $x_k = u_{i_k}(x_{k-1})$, etc. Suppose that to the situation $u = (u_1, \ldots, u_n)$ corresponds, in that sense, the play $x_0, \ldots, x_\ell$. Then we may introduce the notion of the payoff function $K_i : D = D_1 \times \ldots \times D_n \to R^1$,

$$K_i(u) = K_i(u_1, \ldots, u_n) = H_i(x_\ell), \quad (i = 1, \ldots, n)$$

where x_ℓ is the last node in the play $x_0, \ldots, x_\ell$ corresponding to the situation $u = (u_1, \ldots, u_n)$. The collection

$$\Gamma = < D_1, \ldots, D_n, K_1, \ldots, K_n > \tag{7.1.1}$$

is called the n–person game with complete information on graph G.

7.2 Nash equilibrium

We first consider a general case. Let arbitrary n sets $A_1, \ldots, A_n$ be given. The set A_i is called the strategy set of the i–th player $(i = 1, \ldots, n)$. Let further n real functions $K_1, \ldots, K_n$ be given on the Cartesian product $A_1 \times \ldots \times A_n$. The function K_i is called the payoff function of the i–th player. Then the collection

$$\Gamma = < A_1, \ldots, A_n; K_1, \ldots, K_n > \tag{7.2.1}$$

is called the n–person game in normal form. The evolution of the game may be represented as follows. Each of players i simultaneously and independent of each other chooses a strategy $u_i \in A_i$ $(i = 1, \ldots, n)$. The players choices result as an n–taple of strategies $u = (u_1, \ldots, u_n)$ called the situation. Thereafter the player i receives the payoff equal to the value of his payoff function in this situation $K_i(u) = K_i(u_1, \dots, u_n)$. Each player is aiming to obtain a maximal payoff. If $n = 2$ and $K_1 = -K_2$, then the game (7.2.1) is zero–sum game.

This chapter consider noncooperative games (7.2.1) where each player acts individuall and receives the payoff determined by his payoff function. Since the function K_i is not exclusively dependent on the strategy u_i, the size of the payoff of the i–th player in the noncooperative game is not exclusively dependent on his behavior. Therefore, in choosing his strategy, each player has to care for possible behaviors of all his partners.

The situation $\bar{u} = (\bar{u}_1, \dots, \bar{u}_n)$ is called the Nash equilibrium in the noncooperative game (7.2.1) if

$$K_i(\bar{u}) \geq K_i(\bar{u} \parallel u_i)$$

for all $u_i \in A_i$ $(i = 1, \dots, n)$. Here by $(\bar{u} \| u_i)$ we denote the situation $(\bar{u}_1, \dots, \bar{u}_{i-1}, u_i, \bar{u}_{i+1}, \dots, \bar{u}_n)$.

The meaning of the optimality of the strategy $\bar{u}_i$ appearing in the Nash equilibrium is that the i–th player deviation from it, with the other players playing the equilibrium strategies $\bar{u}_j$ $(j \neq i)$ from the situation $\bar{u}$, did not increase the payoff of the player i. Thus, the Nash equilibrium is not necessary insensitive to group deviations from it.

For the noncooperative multistage game with complete information (7.1.1), the notion of equilibrium may enhaced by introducing the notion of absolute equilibrium. The subgame Γ_y is related to any subgraph $G_y = (X_y, \gamma)$ of graph G in the following manner: the priority sets for players in the subgame Γ_y are determined by the rule $X_i^y = X_i \bigcap X_y$ $(i = 1, \dots, n)$, the set of final positions is $X_{n+1}^y = X_{n+1} \bigcap X_y$, and the payoff for players H_i^y in the subgame are assumed to be equal: $H_i^y(x) = H_i(x)$ for all $x \in X_{n+1}^y$ $(i = 1, \dots, n)$. Accordingly, the strategy for the i–th player u_i^y in the subgame under study is defined as a truncation of the strategy for the same player u_i in the game (7.1.1), i.e. $u_i^y(x) \rightleftharpoons u_i(x)$, $x \in X_i^y$ $(i = 1, \dots, n)$. The set of all such strategies is denoted by D_i^y. As a result, each subgraph G_y defines the game is normal form

$$\Gamma_y = < D_1^y, \dots, D_n^y; K_1^y, \dots, K_n^y > \tag{7.2.3}$$

where the payoff function K_i^y is defined on $D_1^y \times \dots \times D_n^y$. The Nash equilibrium $\bar{u} = (\bar{u}_1, \dots, \bar{u}_n)$ is called the absolute Nash equilibrium (subgame perfect) in the game (7.1.1) if for any position $y \in X$ the situation $\bar{u}^y = (\bar{u}_1^y, \dots, \bar{u}_n^y)$, where $\bar{u}_i^y$ is a truncation of the strategy $\bar{u}_i$, is the Nash equilibrium in an appropriate subgame Γ_y (7.2.3). The following statement holds true.

Theorem 1 *In any multistage game with complete information (7.1.1) on a finite graph tree G there exists the absolute Nash equilibrium.*

Proof: The length of the game (7.1.1) on the graph G is called the length of the largest path in the graph G. The proof is prvided by induction over the length of the game. If the length of the game (7.1.1) is equal to 1, this means that in the game (7.1.1) only one of the players can make a move. By choosing the next node from the condition of maximization of his payoff, the player is acting under the strategy which forms the absolute Nash equilibrium.

Let us now assume that the theorem holds true for all games whose length does not exceed $k-1$, and prove it for the game (7.1.1) of the length k. Since the subgames Γ_y, $y \in \gamma(x_0)$ are of the length not exceeding $k-1$, then by the indiction assumption the absolute Nash equilibrium exists therein. For each subgame we denote this equilibrium by

$$\bar{u}^y = (\bar{u}_1^y, \dots, \bar{u}_n^y) \tag{7.2.4}$$

Let $\bar{u}^j(x) = \bar{u}_i^y(x)$ for $x \in X_i \bigcap X_y$, $y \in \gamma(x_0)$ $(i = 1, \dots, n)$ and $\bar{u}_i(x_0) = y^*$, where $x_0 \in X_{i_1}$,

$$K_{i_1}^{y^*}(\bar{u}_1^{y^*}, \dots, \bar{u}_n^{y^*}) = \max_{y \in \gamma(x_0)} K_{i_1}^y(\bar{u}_1^y, \dots, \bar{u}_n^y)$$

The function $\bar{u}_i$ is a strategy in the game (7.1.1), its truncation $\bar{u}_i^y$ appearing in (7.2.4) a strategy in the subgame. Thus, to complete the proof of the theorem, it suffices to show that the sytrategy set $\bar{u} = (\bar{u}_1, \dots, \bar{u}_n)$ constitutes the Nash equilibrium in the game (7.1.1). Indeed,

$$K_{i_1}(\bar{u}) = K_{i_1}^{y^*}(\bar{u}^{y^*}) = \max_{y \in \gamma(x_0)} K_{i_1}^y(\bar{u}) \geq K_{i_1}^y(\bar{u}) \geq K_{i_1}^y((\bar{u} \| u_{i_1}^y)) = K_{i_1}(\bar{u} \| u_{i_1}),$$

$$\bar{u}_{i_1}(x_0) = y,$$

for all $u_{i_1} \in D_{i_1}$; if $i \neq i_1$, then

$$K_i(\bar{u}) = K_i^{y^*}\left(\bar{u}^{y^*}\right) \geq K_i^{y^*}\left((\bar{u} \| u_i)^{y^*}\right) = K_i(\bar{u} \| u_i)$$

for all $u_i \in D_i$. This completes the proof of the theorem. ■

Example 4. Suppose the two–person game (7.1.1), $n = 2$, is played on the graph directed in Fig.37; the structure of the game is presented in Fig.38: the nodes from the priority set X_1 are represented by circles and these from the set X_2 by brocks, wuith the players' payoff written in final positions.

The position from the sets X_i $(i = 1, 2)$ and the arcs emanating from each node $x \in X_i$ are numbered by double indices (i, j) where i is the index of the player, and j–the index of the node x in the set X_i. Since the choice in the node x is equivalent to the choice of the next node $y \in \gamma(x)$, then, for simplicity, we assume that in each node the strategy indicates the number

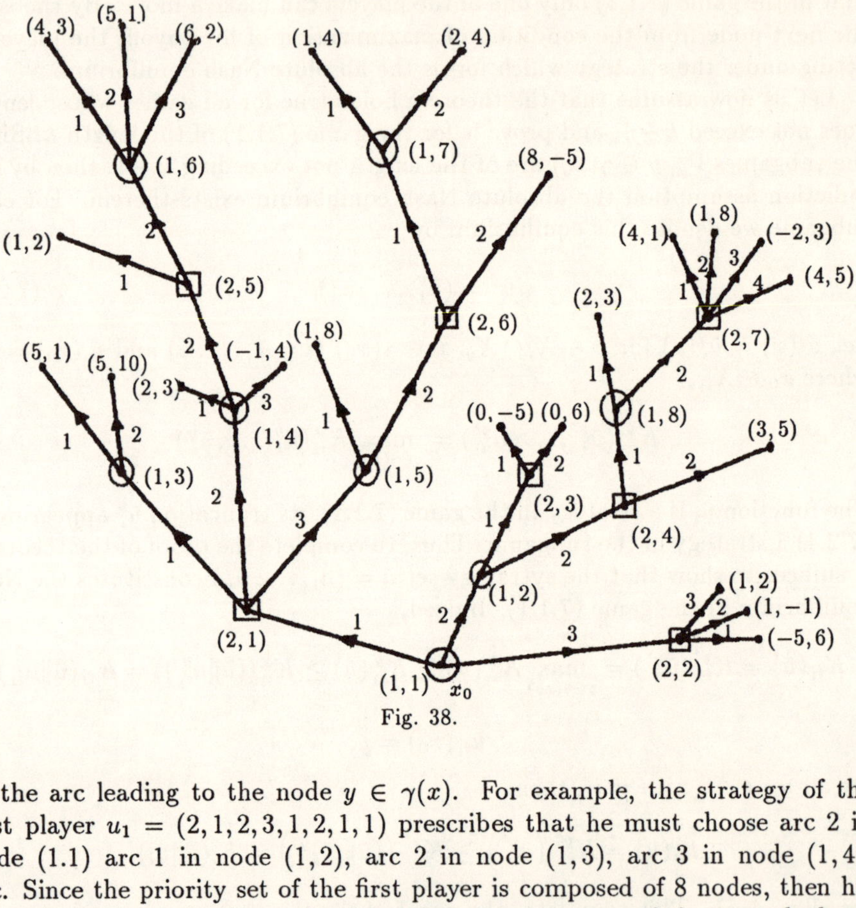

Fig. 38.

of the arc leading to the node $y \in \gamma(x)$. For example, the strategy of the first player $u_1 = (2,1,2,3,1,2,1,1)$ prescribes that he must choose arc 2 in node (1.1) arc 1 in node (1,2), arc 2 in node (1,3), arc 3 in node (1,4), etc. Since the priority set of the first player is composed of 8 nodes, then his strategy in the eight–dimensional vector. Any strategy of the second player is the seven–dimentional vector. Altogether there are 864 strategies for the first player, and 576 strategies for the second. The corresponding game in normal form turns out to be the bimatrix game with two payoff matrices of dimension 864×576. The direct solution of such bimatrix games is not only difficult, but also practically impossible. At the same time, the game involved has a suffiently simple structure and can be readily solved by the backward construction of the absolute Nash equilibrium as it was done in the proof of Theorem 1.

We denote by $v_1(y)$ and $v_2(y)$ the payoffs respectively of players 1 and 2 in the subgame Γ_y in a fixed absolute equilibrium. We first solve the subgammes $\Gamma_{(1,6)}$, $\Gamma_{(1,7)}$, $\Gamma_{(2,7)}$. It can be readily seen that $v_1(1,6) = 6$, $v_2(1,6) = 2$, $v_1(1,7) = 2$, $v_2(1,7) = 4$, $v_1(2,7) = 1$, $v_2(2,7) = 8$. Then we solve the subgames $\Gamma_{(2,5)}$, $\Gamma_{(2,6)}$, $\Gamma_{(2,7)}$. In the subgame $\Gamma_{(2,5)}$ there are two Nash equilibrium, since the second player does not care which of the alternatives to choose. His choice, however, is essential for Player 1 since, with Player 2's choice of the left–hand arc, the first player scores +1, whereas with Player 2's choice of the other arc Player 1 scores +6. Note this and assume that the second player "favours" and chooses the right–hand arc in position (2,5). Then $v_1(2,5) = v_1(1,6) = 6$, $v_2(2,5) = v_2(1,6) = 2$, $v_1(2,6) = v_1(1,7) = 2$, $v_2(2,6) = v_1(1,7) = 4$, $v_1(1,8) = 2$, $v_2(1,8) = 3$. Next we solve the games $\Gamma_{(1,3)}$, $\Gamma_{(1,4)}$, $\Gamma_{(2,3)}$, $\Gamma_{(1,5)}$, $\Gamma_{(2,4)}$. In the subgame $\Gamma_{(1,3)}$ there are two Nash equilibria, since the first player does not care which of the alternatives to choose. His choice, however, proves to be essential for Player 2 since with Player 1's choice of the left–hand alternative he (Player 2) scores -10. Assume that he selects the right–hand alternative in position $(1,3)$. Then

$$v_1(1,3) = 5,\ v_2(1,3) = 10,\ v_1(1,4) = v_1(2,5) = 6,$$
$$v_2(1,4) = v_2(2,5) = 2,\ v_1(1,5) = v_1(2,6) = 2,$$
$$v_2(1,5) = v_2(2,6) = 4,\ v_1(2,3) = 0,$$
$$v_2(2,3) = 6,\ v_1(2,4) = 3,\ v_2(2,4) = 5.$$

For the games $\Gamma_{(2,2)}$, $\Gamma_{(1,2)}$, $\Gamma_{(2,2)}$ we have

$$v_1(2,1) = v_1(1,3) = 5,\ v_2(2,1) = v_2(1,3) = 10,\ v_1(1,2) = v_1(2,4) = 3,$$
$$v_2(1,2) = v_2(2,4) = 3,\ v_1(2,2) = -5,\ v_2(2,2) = 6.$$

Now it is possible to solve the game $\Gamma = \Gamma_{(1,1)}$. It can be readily seen that $v_1(1,1) = v_1(2,1) = 5$, $v_2(1,1) = v_2(2,1) = 10$. As a result, we get the following absolute Nash equilibrium

$$(\bar{u}_1, \bar{u}_2) = ((1,2,2,2,2,3,2,1),(1,3,2,2,2,1,2)) \qquad (7.2.5)$$

In the equilibrium (7.2.5) the game follows the path $((1,1),(2,1),(1,3))$. By construction, the strategies $\bar{u}_i$ $(i = 1,2)$ are "favourable" in that the player i making his move and being equally interested in the choice of subsequent alternatives chooses the strategy that is favourable for the player $3 - i$.

In the game under study there are equilibrium where the payoffs are different. To construct one of such equilibria, it suffices to replace the "favourableness" condition of the players with the inverse condition of "unfavourableness". Then we have a new Nash equilibrium

$$(\tilde{u}_1, \tilde{u}_2) = ((2,2,1,1,2,3,2,1),(3,3,2,2,1,1,2)) \qquad (7.2.6)$$

The payof for both players in (7.2.6) are less then those in the equilibrium (7.2.5). The equilibrium (7.2.6), just as the equilibrium (7.2.5), is the absolute equilibrium. Evidently, apart from "favourable" and "unfavourable" absolute Nash equilibria, there is an entire family of intermediate absolute equilibria. Of interest is the question: when they are not differing in the players' payoffs.

Theorem 2 *Suppose the payoff of players H_i $(i = 1,\dots,n)$ in the game (7.2.1) are such that if there are such k and such x, y $x \neq y$, that $H_k(x) = H_k(y)$, then $H_i(x) = H_i(y)$ for all $i = 1,\dots,n$. Then, in the game (7.1.1) the players' payoff coincide in all absolute equilibria.*

Proof: Consider a family of subgames of the game (7.1.1), the proof is given by induction over their length. Suppose the subgame Γ_x is of length 1 and the player i_1 makes a move in a unique position which is not final. In the equilibrium he then makes a choice $\bar{x} \in \gamma(x)$ from the condition

$$H_{i_1}(\bar{x}) = \max_{y\in\gamma(x)} H_{i_1}(y)$$

If the point $\bar{x}$ is unique, then the players' payoff vector is Nash equilibria is also unique and here equals to $H(\bar{x}) = (H_1(\bar{x}),\dots,H_n(\bar{x}))$. If there exists a point $\tilde{x} \neq \bar{x}$ sach that $H_{i_1}(\tilde{x}) = H_{i_1}(\bar{x})$, then there is one more equilibrium with the payoffs $H(\tilde{x}) = (H_1(\tilde{x}),\dots,H_n(\tilde{x}))$. From the condition of the theorem, however, it follows that $H(x) = H(\tilde{x})$

Let $v(x) = (v_1(x),\dots,v_n(x))$ be the payoff vector in the equilibrium in a single stage subgame Γ_x which, as is previously seen, is defined in a unique mannaer. We will prove that if $v_{i_1}(x') = v_{i_1}(x'')$ for same i_1, where x' and x'' are sach that the length of subgames $\Gamma_{x'}$ $\Gamma_{x''}$ are equal to I, then $v_i(x') = v_i(x'')$ for all $i = 1,\dots,n$. Indeed, let $x' \in X_{i_k}$, $x'' \in X_{i_\ell}$

$$v_{i_k}(x') = H_{i_k}(\bar{x}') = \max_{y\in\gamma(x')} H_{i_k}(y)$$

$$v_{i_\ell}(x'') = H_{i_\ell}(\bar{x}'') = \max_{y\in\gamma(x'')} H_{i_\ell}(y) \tag{7.2.7}$$

Then $v_i(x') = H_i(\bar{x}')$, $v_i(x'') = H_i(\bar{x}'')$, $i = 1,\dots,n$. From (7.2.7) we have that $H_{i_1}(\bar{x}') = H_{i_1}(\bar{x}'')$ and hence using the condition of the theorem we get $v_i(x') = H_i(\bar{x}') = H_i(\bar{x}'') = v_i(x'')$, $i = 1,\dots,n$.

We now assume that in all subgames Γ_x of the length not exceeding $k-1$, the payoff $v_i(x)$ in Nash equilibrium for all players are uniquely defined and if for some two subgames $\Gamma_{x'}$ and $\Gamma_{x''}$ the equality $v_{i_1} = v_{i_1}(x'')$ holds for one index i_1, then $v_i(x') = v_i(x'')$ for all indexes $i \neq i_1$. Let Γ_x be a subgame of length k. If $x \in X_{i_1}$, then in the position x the player i_1, chooses the next position $\bar{x} \in \gamma(x)$ from the condition

$$v_{i_1}(\bar{x}) = \max_{y\in\gamma(x)} v_{i_1}(y).$$

In the case where the point $\bar{x}$ determined by this condition is unique, here is a unique vector $(v_1(\bar{x}), \ldots, v_n(\bar{x}))$ of the players payoffs in the Nash equilibrium in the game Γ_x. If, however, there are two positions $\bar{x}$ and $\tilde{x}$ for which the equality $v_{i_1}(\bar{x}) = v_{i_1}(\tilde{x})$ holds true, then, by the induction assumption we have $v_i(\bar{x}) = v_i(\tilde{x})$ for all $i \neq i_1$. Consequently, the players payoffs $\{v_i(x)\}$ in the Nash equilibrium in the game Γ_x are defined in a unique manner: $v_i(x) = v_i(\bar{x}) = v_i(\tilde{x})$, $i = 1, \ldots, n$.. This completes the proof of the theorem. ■

We will derive functional equations for nonzero–sum games with complete information.It follows from Theorem 1 that in any nonzero–sum game with complete information there exists an absolute Nash equilibrium, i.e. there exist such a situation $(\bar{u}_1, \ldots, \bar{u}_n)$ whose truncation in any of the subgames Γ_x of game Γ forms therein the Nash equilibrium. Denote by $v(x) = \{v_1(x), \ldots, v_n(x)\}$ the payoff vector in the Nash equilibrium of subgame Γ_x which is a truncation of the absolute equilibrium $(\bar{u}_1, \ldots, \bar{u}_n)$. In contrast to the zero–sum case, where the value of the game is defined in a unique manner, here the vector $v(x)$ takes different values for different Nash equilibrium, since in different absolute equilibria the players payoff vectors may be different. It was shown that $v(x)$ may be constructed as single–valued vector function which sets up a correspondence between each x and the payoff vector in one of the absolute equilibria of subgames Γ_x.

We derive a functional equations for the computation of the vector function $v(x)$. We have $v(x) = (v_1(x), \ldots, v_n(x))$, $v_i(x) = K_i^x(\bar{u}_1^x, \ldots, \bar{u}_n^x)$, where $(\bar{u}_1^x, \ldots, \bar{u}_n^x)$ are absolute equilibria in subgames Γ_x. Let $y \in \gamma(x)$ and $x \in X_1$, then

$$v_1(x) = K_1^{y^*}(\bar{u}_1^{y^*}, \ldots, \bar{u}_n^{y^*}) = \max_{y \in \gamma(x)} K_1^y(\bar{u}_1^y, \ldots, \bar{u}_n^y) = \max_{y \in \gamma(x)} v_1(y)$$

and $v_i(x) = v_i(y^*)$ for $i \neq 1$. For $x \in X_i$ we similarly obtain

$$v_i(x) = K_i^{y^*}(\bar{u}_1^{y^*}, \ldots, \bar{u}_n^{y^*}) = \max_{y \in \gamma(x)} K_i^y(\bar{u}_1^y, \ldots, \bar{u}_n^y) = \max_{y \in \gamma(x)} v_i(y)$$

and $v_j(x) = v_j(y^*)$ for $j \neq i$. Thus,

$$v_i(x) = v_i(y^*) = \max_{y \in \gamma(x)} v_i(y) \tag{7.2.8}$$

for all $x \in X_i$ and

$$v_j(x) = v_j(y^*) \text{ for } j \neq i \ (i, j = 1, \ldots, n). \tag{7.2.9}$$

Equations (7.2.8), (7.2.9) are solved under the boundary condition

$$v(x) = H(x), \quad x \in X_{n+1} \tag{7.2.10}$$

The system of equations (7.2.8),(7.2.9) with the boundary condition (7.2.10) makes it possible to construct the backward recurrent procedure for the computation of the vector function $v(x)$ and the corresponding absolute Nash equilibrium. Indeed, let the vector function $v(y)$ be computed for all subgames Γ_y of length $\ell \leq k-1$ and Γ_x be a subgame of length k. In this case, if $x \in X_i$, then $v_i(x)$ is determined from (7.2.8), and $v_j(x)$ $(j \neq i)$ from (7.2.9), i.e.

$$v(x) = (v_1(y^*), \ldots, v_{i-1}(y^*), \max_{y \in \gamma(x)} v_i(y), v_{i+1}(y^*), \ldots, v_n(y^*)).$$

The same formulas indicate the initial optimal choice in position $x \in X_i$ of subgame Γ_x. Indeed, in the position $x \in X_i$ the player i has to choose the node $y^* \in \gamma(x)$ for which the value of $v_i(y^*)$ is maximal.

Evidently, the solution of the system of recurrent equation (7.2.8), (7.2.9) under the boundary condition (7.2.10) is not unique, since a maximum in (7.2.8) may be achieved on different y_1^*, y_2^* nodes although $v_i(y_1^*) = v_i(y_2^*)$, but $v_j(y_1^*)$ for some $j \neq i$ may not coincide with $v_j(y_2^*)$, which essentially affects further solution. As noted in Example 4, this exactly leads to different payoffs in different absolute equilibria.

Although we have proved the existence of an absolute. Nash equilibrium in the nonzero–sum multistage game with complete information, there may exist not necessary absolute equilibria in the basic game (7.1.1). Among such equilibria a special role play the equilibria in penality strategies. This will be illustrated by referring to examples.

Example 5. Consider the nonzero–sum two–person game from Example 4. We compare the payoffs in "favourable" and "unfavourable" equilibria. In the "favourable" equilibrium (7.2.5) the players' payoffs are (5,10), this being (3,5) in the equilibrium (7.2.6). In the equilibrium (7.2.5) a large payoff for Player 2 is fully dependent upon the fancies of Player 1, since in the node (1,3) the choice of the left–hand or right–hand arc makes no difference to Player 1, and if for some reason he decides on the left–hand arc, Player 2's payoff may go down to 1. However, by employing a particular behavior here called a penality strategy, Player 2 may always ensure payoff 5. Indeed, if Player 2 adopts the strategy

$$u_2^* = (2, 3, 2, 2, 1, 1, 2), \tag{7.2.11}$$

the best responce of Player 1 may only come from the class of strategies

$$u_1^*(2, 2, a_3, 1, a_5, a_6, a_7, a_8), \tag{7.2.12}$$

where a_i $(i = 1, 2, 4)$ are the arbitrary acceptable choices of Player 1 in the corresponding nodes. The payoff of Players in the equilibrium (7.2.11) are equal to $K_1(u_1^*, u_2^*) = 3$, $K_2(u_1^*, u_2^*) = 5$ and coincide with those in (7.2.6). It is fairly easy to show that the pair of strategies (7.2.11), (7.2.12) forms the

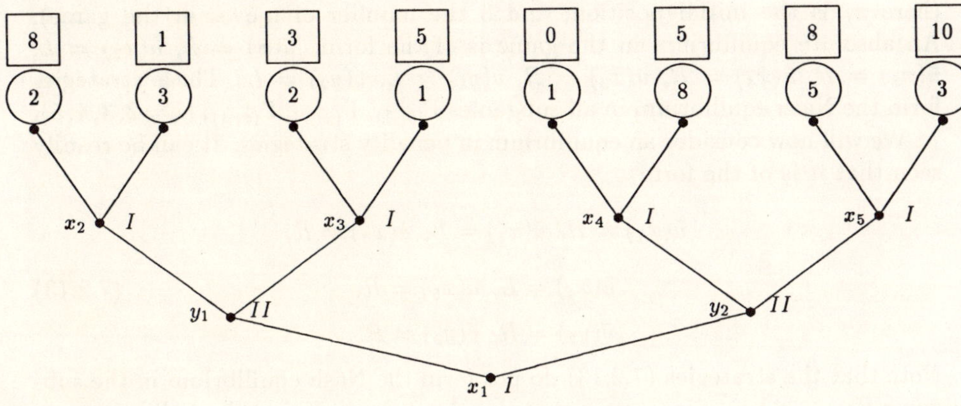

Fig. 39.

Nash equilibrium. At the same time, in this equilibrium Player 2 ensures the payoff 5, but this is not true for the situation (7.2.6) in which Player 1 may choose the strategy $u_1 = (1, a_2, a_3, 3, a_5, a_6, 2, a_8)$, then $K_2(u_1, \tilde{u}_2) = 4$.

Note that the strategy (7.2.11) is a penelty strategy in that, when it is adopted by Player 2, the departure of Player 1 from the strategy (7.2.12) in the node x_0, say by choosing the left–hand arc, involves Player 2's choice in the node (2,1) of the central arc with the quaranteed subsequent payoff for Player 1 equal to 2. In this case, Player 2's choice in the subgame $\Gamma_{(2,1)}$ is far from being optimal (Player 2 in this position should have chosen the right–hand arc) but he "penalizes" Player 1 thereby preventing his departure from the strategy u_1^* in the position (1,1). So the truncation of the Nash equilibrium (7.2.11), (7.2.12) in the subgame $\Gamma_{(2,1)}$ is not a Nash equilibrium, i.e. the equilibrium is not an absolute equilibrium (subgame perfect), although it guarantees Player 2's payoff equal to 5.

Example 6. We now provide a simpler example illustrating the above–mentioned property of equilibrium in penalty strategies. The structure of the game is shown in Fig.39. At the first step Player 1 chooses one of the two alternatives: "right"(R) or "left" (L); at the second step Player 2 knows Player 1's move makes his move and chooses R or L alternative, at the third step Player 1 chooses R or L alternative, whereupon the game terminates and the players receive their payoff in final positions. In Fig.39 the payoff for Player 1 are indicated by blocks, and those for Player 2 by circles. Denote the nodes, in which Player 1 makes a move, by x_1, x_2, x_3, x_4, x_5, and the nodes, in which Player 2 makes a move, by y_1, y_2. In each position $x_i \in X_1$ (respectively $y_i \in X_2$) the strategy for the first (respectively the second) player u (v) may assume one of two values: R or L. Denote the game in question by $\Gamma_{(x_1,3)}$

(here x_1 is the initial position, and 3 the number of moves in the game). An absolute equilibrium in the game is of the form: $\bar{u}(x) = L$, $\bar{u}(x_2) = L$, $\bar{u}(x_3) = R$, $\bar{u}(x_4) = R$, $\bar{u}(x_3) = R$, $\bar{v}(y_1) = L$, $\bar{v}(y_2) = L$. These strategies form the Nash equilibrium in all subgames $\Gamma_{(y_1,7)}$, $\Gamma_{(y_2,2)}$, $\Gamma_{(x_i,1)}$, $(i = 2,3,4,5)$.

We will now consider an equilibrium in penality strategies. It can be readily seen that it is of the form

$$\tilde{u}(x_1) = R,\ \tilde{u}(x_2) = L,\ \tilde{u}(x_3) = R,$$
$$\tilde{u}(x_4) = L,\ \tilde{u}(x_5) = R, \tag{7.2.13}$$
$$\tilde{v}(y_1) = R,\ \tilde{v}(y_2) = R.$$

Note that the strategies (7.2.13) do not form the Nash equilibrium in the subgame $\Gamma_{(x_4,1)}$. Also, of interest is that the players' payoffs in this equilibrium are higher than those in the absolute equilibrium ((10,3) and (8,2), respectively). If Player 2 does not pursue the strategy $\tilde{v}$ (in particular, of importance is the case where $\dot{v}(y_2) = L \neq \tilde{v}(y_2) = R$), Player 1 in the node x_4 choses the alternative $\tilde{u}(x_4) = L$, which minimizes not only the payoff for the "penalized" Player 2, but also his own payoff. Because of this, the truncation of strategies (7.2.13) in the subgame do not form in the subgame $\Gamma_{(x,y)}$ the equilibria in this game $\Gamma_{(x_4,1)}$.

Example 7. We have seen in Example 4 that "Pavourableness" of the players give them higher payoffs in the corresponding Nash equilibria, than the "unfavourable" behavior. But this is not always the case. Sometimes the "unfavourable" Nash equilibria gives higher payoff to all the players than the "favourable" one. We shall illustrate this rather nontrival fact on example below. Consider the two–person game on the Fig.40. The nodes from the priority set X_1 are represent by circles and those from X_2 by brocks, with players payoffs written in the final positions. In Example 4 positions from the set X_i $(i = 1,2)$ are numbered by double indexes where i is the indexes of the player and j–the index of the node x in the set X_i. One can easily see that he "favourable" equilibria has the form $(2,2,1,1,1)$, $(2,1)$ with payoffs (2,1).

The "unfavourable" equilibria has the form $(1,1,2,1,1)$, $(1,1)$ with payoffs (5,3).

We will now provide a precise definition for penality strategies. For simplicity, we restrict ourselves to the case of nonzero–sum two–person multistage game with complete information

$$\Gamma = < D_1, D_2, K_1, K_2 > \tag{7.2.14}$$

on graph G. This game is related to two zero–sum games

$$\Gamma_1 = < D_1, D_2, K_1, -K_1 >, \tag{7.2.15}$$

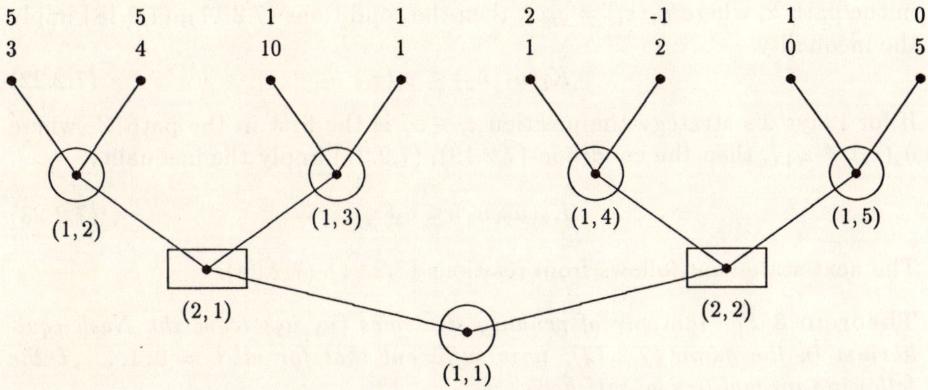

Fig. 40.

$$\Gamma_2 =< D_1, D_2, -K_2, K_2 > \tag{7.2.16}$$

Let Γ_{1x}, Γ_{2x} be subgames of games (7.2.15),(7.2.16); $v_1(x)$, $v_2(x)$ the values of these subgames;$(\bar{u}_{11}^x, \bar{u}_{21}^x)$ and $(\bar{u}_{12}^x, \bar{u}_{22}^x)$ the equilibria in the games Γ_{1x} and Γ_{2x}, respectively. Consider a fixed path $z = (x_0 = z_0, z_1, \ldots, z_\ell)$ on the graph G realized in the games (7.2.14)-(7.2.16).

The strategy $\tilde{u}_1$ is called a penality strategy for Player 1 if

$$\tilde{u}_1(z_i) = z_{i+1} \tag{7.2.17}$$

for positions $z_i \in Z \bigcap X_1$, and

$$\tilde{u}_1(y) = \bar{u}_{12}^x(y) \tag{7.2.18}$$

for positions $y \in X_1 \backslash Z$ such that $(\gamma^{-1})^m(y) = x \in Z$ for some m. The penality strategy $\tilde{u}_2$ for Player 2 is defined in much the same way:

$$\tilde{u}_2(z_i) = z_{i+1} \tag{7.2.19}$$

if $z_i \in Z \bigcap X_2$;

$$\tilde{u}_2(y) = \bar{u}_{21}^x(y) \tag{7.2.20}$$

if $y \in X_2 \setminus Z$ and such that $(\gamma^{-1})^m(y) = x \in Z$ for some m.

Note that

$$K_i(\tilde{u}_1, \tilde{u}_2) = H_i(z_\ell) \ \ (i = 1,2) \tag{7.2.21}$$

If Player 1 employs the strategy u_1, for which the position $z_i \in Z$ is the first in the path z, where $u_1(z_i) \neq z_{i+1}$, then the conditions (7.2.17), (7.2.18) imply the inequality.

$$K_1(u_1, \tilde{u}_2) \leq v_1(z_i) \tag{7.2.22}$$

If for Playr 2's strategy the position $z_i \in Z$ is the first in the path Z, where $u_2(z_i) \neq z_{i+1}$, then the condition (7.2.19), (7.2.21) imply the inequality

$$K_2(\tilde{u}_1, u_2) \leq v_2(z_i) \tag{7.2.23}$$

The next statement follows from relations (7.2.21)–(7.2.23).

Theorem 3 *For the pair of penality strategies $(\tilde{u}_1, \tilde{u}_2)$ to be the Nash equilibrium in the game (7.2.14), it is sufficient that for all $i = 0, 1, \ldots, \ell$ the following inequalities be satisfied.*

$$K_j(\tilde{u}_1, \tilde{u}_2) \geq v_j(z_i) \ \ (j = 1,2), \tag{7.2.24}$$

where $(z_0, z_1, \ldots, z_\ell)$ is the path realized in the situation $(\tilde{u}_1, \tilde{u}_2)$

Let u_{11}^* and u_{21}^* be the strategies for players 1 and 2 in the games (7.2.15) and (7.2.16), respectively; Z the path in the corresponding situation (u_{11}^*, u_{22}^*); $\tilde{u}_1$ and $\tilde{u}_2$ the penality strategies for the path Z. It may be shown that the condition (7.2.24) is then satisfied. In this manner we get the following theorem.

Theorem 4 *In the game (7.2.14) there is always the Nash equilibrium in penality strategies.*

The implication of penality strategies is that with their help the player makes his opponent to follow a particular path in the game (particular choices) by continuously threatening to switch to a strategy designed against the opponent (the strategy being optimal in the corresponding zero–sum game). The set of equilibria in the class of penalization strategies is sufficiently representational. It should be noted that by penalizing his opponent the player may panaliize himself even to a greater extent.

The strong dynamic stability of Nash equilibrium and K–equilibrium. Let us state the concept of dynamic stability (time consistency) for nonzero–sum multistage game on a graph tree. In this section the game (7.1.1) is denoted by $\Gamma_{(x_0,\ell)}$, where ℓ is the length of the game. Let $U(x_0, \ell)$ be the set of all situation in this game satisfying a given optimality principle (for instance the set of Nash equilibria).

Consider a play of the game $(\bar{x}_0, \bar{x}_1, \ldots, \bar{x}_i)$ in the situation $\bar{u} = (\bar{u}_1, \ldots, \bar{u}_n) \in U(x_0, \ell)$. Let $\Gamma(\bar{x}_i, \ell - i)$ be the corresponding subgames. Denote by $\bar{u}^i$ a truncation of the situation $\bar{u}$ on positions of the subgame $\Gamma(\bar{x}_i, \ell - i)$. The dinamic stability requirement then reduces to fulfilment of the condition $\bar{u}^i \in U(\bar{x}_i, \ell - i)$, where $U(\bar{x}, \ell_{x-i})$ is the set of all situations in the subgame $\Gamma(x_i, \ell - i)$ satisfying a given optimality principle (for instance the set of equilibria). Let $\bar{u}^i \in U(\bar{x}_i, \ell - i)$; and if there exists such situation $w = (w_1, \ldots, w_n) \in U(x_0, \ell)$ that

$$w(x) = u^i(x) \tag{7.2.25}$$

on positions of the subgame $\Gamma(\bar{x}_i, \ell - i)$, $0 \leq i \leq \ell$, then the principle of optimality $U(x_0, \ell)$ is said to be strongly dynamic stable, (strongly time constant). Here by $w(x)$ for $x \in X_j$ $(j = 1, \ldots, n)$ is meant the choice made by the strategy of the j–th player $w_j(x)$ in the position $x \in X_j$, i.e

$$w(x) = w_j(x) \text{ for } x \in X_j \ (j = 1, \ldots, n) \tag{7.2.26}$$

Theorem 5 *The set of all Nash equilibria is a strongly–dynamic stable principle of optimality.*

Proof: Let $\{K_j^{\bar{x}_i}[u^i(x)]\} = v_j(\bar{x}_i)$ $(j = 1, \ldots, n)$ be the payoffs of the players in the subgame $\Gamma(\bar{x}_i, \ell - i)$ in the equilibrium u^i of this subgame. Consider an auxiliary game $\bar{\Gamma}_i(x_0, \ell)$ (a truncation of the game $\Gamma(x_0, \ell)$ in the node $\bar{x}_i$) which is the same as the game $\Gamma(x_0, \ell)$ except that in the gtaph $\bar{G}$ of this game the node $\bar{x}_i$ is declared to be final, the other nodes of the subgraph $G_{\bar{x}_i}$ are eliminated from the set X. Is assumed to be

$$\bar{H}_j(\bar{x}_i) = K_j^{\bar{x}_i}[u^i(x)] = v_j(\bar{x}_i)(j = 1, \ldots, n).$$

In other respects the games $\Gamma(x_0, \ell)$ and $\bar{\Gamma}(x_0, \ell)$ coincide. Let $\bar{w}(x)$ be an equilibrium in the game $\bar{\Gamma}(x_0, \ell)$; construct the situation $w(x)$ in the game $\Gamma(x_0, \ell)$ by the rule.

$$w(x) = \begin{cases} \bar{w}(x) & \text{in positions of the game } \Gamma(x_0, \ell), \\ u^i(x) & \text{in positions of the subgame } \Gamma(\bar{x}_i, \ell - i). \end{cases} \tag{7.2.27}$$

We will show that the situation $w(x)$ is the equilibrium in the game $\Gamma(x_0, \ell)$. Note that $K_j(w) = \bar{K}_j(\bar{w})$ for all $j = 1, \ldots, n$, where $\bar{K}_j$ is the payoff of the player j in the game $\bar{\Gamma}(x_0, \ell)$. Now let $u_j \in D_j$. Consider the situation $(w \| u_j)$. If in this situation the play of the game passes through the position $\bar{x}_i$, then

$$K_j(w \| u_j) = K_j^{\bar{x}_i}(u^i \| u_j) \leq K_j^{\bar{x}_i}(u^i) = \bar{K}_j(\bar{w}) = K_j(w).$$

If in the situation $(w \| u_j)$ the play of the game does not pass through the position $\bar{x}_i$, then

$$K_j(w \| u_j) = K_j(\bar{w} \| u_j) = \bar{K}_j(\bar{w} \| u_j) \leq \bar{K}_j(w) = K_j(w)$$

This completes the proof of the theorem. ■

Consider the concept of K–equilibrium introduced by N.N.Vorobyev. Let K be the set of subsets of all players $N = \{1, \ldots, n\}$. The elements of the system K are called coalitions, the system K being a coalition structure. Let $\bar{u} = (\bar{u}_1, \ldots, \bar{u}_n)$ be a situation in the game (7.1.1), and $u_s = \{u_i\}_{i \in S}$ be a strategy set of the players from the coalition S. Denote by $(\bar{u}\|u_s)$ a situation which is the same as the situation $\bar{u}$ except that the strategies $\bar{u}_i$ replaced therein by the strategies u_i, $i \in S$.

The situation $\bar{u}$ is called the K–equilibrium in the game (7.1.1) if for each coalition $S \in K$ and u_s there is such if the player $i \in S$ that

$$K_i(\bar{u}) \geq K_i(\bar{u}\|u_s)$$

Theorem 6 *Suppose that in the truncated game $\bar{\Gamma}_i(x_0, \ell)$ $(i = 1, 2, \ldots)$ and subgames $\Gamma(x_i, \ell - i)$ there are K–equilibria. Then the set of all K–equilibria is a strongly dynamic stable principle of optimality.*

If the coalition structure K contains all individual players, i.e. the sets $\{i\}$ $(i \in N)$, then it can be readily seen that any K–equilibrium is also the Nash equilibrium. If the coalition structure K is composed of all subsets of the set N, then the K equilibrium is called the strong Nash equilibrium. We shall compare the concepts of Nash equilibrium and K–equilibrium.

The Nash equilibrium $\bar{u} = (\bar{u}_1, \ldots, \bar{u}_n)$ implies that if the player departs from the equilibrium strategy (providing the other players employ their equilibrium strategy in the situation $\bar{u}$), then this will never increas his payoff. The condition that, when the player i deviating, other players employ their equilibrium strategies is very important, since, when two players depart from their equilibrium strategies, they may have an opportunity to improve their payoffs. The Nash equilibrium is a strongly dinamic stable principle of optimality, but the inherent impossibility to unite players into groups in order to increase their payoffs seems to be not quite real.

The K–equilibrium of a situation means that in any coalition S from a given coalition structure K there is necessarily a player who derives no benefit when the coalition departs from the equilibrium strategy. The K–equilibrium of a situation may be interpreted as impossibility, in this situation, for the coalition $S \in K$ to raise the payoff for all its members. Thus, the existence of the K–equilibrium prevents formation of the player coalitions $S \in K$ seeking to depart from it. Unfortunately the existence of K–equilibrium is a very exceptional case.

7.3 Definition of the noncooperative differential games in normal form

Let the state of controlled process at the time instant t be described by the vector

$$x = x(t) = (x_1(t), \ldots, x_m(t)) \in R^m \tag{7.3.1}$$

which varies on the time interval $[t_0, T]$ in accordance with a differential motion equation

$$\dot{x} = f(x, u_1(t), \ldots, u_n(t)), \tag{7.3.2}$$

where $f(x, u_1, \ldots, u_n) = \sum_{i=1}^{n} f_i(x, u_i)$. Its is assumed that the player i can have an influence on realization of the state (7.3.1) of the process in hand by choosing, as a control, the u_i– measurable function $u_i : [t_0, T] \to U_i$, where U_i is a compact set in the Euclidean space R^{m_i}. Denote the set of all such controls by D_i ($i \in N = \{1, \ldots, n\}$). We assume that the vector function f is continuous in thee set $R^m \times U_1 \times \ldots \times U_n$ and satisfies Lipschitz condition for x with a constant K

$$\|f(x_1, u_1, \ldots, u_n) - f(x_2, u_1, \ldots, u_n)\| \leq K\|x_1 - x_2\|$$

uniformly in $u_1 \in U_1, \ldots, u_n \in U_n$.

It is well known that these conditions ensure for each initial state

$$x(t_*) = x_* \tag{7.3.3}$$

and a fixed n–tuple of control of player $u_1 \in D_1, \ldots, u_n \in D_n$ the existence of a unique solution

$$x = x(t) = x(t, t_*, x_*, u_1, \ldots, u_n) \tag{7.3.4}$$

of the problem (7.3.2), (7.3.3) on the time interval $[t_0, T]$ In this case the solution is taken in the sence of Carateodori, i.e. as an absolutely continuous function $x(t)$ satisfying the equation (7.3.2) almost everywhere on the interval $[t_0, T]$. From the continuity of the function f, and from the compactness of the set $W = U_1 \times \ldots \times U_n$ is follows that

$$\max_{w \in W} \|f(0, w)\| = B < \infty.$$

Consequently, for any trajectory (7.3.4) the following inequality holds

$$\|x(t)\| \leq \|x_0\| + \int_{t_0}^{t} \|f(x(\tau), u_1(\tau), \ldots, u_n(\tau))d\tau \leq$$

$$\leq \|x_0\| + B(t - t_0) + K \int_{t_0}^{t} \|x(\tau)\|d\tau.$$

By using the Gronwall–Bellman inequality we obtain the estimate

$$\|x(t,t_*,u_1,\ldots,u_n)\| \le [\|x_*\| + B(t-t_*)]\, e^{K(t-t_*)} \le$$
$$\le R(\|x_*\|) = [\|x_*\| + B(t-t_*)]\, e^{K(T-t_*)}$$

Thus, the set $\Phi(\sum_1, t_*, x_*)$ of all trajectories (solutions to problems (7.3.2), (7.3.3) for all possible sets of player controls) is uniformly bounded, all of the trajectories (7.3.4) wholly lying inside a closed sphere $S(O, R\|x_*\|))$ with $x=0$ as center and $R(\|x_*\|)$ as radius in the Euclidean space R^m. Next, from the continuity of the function f on the compact set $C(\|x_*\|) = S(O, R\|x_*\|)) \times W$ it follows that

$$\max_{(xw)\in C(\|x_*\|)} \|f(x,w)\| = B_1(\|x_*\|) < \infty$$

This means that all trajectories (7.3.4) satisfy Lipschitz condition for t:

$$\|x(t,t_*,x_*,u_1,\ldots,u_n) - x(t',t_0,u_1,\ldots,u_n,x_*)\| \le B_1(\|x_1\|)|t-t'|$$

Thus the set $\Phi(\sum_1, t_*, x_*)$ is also uniformly continous. Note that the continuity of the function $(t_*, x_*) \to \chi(t,t_*,x_*,u_1,\ldots,u_n)$ uniformly in time t and controls $u_1,\ldots,u_n$ may be proved.

In this chapter we consider differential games in the class of piecewise open–loop strategies. Denote by $D_i[t,\tau)$ the set of restriction of all admissible controls for the i–th player to the interval $[t,\tau) \subset [t_0,T]$ $(i \in N)$, and by $PF[t_0,T]$ the set of all finite partitions

$$\Delta = \{t_0 = t_0^\Delta < t_1^\Delta < \ldots < t_{n(\Delta)}^\Delta = T\}$$

of the time interval $[t_0,T]$. The sequence $\varphi^i_{\Delta,j} = (\varphi^i_{\Delta,1},\ldots,\varphi^i_{\Delta,n(\Delta)})$, where

$$\varphi^i_{\Delta,j} \;:\; R^m \to D_i[t^\Delta_{j-1}, t^\Delta_j) \;\; (j=1,\ldots,n(\Delta)).$$

is called the Δ–strategy for the i–th player. The pair $a_i = (\Delta, \varphi^i_\Delta)$, where $\Delta \in PF[t_0,T]$, and φ^i_Δ is any Δ–strategy for the i–th player, is called the piecewise open–loop strategy (POLS) for the i–th player. Denote by D_i^* the set of all such strategies $(i \in N)$.

Suppose that in the initial state $x(t_0) = x_0$ the players choose piecewise open–loop strategies

$$a_1 = (\Delta_1, \varphi^1_{\Delta_1}), \ldots, a_n = (\Delta_n, \varphi^n_{\Delta_n}) \tag{7.3.5}$$

The numbers $t_j^{\Delta_i}$ $(j=1,\ldots,n(\Delta_i);\ i \in N)$ are arranged in ascending order. Then we have a sequence $t_0 \le t_1 \le t_2 \le \ldots \le t_q = T$. At the initial instant of time t_0 the players choose controls

$$u_i^1 = \varphi^i_{\Delta_i,1}(x_0) \in D_i[t_0, t_1^{\Delta_i}) \tag{7.3.6}$$

whereupon the trajectory

$$x(t) = \chi(t, t_0, x_0, u_1^1, \dots, u_n^1) \tag{7.3.7}$$

is realized on the time interval $[t_0, t_1]$ players from the set $N_1 = \{i \in N \mid t_1^{\Delta_i} = t_1\}$ stick to controls (7.3.6) up to the time instant t_1, then they choose new controls

$$u_i^2 = \varphi^i_{\Delta_i,2}(x(t_1)) \in D_i[t_1, t_2^{\Delta_i}),$$

the trajectory

$$x(t) = \chi(t, t_1, x(t_1), u_1^2, \dots, u_n^2),\ u_i^2 = u_i^1\ (i \in N). \tag{7.3.8}$$

being realized on the time interval $[t_1, t_2]$. Let $N_2 = \{i \in N \mid t_1^{\Delta_i} = t_2\}$ and $N_3 = \{i \in N \mid t_2^{\Delta_i} = t_2\}$ $(N_1 \bigcap N_2 = \emptyset,\ N_3 \subset N_1)$. At the time instant t_2 the players from the set N_2 choose new controls

$$u_i^3 = \varphi^i_{\Delta_i,1}(x(t_2)) \in D_i[t_2, t_2^{\Delta_i}),$$

and the players from the set N_2 choose new controls

$$u_i^3 = \varphi^i_{\Delta_i,2}(x(t_2)) \in D_i[t_2, t_3^{\Delta_i}).$$

Proceeding with these reasonings, we determine in q steps a unique trajectory

$$x(t) = \chi(t, t_0, x_0, a_1, \dots, a_n) \tag{7.3.9}$$

generated by strategies (7.3.5) from the initial state $x(t_0) = x_0$. On the time interval $[t_0, t_1]$ the trajectory (7.3.8) coincides with the trajectory (7.3.7), on the interval $[t_1, t_2]$ with the trajectory (7.3.8), etc.

Consider a conflict decision process where the i–th player seeks to maximize the functional (payoff).

$$H^{(i)}(x_0, T - t_0, x(\cdot))\ \ (i \in N), \tag{7.3.10}$$

defined on the set of all trajectories (7.3.9). Let $u = (u_1, \dots, u_n)$ be a situation in POLS, Then the functional

$$J^{(i)}(x_0, T - t_0; u_1, \dots, u_n) = H^{(i)}(x_0, T - t_0; u). \tag{7.3.11}$$

defined on the set $D_1^* \times \dots \times D_n^*$ is called the payoff function of the i–th player corresponding to the payoff (7.3.10). The collection

$$\Gamma(x_0, T - t_0) =< D_1^*, \dots, D_n^*; J^{(1)}_{t_0,x_0}, \dots, J^{(n)}_{t_0,x_0} > \tag{7.3.12}$$

is called the noncooperative differential n–person game described by equations (7.3.2) with payoff functions (7.3.11).

In the games (7.3.12) with prescribed duration $T - t_0$ a payoff for the i–th player is of the form

$$H_i(x_0, T - t_0, x(\cdot)) = \int_{t_0}^{T} h_i(x(\tau))d\tau + H_i(x(T)),$$

$$h_i > 0, \quad H_i > 0, \tag{7.3.13}$$

where h_i and H_i are continuous functions on R^m. The addend on the right-hand side (7.3.13) is called the integral payoff, the addend being the terminal payoff. Basically, we will consider the games (7.3.12) with the payoff (7.3.13) for which a special designation $\Gamma(x_0, T - t_0)$ is used.

Let N be the set of all players. Then every subset $S \subset N$ is called a coalition is the game $\Gamma(x_0, T - t_0)$.

In our book we shall pay main attention to the cooperative differential games and corresponding optimality principles. For the noncooperative case we shall give only the definition of main optimality principles, which are generalization of the multimove case.

Nash–equilibrium. The n–tuple of strategies $u^* = (u_1^*, \dots, u_i^n, \dots, u_n^*)$ constitute a Nash equilibrium in the game $\Gamma(x_0, T - t_0)$ if

$$J^{(i)}(x_0, T - t_0; u^*(\cdot)) \geq J^{(i)}(x_0, T - t_0; u^*(\cdot)\|u_i(\cdot))$$

for all $i = 1, \dots, n$, $u_i(t) \in D_i^*$. The situation $u_\epsilon^*(\cdot) = (u_{\epsilon_1}^*(\cdot), \dots, u_{\epsilon_i}^*(\cdot), \dots, u_{\epsilon_n}^*(\cdot))$ is called Nash ϵ–equilibrium in the game $\Gamma(x_0, T - t_0)$ if

$$J^{(i)}(x_0, T - t_0; u_\epsilon^*(\cdot)) + \epsilon \geq J^{(i)}(x_0, T - t_0; (u_\epsilon^*(\cdot)\|u_i(\cdot)))$$

for all $i = 1, \dots, n, u_i(\cdot) \in D_i^*$. The following statement holds. The game $\Gamma(x_0, T - t_0)$ for any $\epsilon > 0$ has an Nash ϵ–equilibrium in the class of piecewise open–loop strategies (POLS). We shall not give the proof of this statement in our book, which is a generalization of the correspong theorem for the descrete time case.

Pareto–optimality. The situation $\bar{u}(\cdot) = (\bar{u}_1(\cdot), \dots, \bar{u}_i(\cdot), \dots, \bar{u}_n(\cdot))$, $\bar{u}_i() \in D_i^*$ is called Pareto–optimal in the game $\Gamma(x_0, T - t_0)$ if from the existence of the situation $u'(\cdot) = (u_1'(\cdot), \dots, u_i'(\cdot), \dots, u_n'(\cdot))$, $u_i'(\cdot) \in D_i^*$ such that

$$J^{(i)}(x_0, T - t_0; u'(\cdot)) \geq J^{(i)}(x_0, T - t_0; \bar{u}(\cdot)), \ i =, \dots, n$$

follows that

$$J^{(i)}(x_0, T - t_0; u'(\cdot)) = J^{(i)}(x_0, T - t_0; \bar{u}(\cdot)).$$

7.4 Definition of cooperative differential game in the form of characteristic sets

Consider a previously defined differential game $\Gamma(x_0, T - t_0)$ assuming that the players' payoffs are nontransferable. Recall that in the nontransferable payoff case, the players from the coalition $S \subset N$ aiming to maximize each component of the vector $J_S = \{J^i, i \in S\}$, where J^i is the payoff function of the player i (see (7.3.11). Therefore the cooperation between the members of the coalition S is only the choice of joint strategies $u_S(\cdot) \in \mathcal{D}_S$ without side payments.

Definition 1. The game $\gamma(x_0, T - t_0)$ is called a cooperative differential game without side payments if

1. its rules provide the formation of any coalition;
2. the players' payoffs are untransferable and they may jointly choose strategies but without making any side payments;
3. the players' objective is to maximize, by using joint efforts, their own payoffs.

For each coalition $S \subset N$, we consider a family of zero-sum games

$$\{\Gamma_S^y(x_0, T - t_0) \mid y \in C^{T-t_0}(x_0)\},$$

where $C^{T-t_0}(x_0)$ is the reachability set in the game $\Gamma(x_0, T - t_0)$. The zero-sum game $\Gamma_S^y(x_0, T - t_0)$ between coalition S and $N \setminus S$, acting as one player, starts at the instant t_0 from the state x_0 and proceeds until the instant T. Its dynamics is described by equations (7.4.2). The objective of Player S in the game $\Gamma_S^y(x_0, T - t_0)$ is to minimize the distance $\rho(x(T), y) = \|x(T) - y\|$ at the game termination instant, where $x(T) = x(T; x_0, u_S(\cdot), u_{N\setminus S}(\cdot))$ is the end point of the trajectory $x(\cdot) = x(\cdot; x_0, u_S(\cdot), u_{N\setminus S}(\cdot))$ which is a solution of the system (7.3.2), $y \in C^{T-T_0}(x_0)$ is a point the coalition $S \subset N$ seeks to reach at the game termination instant T, and

$$u_S(\cdot) \in \mathcal{D}_S = \prod_{i \in S} \mathcal{D}_i, \; u_{N\setminus S}(\cdot) \in \mathcal{D}_{N\setminus S} = \prod_{i \in N\setminus S} \mathcal{D}_i$$

($\mathcal{D}_i$ is the admissible strategy set for the player i in the game $\Gamma(x_0, T - t_0)$).

Using the Theorem 3 from the Ch. 2 we get that in the game $\Gamma_S^y(x_0, T - t_0)$ for any $\epsilon > 0$ there are the ϵ-optimal piecewise open-loop strategies $u_S^\epsilon(\cdot)$, $u_{N\setminus S}^\epsilon(\cdot)$ and the value of this game is equal to

$$val\Gamma_S^y(x_0, T - t_0) = \lim_{\epsilon \to 0} \|x(T; x_0, u_S^\epsilon(\cdot), u_{N\setminus S}^\epsilon(\cdot)) - y\|.$$

Introduce the sets

$$Y_S^{T-t_0}(x_0) = \{z \in C^{T-t_0}(x_0) \mid val\Gamma_S^z(x_0, T-t_0) = 0\},$$

$$X_S^{T-t_0}(x_0) = \{x(\cdot) \mid x(t_0) = x_0,\ x(T) \in Y_S^{T-t_0}(x_0)\}.$$

Purposefully, the set $Y^{T-t_0}(x_0)$ is a subset of the reachability set $C^{T-t_0}(x_0)$ for each point z of which there is the strategy $u_S^\epsilon(\cdot)$ ensuring the arrival of the coalition S in the ϵ–neighborhood of the point z at the instant T irrespective of the behavior of the coalition $N \setminus S$; $X_S^{T-t_0}(x_0)$ is the set of all trajectories of the system (7.3.1.)–(7.3.2.) leading to the set $Y_S^{T-t_0}(x_0)$. Let V be a mapping $V\ :\ 2^N \times R^m \times R^1 \to R^n$ such that for each triplet $(S, x_0, T-t_0) \in 2^N \times R^m \times R^1$

$$V(S; x_0, T-t_0) = \{J(x(\cdot)) \mid x(\cdot) \in X_S^{T-t_0}(x_0)\}, \tag{7.4.1}$$

where $J(x(\cdot)) = (J_1(x(\cdot)), \ldots, J_n(x(\cdot)))$,

$$J_i(x(\cdot)) = \int_{t_0}^{T} h_i(x(t))dt + H_i(x(T)),\ i = 1, \ldots, n$$

Lemma 1 *The mapping V defined by (7.4.1) has the following properties:*

1. *for an empty coalition $\emptyset$ and any $x_0 \in R^m$ and $T \geq t_0$ $V(\emptyset; x_0, T-t_0) = \emptyset$;*

2. *for the maximal coalition N and any $x_0 \in R^m$ and $T \geq t_0$ $V(N; x_0, T-t_0) = \{J(x(\cdot)) \mid x(\cdot) \in X_N^{T-t_0}(x_0)\}$;*

3. *for any fixed $x_0 \in R_m$ and $T \geq t_0$ the mapping is superadditive in S, i.e. for the arbitrary nonintersecting coalitions S and R.*

$$V(S \bigcup R; x_0, T-t_0) \supset V(S; x_0, T-t_0) \bigcup V(R; x_0, T-t_0)$$

Proof: Since $\mathcal{D}_\emptyset = \emptyset$, then we get sequentially $Y_\emptyset^{T-t_0}(x_0) = \emptyset$, $X_\emptyset^{T-t_0}(x_0) = \emptyset$, $V(\emptyset; x_0, T-t_0) = \emptyset$.

By the definition of the reachability region, for each $y \in C^{T-t_0}(x_0)$ there is a strategy $\bar{u}_N(\cdot) \in \mathcal{D}_N$ such that $\rho(x(T; x_0, \bar{u}_N(\cdot)), y) = 0$, hence $Y_N^{T-t_0}(x_0) = C^{T-t_0}(x_0)$ and, consequently, $V(N; x_0, T-t_0) = \{J(x(\cdot)) \mid x(\cdot) \in X_N^{T-t_0}(x_0)\}$ where $X_N^{T-t_0}(x_0) = \{x(\cdot) \mid x(t_0) = x_0,\ x(T) \in C^{T-t_0}(x_0)\}$ is the set of all trajectories to be realized in the game $\Gamma(x_0, T-t_0)$

We will now prove the superadditivity. Show that $S' \subset S$ implies $X_{S'}^{T-t_0}(x_0) \subset X_S^{T-t_0}(x_0)$. Let $J \in V(S'; x_0, T-t_0)$. Then there exists a trajectory $x(\cdot) \in X_{S'}^{T-t_0}(x_0)$ for which $J_i = J_i(x(\cdot))$, $i = 1, \ldots, n$. Since $S' \subset S$, then $x(\cdot) \in X_S^{T-t_0}(x_0)$ (the following strategy is always possible: in the game $\Gamma_S^y(x_0, T-t_0)$, the players from S' employ the strategy that is ϵ–optimal in the game

$\Gamma^y_{S'}(x_0, T-t_0)$ whereas the players from $S \setminus S'$ employ the truncation on the set $S \setminus S'$ of the ϵ–optimal strategies of the coalition $N \setminus S$ in the same game). Consequently, $J = J(x(\cdot)) \in V(S; x_0, T-t_0)$ and $V(S'; x_0, T-t_0) \subset V(S; x_0, T-t_0)$. For any $S, R \subset N$, $S \bigcap R = \emptyset$, by what has been proved we have $V(S; x_0, T-t_0) \subset V(S \bigcup R; x_0, T-t_0)$, $V(R; x_0, T-t_0) \subset V(S \bigcup R; x_0, T-t_0)$. But then $V(S; x_0, T-t_0) \bigcup V(R; x_0, T-t_0) \subset V(S \bigcup R; x_0, T-t_0)$. This completes the proof of the lemma. ■

Using the set $\mathcal{D}_N$ of all admissible situations in the game $\Gamma(x_0, T-t_0)$ one can write $V(N; x_0, T-t_0) = \{J(x_0, u_N(\cdot)) = \{J_i(x_0, u_N(\cdot))\},\ i \in N \mid u_N(\cdot) \in \mathcal{D}_N\}$. For each coalition $S \subset N$, the set $V(S; x_0, T-t_0)$ may be interpreted as a set of all payoff vectors which can be realized in the game $\Gamma(x_0, T-t_0)$ only by the efforts of the members of the coalition S irrespective of the behavior of the players from the set $N \setminus S$. That is, this is the set of the payoff vectors $J \in V(N; x_0, T-t_0)$ whose corresponding components may be guaranteed by the coalition S to its members. From the superadditivity of the mapping V in S it follows that for $S \subset S'$, $V(S; x_0, T-t_0) \subset V(S'; x_0, T-t_0)$,i.e. the larger the coalition the wider its opportunities. Thus, the players can obtain the largest–benefit by uniting in the maximal coalition N.

Definition 2. The set $V(S; x_0, T-t_0)$ defined by (7.4.1) is called a characteristic set of the coalition S in the game $\Gamma(x_0, T-t_0)$

The mapping V is the analog of the characteristic function in the cooperative game with side payments and can be used in the definition of solutions of the cooperative differential game without side payments.

The pair $< N, V(S; x_0, T-t_0) >$,where N is the set of players, and $V$the mapping defined by (7.4.1.), is called the cooperative differential game in the form of characteristic sets. For short,we denote it by $\Gamma_V(x_0, T-t_0)$.

Proceeding from the minimax principle, we constructed a characteristic set. As is shown by Lemma, it is superadditive in S at all times. Note that the characteristic set may also be constructed by employing some other methods. However, it should always have the meaning of the guaranteed payoff set for coalition and possess superadditivity in S and $V(\emptyset; x_0, T-t_0) = \emptyset$.

In the game $\Gamma_V(x_0, T-t_0)$ the players are facing the problem of realization of (in a sense) optimal payoff vector $J = (J_1, \ldots, J_n)$ from the set $V(N; x_0, T-t_0)$. The set of payoff vectors satisfying some optimality principle is called the solution of the game $\Gamma_V(x_0, T-t_0)$ and is denoted by $W_V(x_0, T-t_0) \subset V(N; x_0, T-t_0)$. Specify now the form of the set $W_V(x_0, T-t_0)$.

Definition 3. The vector $J \in V(N; x_0, T-t_0)$ $(J = (J_1, \ldots, J_n))$ is called the imputation in the game $\Gamma_V(x_0, T-t_0)$ if

$$J_i \geq \max\{J'_i \mid J' \in V(\{i\}; x_0, T-t_0)\},\ i = 1, \ldots, n \qquad (7.4.3)$$

The condition (7.4.3) reflects the individual rationality of the imputations vector.The set of all imputations in the game $\Gamma_V(x_0, T-t_0)$ is denoted by $E_V(x_0, T-t_0)$. Each imputation is a payoff vector and $E_V(x_0, T-t_0) \subset V(N; x_0, T-t_0)$.

Definition 4 Let $J', J'' \in E_V(x_0, T-t_0)$. We say that the imputationJ' dominates the imputation J'' for the coalition $S(J' \succ_S J'')$ if

1. $J'_i > J''_i$ for all $i \in S$;

2. $J' \in V(S; x_0, T-t_0)$.

The imputation J' is said to dominate the imputation J'' $(J' > J'')$if there is a nonempty coalition $S \subset N$ such that $J' \succ_S J''$.

Note that in the untransferable payoff games the domination is possible for all coalitions containing more than one player, and for the coalition N.

Definition 5. By the c–kernel in the game $\Gamma_V(x_0, T-t_0)$ without side payments is meant the set of nondominated imputation and is denoted by $C_V(x_0, T-t_0)$

Definition 6. By the NM–solution in the game $\Gamma_V(x_0, T-t_0)$ without side payments is meant a subset $L_V(x_0, T-t_0)$ of the imputation set $E_V(x_0, T-t_0)$ for which the following conditions are satisfied:

1. if $J', J'' \in L_V(x_0, T-t_0)$ it follows that J' and J'' do not dominate one another;

2. for any $J \in [E_V(x_0, T-t_0) \setminus L_V(x_0, T-t_0)]$ there is an imputation $\bar{J} \in L_V(x_0, T-t_0)$ such that $\bar{J}$ dominates J.

Theorem 7 *The imputation J belongs to the c–kernel $C_V(x_0, T-t_0)$ of the game $\Gamma_V(x_0, T-t_0)$ if and only if, for each coalition $S \subset N$ and each imputation $J' \in V(S; x_0, T-t_0)$ there exists such $i \in S$ that $J_i \geq J'_i$.*

Proof: If $J \in C_V(x_0, T-t_0)$ and for a coalition $S \subset N$ there exist such an imputation $J' \in V(S; x_0, T-t_0)$ that $J'_i > J_i$ for all $i \in S$, then$J' \succ_S J$. This disagrees with the belonging of the imputation J to the c–kernel. If, however the imputation J has the property indicated in the condition of the theorem, then it is nondominated, i.e. it belongs to the c–kernel. ■

Theorem 8 *If there exist the c–kernel $C_V(x_0, T-t_0)$ and the NM–solution $L_V(x_0, T-t_0)$ in the game $\Gamma_V(x_0, T-t_0)$, then $C_V(x_0, T-t_0) \subset L_V(x_0, T-t_0)$.*

The proof follows from the definition of domination and from that of the sets $C_V(x_0, T-t_0)$, $L_V(x_0, T-t_0)$.

Definition 7. The situation $u_N^p(\cdot) \in \mathcal{D}_N$ is called Pareto optimal in the game $\Gamma_V(x_0, T-t_0)$ if for any situation $u_N(\cdot) \in \mathcal{D}_N$ the condition $J_i(x_0, u_N(\cdot)) \geq J_i(x_0, u_N^p(\cdot))$ for all $i = 1, \ldots, n$ implies $J_i(x_0, u_N(\cdot)) = J_i(x_0, u_N^p(\cdot))$ for all $i = 1, \ldots, n$.

As is seen from the latter definition, any deviation from the Pareto optimal situation involves a decrease in the payoff for at least one player (at the same time, the other players' payoffs may increase). Let$\mathcal{D}_N^p$ be the set of all Pareto optimal situations in the game $\Gamma_V(x_0, T-t_0)$. Introduce the set

$$P(x_0, T-t_0) = \{J(x_0, u_N(\cdot)) \mid u_N(\cdot) \in \mathcal{D}_N^p\}$$

which is called the Pareto optimal set of payoff vectors in the game $\Gamma_V(x_0, T-t_0)$. Evidently, $P(x_0, T-t_0) \subset V(N; x_0, T-t_0)$. The vectors from the set $P(x_0, T-t_0)$ do not dominate one another for the coalition N.

7.5 Classification of dinamic stable (time – consistant) solutions

As is seen from Definition 3, any imputation $K \in E_V(x_0, T-t_0)$ is a payoff vector in the game $\Gamma_V(x_0, T-t_0)$ and hence is realized along a trajectory $x(\cdot)$, i.e. $K = J(x(\cdot))$. Along the trajectory $x(\cdot)$, the imputation K is unique because of the lack of side payments. And each imputation from the set $E_V(x_0, T-t_0)$ corresponds to its own trajectory. Hence the conditionally optimal trajectory in the game $\Gamma_V(x_0, T-t_0)$ has to be defined as a trajectory along which one of the imputations from the solution $W_V(x_0, T-t_0)$ is realized.

Definition 8. Let $W_V(x_0, T-t_0) \neq \emptyset$. Any trajectory $\bar{x}(\cdot)$ of the system (7.3.1.)–(7.3.2.) such that $J(\bar{x}(\cdot)) \in W_V(x_0, T-t_0)$ will be called the conditionally optimal trajectory in the cooperative differential game $\Gamma_V(x_0, T-t_0)$ without side payments.

In Definition 8, $J(\bar{x}(\cdot)) = \{J_i(\bar{x}(\cdot)),\ i = 1, \ldots, n\}$,where $J_i(\bar{x}(\cdot)) = \int_{t_0}^{T} h_i(\bar{x}(t))dt + H_i(\bar{x}(T))$, $i = 1, \ldots, n$, are the players' payoffs along the trajectory $\bar{x}(\cdot)$ (see 7.3.13).

Suppose that in the initial state of the game $\Gamma_V(x_0, T-t_0)$ an optimality principle is chosen and the solution $W_V(x_0, T-t_0)$ is constructed in compliance with it. Let $\bar{x}(\cdot)$ be a conditionally optimal trajectory, i.e. $J(\bar{x}(\cdot)) \in W_V(x_0, T-t_0)$. Assume that the game $\Gamma_V(x_0, T-t_0)$ proceeds along the trajectory $\bar{x}(\cdot)$, i.e. consider the current family of subgames $\{\Gamma_V(\bar{x}(t), T-t),\ t_0 \leq t \leq T\}$ defined along the trajectory $\bar{x}(\cdot)$. The current game $\Gamma_V(\bar{x}(t), T-t)$ is defined as follows. The following sets are constructed for each coalition $S \subset N$:

$$Y_S^{T-t}(\bar{x}(t)) = \{z \in C^{T-t}(\bar{x}(t)) \mid val\Gamma_S^z(\bar{x}(t), T-t) = 0\},$$

$$X_S^{T-t}(\bar{x}(t)) = \{y(\cdot) \mid y(t) = \bar{x}(t),\ y(T) \in Y_S^{T-t}(\bar{x}(t))\}.$$

Here $C^{T-t}(\bar{x}(t))$ is the reachability set of system from the initial state $\bar{x}(t)$ at the time instant $T-t$; $\Gamma_S^z(\bar{x}(t), T-t)$ is a zero–sum game between the coalition S and $N\backslash S$ described by the equation $\dot{y} = f(y, v_S(\cdot), v_{N\backslash S}(\cdot))$, $v_S(\cdot) \in \mathcal{D}_S[t,T]$, $v_{N\backslash S}(\cdot) \in \mathcal{D}_{N\backslash S}[t,T]$, $(\mathcal{D}_S[t,T] \prod_{i\in S} \mathcal{D}_i[t,T]$ is the set of admissible strategies for the coalition S defined in the restricted region $[t,T])$, the game duration being $T-t$ and the initial state $y(t) = \bar{x}(t)$. In this game, the objective of Player S is to minimize the distance $\|y(T) - z\|$ by the instant T (the objective of Player $N \setminus S$ is he opposite one), where $z \in C^{T-t}(\bar{x}(t))$is a fixed point of the reachability set in the current game $\Gamma_V(\bar{x}(t), T-t)$.

We will now construct the set

$$V(S; \bar{x}(t), T-t) = \{J(y(\cdot)) \mid y(\cdot) \in X_S^{T-t}(\bar{x}(t))\},$$

where $J(y(\cdot)) = \{J_i(y(\cdot)),\ i = 1, \ldots, n\}$,

$$J_i(y(\cdot)) = \int_t^T h_i(y(\tau))d\tau + H_i(y(T)),\ i = 1, \ldots, n.$$

The construction of the characteristic sets $V(S; \bar{x}(t), T-t)$ for each $S \in 2^N$ exactly defines the current cooperative game $\Gamma_V(\bar{x}(t), T-t) =< N, V(S; \bar{x}(t), T-t) >$.

Let $E_V(\bar{x}(t), T-t)$ be a imputation set, i.e. $E_V(\bar{x}(t), T-t) = \{J \in V(N; \bar{x}(t), T-t) \mid J_i \geq \max\{J_i' \mid J' \in V(\{i\}; \bar{x}(t), T-t)\},\ i = 1, \ldots, n\}$ and $W_V(\bar{x}(t), T-t) \subset E_V(\bar{x}(t), T-t)$ be a solution to the game $\Gamma_V(\bar{x}(t), T-t)$ generated by the same optimality principle as the solution $W_V(x_0, T-t_0)$ of the game $\Gamma_V(x_0, T-t_0)$. The current game $\Gamma_V(\bar{x}(T), 0)$ is of zero duration, and hence a unique payoff vector defined in the initial state $\bar{x}(T)$ exists therein. For this reason, we assume that the set $W_V(\bar{x}(T), 0)$, which is a solution of the current game $\Gamma_V(\bar{x}(T), 0)$, is composed of a unique imputation $H(\bar{x}(T)) = \{H_i(\bar{x}(T)),\ i = 1, \ldots, n\}$, where $H_i(\bar{x}(T))$ is a terminal part of the player i's payoff (see (7.3.13)).

For the trajectory $\bar{x}(\cdot)$ to be realized in the game $\Gamma_V(x_0, T-t_0)$, it should be conditionally optimal in the initial sense in all current subgames (otherwise the players have no reason to keep to this trajectory until the game terminates). In other words, the conditionally optimal trajectory $\bar{x}(\cdot)$ in the game $\Gamma_V(x_0, T-t_0)$ can be realized only when the imputation $J(\bar{x}(\cdot)) \in W_V(x_0, T-t_0)$, has the property of dynamic stability.

Dynamic stable solution. Let $\bar{x}(\cdot)$ be a conditionally optimal trajectory in the game $\Gamma_V(x_0, T-t_0)$. For $t \in [t_0, T]$ it is denoted by $\bar{x}(\cdot)[t, T]$. If the game $\Gamma_V(x_0, T-t_0)$ proceeds along the trajectory $\bar{x}(\cdot)$ until the instant t, then the player i's payoff on the time interval $[t_0, t]$ is equal to

$$J_i(\bar{x}(\cdot)[t_0, t]) = \int_{t_0}^{t} h_i(\bar{x}(\tau))d\tau.$$

for the imputation $J(\bar{x}(\cdot)) \in W_V(x_0, T-t_0)$ to be realized in the game $\Gamma_V(x_0, T-t_0)$, it is essential that each instant $t_0 \le t \le T$ the vector

$$J(\bar{x}(\cdot)[t, T]) = \int_{t}^{T} h(\bar{x}(\tau))d\tau + H(\bar{x}(T))$$

representing the players' payoffs in the game $\Gamma_V(\bar{x}(t), T-t)$ should be the imputation in the game $\Gamma_V(\bar{x}(t), T-t)$ belong to its solution $W_V(\bar{x}(t), T-t)$. In this case, at each instant the part $J(\bar{x}(\cdot)[t, T]) \in W_V(\bar{x}(t), T-t)$ of the imputation $J(\bar{x}(\cdot)[t_0, T]) \in W_V(x_0, T-t_0)$ is optimal in the initial sense and hence the players have no reason to reject orientation towards its realization in the current game $\Gamma_V(\bar{x}(t), T-t)$, $t_0 \le t \le T$. And this means that during the game the players have no intention to reject the conditionally optimal trajectory $\bar{x}(\cdot)$ and its corresponding imputation $J(\bar{x}(\cdot)) \in W_V(x_0, T-t_0)$.

Definition 9. The imputation $J(\bar{x}(\cdot)) \in W_V(x_0, T-t_0)$ is called dynamic stable in the game $\Gamma_V(x_0, T-t_0)$ if

1. along the conditionally optimal trajectory $\bar{x}(\cdot)$ $W_V(\bar{x}(t), T-t) \neq \emptyset$, $t_0 \le t \le T$;
2. $$J(\bar{x}(\cdot)) \in \bigcap_{t_0 \le t \le T} [J(\bar{x}(\cdot)[t_0, t]) \oplus W_V(\bar{x}(t), T-t)].$$

We say that the cooperative differential game $\Gamma_V(x_0, T-t_0)$ without side payments has a dynamic stable solution $W_V(x_0, T-t_0)$ if all the imputations from $W_V(x_0, T-t_0)$ are dynamic stable. Any conditionally optimal trajectory, along which there exists a dynamic stable imputation entering into the solution, will be called an optimal trajectory.

The dynamic stable solution can be placed in correspondence with the optimal trajectory set $\bar{X}$: each imputation from the dynamic stable solution $W_V(x_0, T-t_0)$ corresponds to at least one optimal trajectory, so that

$$W_V(x_0, T-t_0) = \{J(\bar{x}(\cdot)) \mid \bar{x}(\cdot) \in \bar{X}\}$$

Let $J', J'' \in W_V(x_0, T-t_0)$ be dynamic stable imputations. Then there exist such trajectories $\bar{x}(\cdot), \bar{\bar{x}}(\cdot) \in \bar{X}$ that $J' = J(\bar{x}(\cdot))$, $J'' = J(\bar{\bar{x}}(\cdot))$ and at any instant $t_0 \leq t \leq T$

$$J(\bar{x}(\cdot)) \in [J(\bar{x}(\cdot))[t_0, t]) \oplus W_V(\bar{x}(\cdot), T - t)],$$

$$J(\bar{\bar{x}}(\cdot)) \in [J(\bar{\bar{x}}(\cdot))[t_0, t]) \oplus W_V(\bar{\bar{x}}(t), T - t)].$$

Let $a \in R^n$, $A \subset R^n$, then by $a \oplus A$ we understand the set of all vectors ofthe form $a + z$, where $z \in A$ ($a \oplus A = a + z : z \in A$). Here, in general, $J(\bar{x}(\cdot)[t_0, t]) \neq J(\bar{\bar{x}}(\cdot)[t_0, t])$, and the sets $W_V(\bar{x}(t), T - t)$ and $W_V(\bar{\bar{x}}(t), T - t)$, as solutions of different games $\Gamma_V(\bar{x}(t), T - t)$ and $\Gamma_V(\bar{\bar{x}}(t), T - t)$, are not related to one another.

Evidently, the imputation $J(\bar{x}(\cdot))$ is dynamic stable in the game $\Gamma_V(x_0, T - t_0)$ if, at each instant $t_0 \leq t \leq T$ the vector $J(\bar{x}(\cdot)[t, T])$ is a dynamic stable imputation in the current game $\Gamma_V(\bar{x}(t), T - t)$.

Strongly–dynamic stable solution. The dynamic stability of the imputation $J(\bar{x}(\cdot)) \in W_V(x_0, T - t_0)$ implies that the inclusion

$$J(\bar{x}(\cdot), [t, T]) \in W_V(\bar{x}(t), T - t) \tag{7.5.1}$$

is satisfied at each instant $t_0 \leq t \leq T$, i.e.on each interval $[t, T]$ the remaining portion $\bar{x}(\cdot)[t, T]$ of the trajectory $\bar{x}(\cdot)$ is to be optimal in the game $\Gamma_V(\bar{x}(t), T - t)$. If, together with (7.5.1), at very instant $t_0 \leq t \leq T$ for all conditionally optimal trajectories $\tilde{x}_t(\cdot)$ in the game $\Gamma_V(\bar{x}(t), T - t)$ there is the inclusion

$$J(\bar{x}(\cdot)[t_0, t]) + J(\tilde{x}_t(\cdot)) \in W_V(x_0, T - t_0)$$

then a "good" continuation of the game after the instant t is not only the imputation $J(\bar{x}(\cdot)[t, T]) \in W_V(\bar{x}(t), T - t)$ but also any imputation $J(\tilde{x}_t(\cdot)) \in W_V(\bar{x}(t), T - t)$. By slightly strengthening this requirement, we obtain new concept of dynamic stability which is called the strongly dynamic stability.

Definition 10. The imputation $J(\bar{x}(\cdot)) \in W_V(x_0, T - t_0)$ is called strongly dynamic stable in the game $\Gamma_V(x_0, T - t_0)$ if:

1. the imputation $J(\bar{x}(\cdot))$ is dynamic stable;
2. for any $t_0 \leq t_1 \leq t_2 \leq T$

$$[J(\bar{x}(\cdot)[t_0, t_2]) \oplus W_V(\bar{x}(t_2), T - t_2)] \subset$$

$$\subset [J(\bar{x}(\cdot)[t_0, t_1]) \oplus W_V(\bar{x}(t_1), T - t_1)] \tag{7.5.2}$$

We say that the cooperative differential game $\Gamma_V(\bar{x}(t), T - t)$ without side payments has a strongly dynamic stable solution $W_V(x_0, T - t_0)$ if all the imputations from $W_V(x_0, T - t_0)$ are strongly–dynamic stable. The conditionally optimal trajectory, along which a strongly–dynamic stable imputation from the solution exists, is called the strongly optimal trajectory.

From (7.5.2), for $t_2 = t$, $t_1 = t_0$ we have

$$J(\bar{x}(\cdot)[t_0, t]) \oplus W_V(\bar{x}(t), T - t) \subset W_V(x_0, T - t_0), \; t_0 \leq t \leq T;$$

for $t_2 = T$, $t_1 = t$ we obtain

$$J(\bar{x}(\cdot)[t_0, T]) \oplus W_V(\bar{x}(T), 0) \subset J(\bar{x}(\cdot)[t_0, t]) \oplus W_V(\bar{x}(t), T - t), \; t_0 \leq t \leq T,$$

or

$$\left[\int_{t_0}^{T} h(\bar{x}(t))dt + H(\bar{x}(T))\right] \in [J(\bar{x}(\cdot)[t_0, t]) \oplus W_V(\bar{x}(t), T - t), \; t_0 \leq t \leq T.$$

Remark. Let $J(\bar{x}(\cdot)) \in W_V(x_0, T - t_0)$ be a strongly–dynamic stable imputation, and $\tilde{x}_t(\cdot)$ an arbitrary conditionally optimal trajectory in the current game $\Gamma_V(\bar{x}(t), T - t)$. From the definition of the strongly dynamic stability it follows that $J(\bar{x}(\cdot)[t_0, t]) + J(\tilde{x}_t(\cdot)) \in W_V(x_0, T - t_0)$. The vector $J(\tilde{x}_t(\cdot))$ is realized in the game $\Gamma_V(\bar{x}(t), T - t)$ along the trajectory $\tilde{x}_t(\cdot)$, hence the vector $J(\bar{x}(\cdot)[t_0, t]) + J(\tilde{x}_t(\cdot))$ will be realized in the game $\Gamma_V(x_0, T - t_0)$ if the following trajectory is realized there

$$\hat{x}(\cdot) = \begin{cases} \bar{x}(\cdot) & \text{on } [t_0, t], \\ \tilde{x}_t(\cdot) & \text{on } [t, T]. \end{cases}$$

The existence of a strongly dynamically stable solution in the cooperative differential game without side payments involves fulfilment of rather rigid conditions and hence is often empty.

Terminal payoffs. Consider the case when in (7.3.13) $h_i \equiv 0$, $i = 1, \ldots, n$. Then it follows from the nontransferability of the players' payoffs that to each point $x \in C^{T-t_0}(x_0)$ corresponds one vector of the payoffs $H(x) = \{H_i(x),\; i = 1, \ldots, n\}$, where $C^{T-t_0}(x_0)$ is the reachability set of system (7.3.2)–(7.3.3) on the interval $[t_0, T]$. Therefore the imputations in the game $\Gamma_V(x_0, T - t_0)$ with terminal payoffs will be realized in the final states (points) rather than along the trajectories, accordingly.

$$V(N; x_0, T - t_0) = \{H(x) \mid x \in C^{T-t_0}(x_0)\}.$$

The conditionally optimal trajectory is defined as follows.

Definition 11. Let $W_V(x_0, T - t_0) \neq \emptyset$. Any trajectory of system (7.3.2)–(7.3.3) such that $H(\bar{x}(T)) \in W_V(x_0, T - t_0)$ is called the conditionally optimal trajectory in the game $\Gamma_V((x_0, T - t_0)$ with terminal payoffs.

Let $\bar{x}(\cdot)$be the conditionally optimal trajectory. Consider the current games $\Gamma_V(\bar{x}(t), T-t)$, $t_0 \leq t \leq T$ along this trajectory. The game $\Gamma_V(\bar{x}(t), T-t)$ differes from the game $\Gamma_V(x_0, T-t_0)$ in the initial state and duration, although the players' payoff functions therein are the same as in the game $\Gamma_V(x_0, T-t_0)$, i.e. $J_i = H_i(\bar{x}(T)), i = 1, \ldots, n$.

As before, the solution of the current game $\Gamma_V(\bar{x}(t), T-t)$ is denoted as $W_V(\bar{x}(t), T-t)$.

Definition 12. The imputation $H(y) \in W_V(x_0, T-t_0)$ is called dynamic stable in the game $\Gamma_V(x_0, T-t_0)$ with terminal payoffs if

1. there exists such conditionally optimal trajectory $\bar{x}(\cdot)$ that $\bar{x}(T) = y$ and $W_V(\bar{x}(t), T-t) \neq \emptyset$, $t_0 \leq t \leq T$;

2. $$H(y) \in \bigcap_{t_0 \leq t \leq T} W_V(\bar{x}(t), T-t).$$

Definitions of the dynamic stable solution and optimal trajectory in the game $\Gamma_V(x_0, T-t_0)$ with terminal payoffs are the same as in general case.

Let
$$\bar{X} = \{\bar{x}(\cdot) \mid H(\bar{x}(T)) \in W_V(\bar{x}(t), T-t)\ , t_0 \leq t \leq T\}$$
be a set of optimal trajectories in the game $\Gamma_V(x_0, T-t_0)$. The dynamic stability of the solution $W_V(x_0, T-t_0)$ means that

$$W_V(x_0, T-t_0) \subset \bigcup_{\bar{x}(\cdot) \in X} \left[\bigcap_{t_0 \leq t \leq T} W_V(\bar{x}(t), T-t) \right].$$

Definition 13. We say that the game with terminal payoff $\Gamma_V(x_0, T-t_0)$ has a strongly dynamic stable solution $W_V(x_0, T-t_0)$if:

1. it is dynamic stable;

2. for any optimal trajectory $\bar{x}(\cdot)$ and any $t_0 \leq t_1 \leq t_2 \leq T$
$$W_V(\bar{x}(t_2), T-t_2) \subset W_V(\bar{x}(t_1), T-t_1).$$

In this case the conditionally optimal trajectory is called a strongly optimal trajectory.

Note that if $W_V(x_0, T-t_0)$ is strongly–dynamic stable, and $H(\bar{x}(T))$, $H(\bar{\bar{x}}(T)) \in W_V(x_0, T-t_0)$, then at any instant $t_0 < t \leq T$ the sets $W_V(\bar{x}(t), T-t)$ and $W_V(\bar{\bar{x}}(t), T-t)$ may be arbitrarily with respect to one another, but

$$W_V(\bar{x}(t), T-t) \subset W_V(\bar{x}(\tau), T-\tau),$$

$$W_V(\bar{\bar{x}}(t), T-t) \subset W_V(\bar{\bar{x}}(\tau), T-\tau)$$

for all $\tau < t$ and, in particular, for $\tau = t_0$.

Regularization of the game. The players are actually unable to affect the dynamic stability of the solution and restrict themselves to the choice of a conditionally optimal trajectory only. If along a conditionally optimal trajectory there is no dynamic stability of the solution,then because nontransferability the players cannot change this circumstances by effecting mutual (side) payments. To solve this problem, we perform regularization of the game $\Gamma_V(x_0, T-t_0)$ by providing somewhatwider opportunities for the players.

Consider the integral payoff game, i.e. in (7.3.13.) we set $H_i \equiv 0$, $i = 1, \dots, n$.

Let $\bar{x}^0(\cdot)$ be such conditionally optimal trajectory in the game $\Gamma_V(x_0, T-t_0)$ that

$$W_V(\bar{x}^0(t), T-t) \neq \emptyset, \; t_0 \leq t \leq T \tag{7.5.3}$$

Assume that the imputation $J(\bar{x}^0(\cdot)) \in W_V(x_0, T-t_0)$ is dynamic unstable, and there exists the first instant $t_1 > t_0$ when the dynamic stability of this imputation is disturbed, i.e.

$$\int_{t_1}^{T} h(\bar{x}^0(t))dt \notin W_V(\bar{x}^0(t_1), T-t_1).$$

Let us choose an arbitrary imputation

$$\int_{t_1}^{T} h(\bar{x}^1(t))dt \in W_V(\bar{x}^0(t_1), T-t_1) \tag{7.5.4}$$

where $\bar{x}^1(\cdot)$ is conditionally optimal trajectory in the current game $\Gamma_V(\bar{x}^0(t_1), T-t_1)$ $(\bar{x}^1(t_1) = \bar{x}^0(t_1)$, $\bar{x}^1(t) \neq \bar{x}^0(t)$, $t > t_1)$. Denote

$$\eta_{t_1} = \int_{t_1}^{T} [h(\bar{x}^1(t)) - h(\bar{x}^0(t))]dt \tag{7.5.5}$$

From (7.5.4) and (7.5.5) it follows that

$$\eta_{t_1} + \int_{t_1}^{T} h(\bar{x}^0(t))dt \in W_V(\bar{x}^0(t_1), T-t_1).$$

Note that there are no restriction for the sign of the components of the vector η_{t_1}. Set

$$J^1(\bar{x}^0(\cdot)[t_0, t_1]) = \int_{t_0}^{t_1} h(\bar{x}^0(t))dt - \eta_{t_1},$$

$$J^1(\bar{x}^0(\cdot)(t_1, T]) = \int_{t_1}^{T} h(\bar{x}^1(t))dt.$$

Evidently, $J^1(\bar{x}^0(\cdot)) = J(\bar{x}^0(\cdot))$. Let

$$J^1(\bar{x}^0(\cdot)[t,T]) = \int_t^T h(\bar{x}^1(\tau))d\tau,\ t \in (t_1, T],$$

and let there be the first instant of time $t_2 > t_1$ such that

$$J^1(\bar{x}^0(\cdot)[t,T]) \in W_V(\bar{x}^0(t), T-t),\ t \in (t_1, t_2),$$

and

$$J^1(\bar{x}^0(\cdot)[t_2,T]) \notin W_V(\bar{x}^0(t_2), T-t_2).$$

We choose an arbitrary imputation

$$\int_{t_2}^T h(\bar{x}^2(t))dt \in W_V(\bar{x}^0(t_2), T-t_2) \tag{7.5.6}$$

where $\bar{x}^2(\cdot)$ is a conditionally optimal trajectory in the current game $\Gamma_V(\bar{x}^0(t_2), T-t_2)$ $(\bar{x}^2(t_2) = \bar{x}^0(t_2),\ \bar{x}^2(t) \neq \bar{x}^0(t), t > t_2)$. Denote

$$\eta_{t_2} = \int_{t_2}^T [h(\bar{x}^2(t)) - h(\bar{x}^1(t))]dt \tag{7.5.7}$$

From (7.5.6) and(7.5.7) it follows that

$$\eta_{t_2} + \int_{t_2}^T h(\bar{x}^1(t))dt \in W_V(\bar{x}^0(t_2), T-t_2).$$

Note that there are no restrictions for the sign of components of the vector η_{t_2}. Set

$$J^2(\bar{x}^0(\cdot)[t_0,t_2]) = J^1(\bar{x}^0(\cdot)[t_0,t_1]) + \int_{t_1}^{t_2} h(\bar{x}^1(t))dt - \eta_{t_2},$$

$$J^2(\bar{x}^0(\cdot)(t_2,T]) = \int_{t_2}^T h(\bar{x}^2(t))dt.$$

It can be readily seen that $J^2(\bar{x}^0(\cdot)) = J(\bar{x}^0(\cdot))$. Let

$$J^2(\bar{x}^0(\cdot)[t,T]) = \int_t^T h(\bar{x}^2(\tau))d\tau,\ t \in (t_2, T],$$

and let there be the first instant of time $t_3 > t_2$ such that

$$J^2(\bar{x}^0(\cdot)[t,T]) \in W_V(\bar{x}^0(t), T-t),\ t \in (t_2, t_3),$$

and

$$J^2(\bar{x}^0(\cdot)[t_3,T]) \notin W_V(\bar{x}^0(t_3), T-t_3).$$

Assume that the further continuation of this process makes it possible to obtain a sequence of time instants $t_1 < t_2 < t_3 < \dots$ and imputations $J(\bar{x}^0(\cdot)[t,T])$ for $t \in [t_0, t_1)$, $J^1(\bar{x}^0(\cdot)[t,T])$ for $t \in (t_1, t_2)$, $J^2(\bar{x}^0(\cdot)[t,T])$ for $t \in (t_2, t_3), \dots$, $J^k(\bar{x}^0(\cdot)[t,T])$ for $t \in (t_k, t_{k+1}), \dots$, for which we have the inclusion $J^k(\bar{x}^0(\cdot)[t,T]) \in W_V(\bar{x}^0(t), T-t)$ for $t \in (t_k, t_{k+1})$, $k = 0, 1, \dots$, and

$$J^k(\bar{x}^0(\cdot)[t_{k+1},T]) \notin W_V(\bar{x}^0(t_{k+1}), T - t_{k+1}).$$

Also, assume that there exists the time instant $t_\ell < T$ for which $J^\ell(\bar{x}^0(\cdot),[t,T]) \in W_V(\bar{x}^0(t), T-t)$ for all $t \in [t_\ell, T]$.

Denote

$$\eta_{t_k} = \int_{t_k}^{T} [h(\bar{x}^k)) - h(\bar{x}^{k-1}(t))]dt$$

and define the function γ^t by the formula

$$\gamma^t = \sum_{r=0}^{k-1} \int_{t_r}^{t_{r+1}} h(\bar{x}^r(\tau))d\tau - \sum_{r=1}^{k} \eta_{t_r} + \int_{t_k}^{t} h(\bar{x}^k(\tau))d\tau,$$

$$t \in [t_k, t_{k+1}),\ k = 0, 1, \dots, \ell\ (t_{\ell+1} = T).$$

Let us introduce a vector function $\beta(t) = (\beta_1(t), \dots, \beta_n(t))$, defined on the intervals $[t_k, t_{k+1})$, $k = 0, 1, \dots, \ell$ $(t_{\ell+1} = T)$ by the formulas

$$\begin{aligned}
&\beta(t) = 1, && t \in [t_0, t_1);\\
&\beta(t) = \tfrac{h(\bar{x}^1(t))}{h(\bar{x}^0(t))} > 0, && t \in [t_1, t_2);\\
&\dots\dots && \dots\\
&\beta(t) = \tfrac{h(\bar{x}^k(t))}{h(\bar{x}^0(t))} > 0, && t \in [t_k, t_{k+1});\\
&\dots\dots && \dots
\end{aligned}$$

Then the expression for γ^t may be written in the form

$$\gamma^t = \sum_{r=0}^{k-1} \int_{t_r}^{t^{r+1}} \beta(\tau)h(\bar{x}^0(\tau))d\tau - \sum_{r=1}^{k} \eta_{t_r} + \int_{t_k}^{t} \beta(\tau)h(\bar{x}^0(\tau))d\tau,$$

$$t \in [t_k, t_{k+1}),\ k = 0, 1, \dots, \ell\ (t_{\ell+1} = T).$$

We may show that for any $t \in [t_0, T]$

$$J(\bar{x}^0(\cdot)) - \gamma^t \in W_V(\bar{x}_0(t), T-t).$$

The proof of this assertion is left to the reader.

Now, because of the players payoffs being corrected at discrete time instants $t_1, \dots, t_\ell$ by the quantities $\eta_{t_1}, \dots, \eta_{t_\ell}$ and regularized with the use of multipliers $\beta(t)$ for $t \in [t_k, t_{k+1})$, the conditions for the belonging of the remaining part of the imputation $J(\bar{x}^0(\cdot)) - \gamma^t$ to the current game solution are satisfied. The regularization of the game is such that at the start of the game the players agree to distribute the imputation $J(\bar{x}^0(\cdot))$ on the time interval $[t_0, T]$ in such a way that the vector of the obtained payoffs on each interval $[t_0, t]$ is equal to the vector γ^t, $t \in [t_0, T]$. Then the remaining imputation $J(\bar{x}^0(\cdot)) - \gamma^t$ always belongs to the solution of the current game $\Gamma_V(\bar{x}^0(t), T-t)$ and the players have no reason to depart from the trajectory $\bar{x}^0(\cdot)$ chosen at the start of the game.

Remark. Regularization of the game is performed providing that $\bar{x}^0(t)$ $W_V(x^0(t), T-t) \neq \emptyset$, $t_0 \leq t \leq T$. If, this condition is not satisfied, then at the instant t, when $W_V(\bar{x}^0(t), T-t) = \emptyset$, in the current game $\Gamma_V(\bar{x}^0(t), T-t)$ there are no imputation satisfying the initial optimality principle (the optimality principle which has generated the solution $W_V(x_0, T-t_0)$ of the game $\Gamma_V(x_0, T-t_0)$). Therefore there is no reason to develop further the game along the trajectory $\bar{x}^0(\cdot)$. In this case we say that the game $\Gamma_V(x_0, T-t)$ has no dynamic stable solution.

7.6 Structure and dynamic stability of pareto optimal solution in the game of approaching

A typical example of untransferable payoffs in differential games accures when this payoff are the distances. Here, and in subsequent sections of this chapter, we will show an application of the above–developed approach to the differential games with such type of payoffs.

We assume that the dynamics of the n–person game is described by equations (7.3.2)–(7.3.3), and the players' objective is to obtain the closest approach to specified stationary points $M_i \in R^m$, $i = 1, \ldots, n$ by the game termination instant T. That is, the player i's payoff is of the form

$$H_i(x(T)) = -\rho(x(T), M_i), \tag{7.6.1}$$

where ρ is the Euclidean distance. For each coalition $S \subset N$ $(N = \{1, \ldots, n\})$ we construct a characteristic set $V(S; x_0, T-t_0)$ by the rule (7.4.1) setting $J_i(x(\cdot)) = H_i(x(T))$, $i = 1, \ldots, n$. The obtained cooperative game is denoted by $\Gamma_V^c(x_0, T-t_0)$ and is called the cooperative game of approaching. The game $\Gamma_V^c(x_0, T-t_0)$ is an example of cooperative differential games without side payments defined in 4 of this chapter.

Here, and in the subsequent sections, the reachability set $C^{T-t_0}(x_0)$ of the system (7.3.2)–(7.3.3) is assumed to be convex for all $x_0 \in R^m$, and compact in the sense of Euclidean metric.

In the set $C^{T-t_0}(x_0)$, for each player i there is his best (closest to M_i) point x^*:

$$\rho(x^*, M_i) = \min_{x \in C^{T-t_0}(x_0)} \rho(x, M_i).$$

None of the players, however, can gurantee achievment of his best point. Naturally, because of the dependent motion and as result of discussions at the start of the game, the players restrict themselves to a subset of the set $C^{T-t_0}(x_0)$ confined between the best points for all players. It turns out that in a number of cases exactly this set corresponds to dynamic stable solutions of the game $\Gamma_V^c(x_0, T - t_0)$ in the sense of one or another optimality principle.

We denote by π_B an operator of the orthogonal projection from R^m onto a convex compact set $B \subset R^m$.

By the orthogonal projection of the point $x \in R^m$ onto B is meant the point $\pi_B x \in B$:

$$\rho(x, \pi_B x) = \min_{y \in B} \rho(x, y).$$

We say that the point $\pi_B x$ is the image of the point x for the orthogonal projection π_B. The point $x \in R^m$ is called a pre–image. For $x \in B$, $\pi_B x = x$; for $A \subset R^m$, $\pi_B A = \{\pi_B x \mid x \in A\}$.

Let $A = conv\{A_i,\ i = 1, \ldots, n\}$, $A_i \in R^m$, where $conv\{A_i,\ i = 1, \ldots, n\}$ is a convex hull of the points $A_1, \ldots, A_n$. Recall that A is some r–dimensional polyhedron ($r \leq m$). The latter means that A is contained in some r–dimensional space R^m, but is not wholly contained in any $(r - 1)$ hyperplane. The boundary of any r–dimensional polyhedron is composed of a finite number of $(r - 1)$ dimentional polyhedrons that are called the $(r - 1)$ faces. In what follows, we need the lemmas given below.

Lemma 2 *Let B be a convex compact set in R^m, and x be a point in R^m which does not belong to B. Then the inequality $\rho(\pi_B x, y) < \rho(x, y)$ holds for all $y \in B$.*

Proof: The point $\pi_B x$ exists (since B is compact) and is boundary point. Let us take an arbitrary point $y \in B$, $y \neq \pi_B x$. Since the set B is convex, then $w_\lambda = \lambda y + (1 - \lambda)\pi_B x \in B$, $0 \leq \lambda \leq 1$. Consider the function

$$\rho^2(x, w_\lambda) = \sum_{j=1}^{m} [x^j - \lambda y^j - (1 - \lambda)\pi_B x^j]^2.$$

By the definition of the operator π_B, the point $\pi_B x$ provides a minimum of the function $\rho^2(x, w_\lambda)$. Hence

$$\frac{\partial \rho^2(x, w_\lambda)}{\partial \lambda}\Big|_{\lambda=0} = 2\sum_{j=1}^{m} (\pi_B x^j - y^j)(x^j - \pi_B x^j) \geq 0.$$

From this we have

$$-2(y - \pi_B x, x - \pi_B x) \geq 0,$$

where $(\cdot, \cdot)$ is the scalar product of vectors. Cosequently, the angle between vectors $y - \pi_B x$, $x - \pi_B x$ is not less than the right angle and is the largest of all interior angles of a triangle with vertices $x, \pi_B x, y$, i.e. $\rho(x, y) > \rho(\pi_B x, y)$. This completes the proof of the lemma. ■

Let $\bar{x}, \bar{\bar{x}} \notin w_\lambda$, where $w_\lambda = \lambda\bar{y} + (1-\lambda)\bar{\bar{y}}$, $\lambda \in [0,1]$, $\bar{x}, \bar{\bar{x}}, \bar{y}, \bar{\bar{y}} \in R^m$. Introduce the function $F(\lambda) = \rho(\bar{x}, w_\lambda) - \rho(\bar{\bar{x}}, w_\lambda)$, where

$$\rho(x, w_\lambda) = \sqrt{\sum_{j=1}^{m}[x^j - \lambda\bar{y}^j - (1-\lambda)\bar{\bar{y}}^j]^2}.$$

Lemma 3 *From the fact that* $F(\lambda) \geq 0$ *for* $\lambda = 0, 1$ *it follows that* $F(\lambda) \geq 0$ *for all* $0 < \lambda < 1$.

Proof: Since $\rho(\bar{x}, w_\lambda)$ and $\rho(\bar{\bar{x}}, w_\lambda)$ are positive quantities, then the inequality $\rho^2(\bar{x}, w_\lambda) \geq \rho^2(\bar{\bar{x}}, w_\lambda)$ implies the inequality $\rho(\bar{x}, w_\lambda) \geq \rho(\bar{\bar{x}}, w_\lambda)$ for $0 \leq \lambda \leq 1$. Therefore, for simplicity, we consider $F'(\lambda) = \rho^2(\bar{x}, w_\lambda) - \rho^2(\bar{\bar{x}}, w_\lambda)$.

For an arbitrary $\lambda = \lambda_1 \in (0,1)$ $w_{\lambda_1} = \tilde{x} = \lambda_1\bar{y} + (1-\lambda_1)\bar{\bar{y}}$ and

$$F'(\lambda_1) = \sum_{j=1}^{m}[\bar{x}^j - \lambda_1\bar{y}^j - (1-\lambda_1)\bar{\bar{y}}^j]^2 - \sum_{j=1}^{m}[\bar{\bar{x}}^j - \lambda_1\bar{y}^j - (1-\lambda_1)\bar{\bar{y}}^j]^2 =$$

$$= \sum_{j=1}^{m}[(\bar{x}^j)^2 - (\bar{\bar{x}}^j)^2 - 2(1-\lambda_1)\bar{\bar{y}}^j\bar{x}^j - 2\lambda_1\bar{y}^j\bar{x}^j + 2(1-\lambda_1)\bar{\bar{y}}^j\bar{\bar{x}}^j + 2\lambda_1\bar{y}^j\bar{\bar{x}}^j].$$

Considering the equalities

$$F'(0) = \sum_{j=1}^{m}[(\bar{x}^j)^2 - 2\bar{x}^j\bar{\bar{y}}^j - (\bar{\bar{x}}^j)^2 + 2\bar{\bar{x}}^j\bar{\bar{y}}^j],$$

$$F'(1) = \sum_{j=1}^{m}[(\bar{x}^j)^2 - 2\bar{x}^j\bar{y}^j - (\bar{\bar{x}}^j)^2 + 2\bar{\bar{x}}^j\bar{y}^j],$$

we get

$$F'(\lambda_1) = F'(0) + F'(1) +$$

$$+ \sum_{j=1}^{m}\{(\bar{\bar{x}}^j)^2 - (\bar{x}^j)^2 + 2\bar{x}^j[\lambda_1\bar{\bar{y}}^j + (1-\lambda_1)\bar{y}^j] - 2\bar{\bar{x}}^j[\lambda_1\bar{\bar{y}}^j + (1-\lambda_1)\bar{y}^j]\} =$$

$$= F'(0) + F'(1) + \sum_{j=1}^{m}[(\bar{\bar{x}}^j)^2 - (\bar{x}^j)^2 + 2\bar{x}^j\hat{x}^j - 2\bar{\bar{x}}^j\hat{x}^j],$$

where $\hat{x} = \lambda_1\bar{\bar{y}} + (1-\lambda_1)\bar{y}$ is a point that is symmetric to $\tilde{x}$ with respect to the midpoint of the line segment w_λ. Then

$$\sum_{j=1}^{m}[(\bar{\bar{x}}^i)^2 - (\bar{x}^j)^2 + 2\bar{x}^j\hat{x}^j - 2\bar{\bar{x}}^j\hat{x}^j] = \sum_{j=1}^{m}(\bar{\bar{x}}^j - \hat{x}^j)^2 - \sum_{j=1}^{m}(\bar{x}^j - \hat{x}^j)^2 =$$

$$= \rho^2(\bar{\bar{x}}, \hat{x}) - \rho^2(\bar{x}, \hat{x}) = -F'(1-\lambda_1).$$

Finally $F'(\lambda_1) = F'(0) + F'(1) - F'(1-\lambda_1)$. Set $\lambda_1 = 1/2$ and

$$F'\left(\frac{1}{2}\right) = \frac{1}{2}[F'(0) + F'(1)] \geq 0.$$

Similarly, we may show that for all $\lambda_\ell = \ell/2^k$, $k = 1, 2, \ldots$, $\ell = 1, \ldots, 2^k - 1$ $(0 < \lambda_\ell < 1)$

$$F'\left(\frac{\ell}{2^k}\right) = \frac{1}{2}\left[F'\left(\frac{\ell-1}{2^k}\right) + F'\left(\frac{\ell+1}{2^k}\right)\right] \geq 0.$$

The asswertion of the lemma then follows from the continuity of the function $F'(\lambda)$. ■

Lemma 4 *Let B^r be the r-dimensional closed convex polyhedron in R^m with vertices $B_1, \ldots, B_q$. Then, for all $y \in B^r$ and $x \notin B^r$, the inequality $\rho(y, B_k) < \rho(x, B_k)$ is satisfied at least for one $k \in I$ $(I = \{1, \ldots, q\})$.*

Proof: The proof is obtained by induction. For $r = 1$, B^1 is a line segment, and lemma is evedent. Suppose it holds for $r = p - 1$. Since B^p is convex, the line segment $w_\lambda = [\lambda\xi + (1-\lambda)\pi_B^p x]$ belongs to the set B^p for all $0 \leq \lambda \leq 1$, where the point $\xi \in B^{p-1}$ is such that $y \in w_\lambda$. By induction, there exists $k_0 \in I$ such that $\rho(\xi, B_{k_0}) < \rho(x, B_{k_0})$, where $B_{k_0} \in B^{p-1}$. But

$$\rho(yB_{k_0}) = \rho[\bar{\lambda}\xi + (1-\bar{\lambda})\pi_{B^p}x, B_{k_0}] \leq \bar{\lambda}\rho(\xi, B_{k_0}) +$$

$$+(1-\bar{\lambda})\rho(\pi_{B^p}x, B_{k_0}) < \bar{\lambda}\rho(x, B_{k_0}) + (1-\bar{\lambda})\rho(\pi_{B^p}x, B_{k_0}).$$

And it follows from Lemma 2 that $\rho(\pi_{B^p}x, B_{k_0}) \leq \rho(x, B_{k_0})$. Therefore,

$$\bar{\lambda}\rho(x, B_{k_0}) + (1-\bar{\lambda})\rho(\pi_{B_p}x, B_{k_0}) \leq$$

$$\leq \bar{\lambda}\rho(x, B_{k_0}) + (1-\bar{\lambda})\rho(x, B_{k_0}) = \rho(x, B_{k_0}).$$

As a result, we get $\rho(y, B_{k_0}) < \rho(x, B_{k_0})$, $k_0 \in I$, quod erat demonstrandum. ■

Lemma 5 *Let B^r be the r–dimensional closed convex polyhedron in R^m with vertices $B_1, \dots, B_q$ and x, y be some points not belonging to B_r. If there is the point $\bar{\xi} \in B^r$, for which $\rho(y, \bar{\xi}) < \rho(y, \bar{\xi})$ then $\rho(y, B_k) < \rho(x, B_k)$ for all at least one $k \in I$.*

Proof: For $\bar{\xi} \in B^1$, the lemma is evedent. Assume that the lemma holds for $\bar{\xi} \in B^{p-1}$ and assume that $\bar{\xi} \in B^p$. By the definition of B^{p-1} and B^p, there exists such index $k_0 \in I$ that $B_{k_0} \in [B^p \setminus B^{p-1}]$. Then there exists the point $\xi_{k_0} \in B^{p-1}$ for which $\bar{\xi} \in w_\lambda = [\lambda B_{k_0} + (1-\lambda)\xi_{k_0}] \in B^p$, $0 \le \lambda \le 1$.

Suppose $\rho(y, B_{k_0}) \ge \rho(x, B_{k_0})$. If $\rho(y, \xi_{k_0}) < \rho(x, \xi_{k_0})$, then by indiction there is $B_k \in B^{p-1} \subset B^p$ such that $\rho(y, B_k) < \rho(x, B_k)$. If $\rho(y, \xi_{k_0}) \ge \rho(x, \xi_{k_0})$, then by Lemma 3 $\rho(\xi, y) \ge \rho(\xi, x)$ for all $\xi \in w_\lambda = \lambda\xi_{k_0} + (1-\lambda)B_{k_0}$, $0 \le \lambda \le 1$, which contradicts the condition in the lemma. Then $\rho(y, B_{k_0}) < \rho(x, B_{k_0})$, and this completes the proof of the lemma. ∎

Let

$$P^c(x_0, T-t_0) = \{H(x(T; x_0, u_N^P(\cdot))) \mid u_N^P(\cdot) \in \mathcal{D}_N^P\}$$

where $\mathcal{D}_N^P$ is the set of Pareto optimal situations in the game $\Gamma_V^c(x_0, T-t_0)$, and $H(\cdot) = \{H_i(\cdot),\ i = 1, \dots, n\}$. The set $P^c(x_0, T-t_0)$ is called the Pareto optimal set of payoff vectors in the game $\Gamma_V^c(x_0, T-t_0)$ or simply the Pareto optimal set.

In this section, by the solution $W_V^c(x_0, T-t_0)$ of the game $\Gamma_V^c(x_0, T-t_0)$ is meant the Parcto optimal set $P^c(x_0, T-t_0)$. We will study this solution.

Denote $\hat{M} = conv\{M_1, \dots, M_n\}$.

Theorem 9 *If $\hat{M} \bigcap C^{T-t_0}(x_0) = \emptyset$, then*

$$P^c(x_0, T-t_0) = \{H(x) \mid x \in \pi_{C^{T-t_0}(x_0)}\hat{M}\}.$$

Proof: Let

$$y' \in \pi_{C^{T-t_0}(x_0)}\hat{M},\ x \in C^{T-t_0}(x_0),\ x \ne y' \tag{7.6.2}$$

Consider the function $F(\xi) = \rho(y', \xi) - \rho(x, \xi)$, $\xi \in \hat{M}$, and the set

$$Y = \{y \in \hat{M} \mid y' = \pi_{C^{T-t_0}(x_0)} y\}$$

For definiteness, take the point $\bar{y} \in Y$ such that

$$\rho(y', \bar{y}) = \min_{y \in Y} \rho(y', y).$$

Such a point exists, since the set Y is compact. By the definition of the point $\bar{y}$, for all x from (7.6.2) the inequality $F(\bar{y}) < 0$ holds true. Then Lemma 5 implies the inequality $\rho(y', M_i) < \rho(x, M_i)$ for at least one $i = 1, \dots, n$, i.e.

$$H_i(y') > H_i(x) \text{ for at least one } i = 1, \dots, n. \tag{7.6.3}$$

And this means that $H(y') \in P^c(x_0, T-t_0)$. We show that, apart from the points of the set $\pi_{C^{T-t_0}(x_0)}\hat{M}$, in $C^{T-t_0}(x_0)$ there are no points having the property (7.6.3), this is equivalent to the fact that for any $x \in C^{T-t_0}(x_0) \setminus \pi_{C^{T-t_0}(x_0)}\hat{M}$ there is the point $\tilde{y} \in \pi_{C^{T-t_0}(x_0)}\hat{M}$ wherein

$$H_i(\tilde{y}) > H_i(x) \text{ for all } i = 1, \ldots, n. \tag{7.6.4}$$

Consider the sets

$$\hat{M}_1 = conv\{\hat{M}, \pi_{C^{T-t_0}(x_0)}\hat{M}\}, \quad \hat{M}_2 = conv\pi_{C^{T-t_0}(x_0)}\hat{M}.$$

Evidently, $\hat{M}_2 \subset \hat{M}_1$. Denote by $\overline{C^{T-t_0}(x_0)}$ and $\overline{\pi_{C^{T-t_0}(x_0)}\hat{M}}$ the boundaries of the sets $C^{T-t_0}(x_0)$ and $\pi_{C^{T-t_0}(x_0)}\hat{M}$, respectively. Since $C^{T-t_0}(x_0) \bigcap \hat{M} = \emptyset$, then $\pi_{C^{T-t_0}(x_0)}\hat{M} \subset \overline{C^{T-t_0}(x_0)}$ and $\overline{\pi_{C^{T-t_0}(x_0)}\hat{M}} \subset \overline{C^{T-t_0}(x_0)}$.

Let $z \in \overline{\pi_{C^{T-t_0}(x_0)}\hat{M}}$ and Q_z be a hyperplane tangent to $C^{T-t_0}(x_0)$ at the point z. The hyperplane Q_z divides the space R^m into two portions. Since the set $C^{T-t_0}(x_0)$ is convex, it is wholly lying on one side of Q_z. Denote this portion of space by Q_z^+. Let

$$Q^+ = \bigcap_{z \in \overline{\pi_{C^{T-t_0}(x_0)}\hat{M}}} Q_z^+.$$

It can be readily seen that $C^{T-t_0}(x_0) \subset Q^+$. Let further $Q_1^+ = Q^+ \setminus \hat{M}_1$, and $\bar{Q}_1^+$ be the boundary of Q_1^+. By construction, the manifold $\bar{Q}_1^+$ is the locus of the point in R^m whose orthogonal projections onto $\hat{M}_1$ are lying on $\pi_{C^{T-t_0}(x_0)}\hat{M}$. Hence it is clear that for each point $x \in Q_1^+$ its projection

$$\hat{\pi}_{\hat{M}_1} x \in \hat{M}_2 = [C^{T-t_0}(x_0) \bigcap \hat{M}_1]$$

(otherwise, i.e. if $\hat{\pi}_{M_1} x \in \hat{M}_1 \setminus C^{T-t_0}(x_0)$, then on $\bar{Q}_1^+$ there is a point, viz. the point of intersection of the line segment $\lambda x + (1-\lambda)\pi_{\hat{M}_1} x$ and the set $\bar{Q}_1^+$, whose orthogonal projection onto $\hat{M}_1$ does not belong to the set $C^{T-t_0}(x_0)$, and this is inconsistent with the definition of $\bar{Q}_1^+$).

For any $x \in C^{T-t_0}(x_0) \setminus \hat{M}_1$ there is the point $\tilde{y} = \pi_{\hat{M}_1} x$ (since $x \in Q_1^+$ and $\pi_{\hat{M}_1} x \in \hat{M}_2$) which satisfies the inequality (7.6.4) (see Lemma 2).

Now let $x \in \hat{M}_2 \setminus \pi_{C^{T-t_0}(x_0)}\hat{M}$, and $\pi_{\hat{M}} x$ be its image on $\hat{M}$. Since the set $C^{T-t_0}(x_0)$ is, by the assumption, convex and $\pi_{C^{T-t_0}(x_0)}\hat{M} \subset \overline{C^{T-t_0}(x_0)}$, then there exists the point

$$x' \in [\pi_{C^{T-t_0}(x_0)}\hat{M} \bigcap w_\lambda], \; w_\lambda = \lambda x + (1-\lambda)\pi_{\hat{M}} x, \; 0 \leq \lambda \leq 1.$$

This point is the desired one, since by employing Lemma 2 we may easily show that it satisfies the inequality (7.6.4).

Thus, only the points of the set $\pi_{C^{T-t_0}(x_0)}\hat{M}$ have the property (7.6.3). Consequently, $P^c(x_0, T-t_0) = \{H(y) \mid y \in \pi_{C^{T-t_0}(x_0)}\hat{M}\}$. ■

Theorem 10 *If $\hat{M} \subset C^{T-t_0}(x_0)$, then*

$$P^c(x_0, T-t_0) = \{H(x) \mid x \in \pi_{C_{T-t_0}(x_0)}\hat{M} = \hat{M}\}.$$

Proof: As follows from Lemma 2, for every $x \notin \hat{M}$ there is a point $\xi \in \hat{M}(\xi = \pi_{\hat{M}}x)$ such that $H_i(\xi) > H_i(x)$ for all $i, \ldots, n$. Moreover, by Lemma 4, for any $x \notin \hat{M}$ and for all $y \in \hat{M}$ the inequality $H_i(y) > H_i(x)$ is satisfied at least for one $i = 1, \ldots, n$. This means that only the payoff vectors defined on the set $\hat{M}$ can be Pareto optimal. We will show that all vectors in this set are such.

Let $x, y \in \hat{M}$, $x \neq y$. For $r = 1$, $\hat{M}$ is the line segment with the end points M_1 and M_2. We have $\rho(M_1, M_2) = \rho(M_1, y) + \rho(y, M_2) = \rho(M_1, x) + \rho(x, M_2)$. But since $x \neq y$, then either $\rho(M_1, y) < \rho(M_1, x)$, $\rho(y, M_2) > \rho(x, M_2)$ or $\rho(M_1, x) < \rho(M_1, y)$, $\rho(x, M_2) > \rho(y, M_2)$. Hence $H(x), H(y) \in P(x_0, T-t_0)$.

Suppose the theorem holds true for $r = k-1$. Consider the ray Z_x emanating from the point x and passing through the point y. If $r = k$, then there is such point $\bar{z} \in Z_x$ that

$$\rho(x, \bar{z}) = \max_{z \in \hat{M} \bigcap Z_x} \rho(x, z).$$

Evidently, the point $\bar{z}$ is lying on the boundary of the set $\hat{M}$, i.e. it belongs to the $(k-1)$ face of the k–dimensional polyhedron $\hat{M}$. By induction, $H(z) \in P(x_0, T-t_0)$. Then, by the definition of the Pareto optimal set, there exists at least one index $i_0 \in N$ for which $H_{i_0}(\bar{z}) > H_{i_0}(x)$ or, which is the same, $\rho(\bar{z}, M_{i_0}) < \rho(x, M_{i_0})$. Considering that $y \in w_\lambda = \lambda x + (1-\lambda)\bar{z}$, $0 \leq \lambda \leq 1$, we have

$$\rho(y, M_{i_0}) = \rho(\bar{\lambda}x + (1-\bar{\lambda})\bar{z}, M_{i_0}) \leq \bar{\lambda}\rho(x, M_{i_0}) +$$
$$+(1-\bar{\lambda})\rho(\bar{z}, M_{i_0}) < \bar{\lambda}\rho(x, M_{i_0}) + (1-\bar{\lambda})\rho(x, M_{i_0}) = \rho(x, M_{i_0}),$$

i.e. $\rho(y, M_{i_0}) < \rho(x, M_{i_0})$. Consequently, $H(y) \in P^c(x_0, T-t_0)$. By interchanging x and y, we similarly have that $H(x) \in P^c(x_0, T-t_0)$. Thus, for all $y \in \hat{M} = \pi_{C^{T-t_0}(x_0)}\hat{M}$ and only for them $H(y) \in P^c(x_0, T-t_0)$. This completes the proof of the theorem. ■

Theorem 11 *Suppose the points $M_1, \ldots, M_n$ are arranged in such a way that $\hat{M} \supset C^{T-t_0}(x_0)$. Then*

$$P^c(x_0, T-t_0) = \{H(x) \mid x \in \pi_{C^{T-t_0}(x_0)}\hat{M} = C^{T-t_0}(x_0)\}.$$

The proof of Theorem 11 may be obtained from that of Theorem 8 if the sets $\hat{M}$ and $C^{T-t_0}(x_0)$ are interchanged.

We will now consider the general case of arrangement of the points $M_1, \ldots,$ M_n with respect to the reachability set $C^{T-t_0}(x_0)$.

Let

$$N_1 = \{i \in N \mid M_i \in C^{T-t_0}(x_0)\},$$
$$N_2 = \{i \in N \mid M_i \notin C^{T-t_0}(x_0)\}.$$

Evidently, $N_1 \bigcap N_2 = \emptyset$ and $N_1 \bigcup N_2 = N$.

Theorem 12 *In the game of approaching* $\Gamma_V^c(x_0, T-t_0)$, *the set* $P^c(x_0, T-t_0)$ *of Pareto optimal payoff vectors has the following structure:*

$$P^c(x_0, T-t_0) = \{H(x) \mid x \in \pi_{C^{t-t_0}(x_0)}\hat{M}\}.$$

Proof: If $N_1 = \emptyset$ then the conditions of Theorem 6 or Theorem 8 apply when $\hat{M} \bigcap C^{T-t_0}(x_0) = \emptyset$ or $C^{T-t_0}(x_0) \subset \hat{M}$, respectively. There is another case where $C^{T-t_0}(x_0) \not\subset \hat{M}$, $\hat{M} \bigcap C^{T-t_0}(x_0) \neq \emptyset$ when for all $i = 1, \ldots, n$ $M_i \notin C^{T-t_0}(x_0)$. As is readily seen, in this case the theorem is proved in the same way as in the cases 1),2) that are given below.

If $N_2 = \emptyset$, then $\hat{M} \subset C^{T-t_0}(x_0)$ and the conditions of Theorem 7 apply.

Assume that $N_1, N_2 \neq \emptyset$. Introduce the following notation

$$\hat{M}_1 = \hat{M} \bigcap C^{T-t_0}(x_0), \ \hat{M}_2 = \hat{M} \setminus \hat{M}_1.$$

Then $\pi_{C^{T-t_0}(x_0)}\hat{M} = \hat{M}_1 \bigcup \pi_{C^{T-t_0}(x_0)}\hat{M}_2$. Let

$$y' \in \pi_{C^{T-t_0}(x_0)}\hat{M}, \ x \in C^{T-t_0}(x_0), x \neq y'. \tag{7.6.5}$$

We shall show that

$$H_i(y') > H_i(x) \text{ at least for one } i = 1, \ldots, n. \tag{7.6.6}$$

To do this, consider two cases:

1. $y' \in \hat{M}_1$. In this case, the inequality (7.6.6) is proved in much the same way as Theorem 7;

2. $y' \in [\pi_{C^{T-t_0}(x_0)}\hat{M}] \setminus \hat{M}_1 = \pi_{C^{T-t_0}(x_0)}\hat{M}_2$.

In the latter case, for any x from (7.6.5) and $\bar{y} \in Y$, where $Y = \{y \in \hat{M}_2 \mid \pi_{C^{T-t_0}(x_0)}y = y'\}$, the inequality $\rho(y', \bar{y}) < \rho(x, \bar{y})$ is satisfied. Therefore, the inequality (7.6.6) follows from Lemma 5. Thus the set $\pi_{C_{T-t_0}(x_0)}\hat{M}$ is composed only of the points satisfying the condition (7.6.6).

We will show that, apart from the points of the set $\pi_{C^{T-t_0}(x_0)}\hat{M}$ there are no points having the property (7.6.6), i.e. for any $x \in C^{T-t_0}(x_0) \setminus \pi_{C^{T-t_0}(x_0)}\hat{M}$ there is the point $\tilde{y} \in \pi_{C^{T-t_0}(x_0)}\hat{M}$ wherein $\rho(\tilde{y}, M_i) < \rho(x, M_i)$ for all $i = 1, \dots, n$.

Consider the set

$$\hat{M}_3 = conv\pi_{C^{T-t_0}(x_0)}\hat{M}.$$

By Lemma 2, for any $x \in C^{T-t_0}(x_0) \setminus \hat{M}$ there exists such point $\tilde{y} = \pi_{\hat{M}_3} x \in \hat{M}_3$ that $\rho(\tilde{y}, M_i) < \rho(x, M_i)$ for all $i = 1, \dots, n$.

Now let $x \in [\hat{M}_3 \setminus \pi_{C^{T-t_0}(x_0)}\hat{M}]$. If $\tilde{y} = \pi_{\hat{M}} x \in \hat{M}_1$, then the point $\tilde{y}$ is the desired one. If, however, $\pi_{\hat{M}} x \in \hat{M}_2$, then $\pi_{\hat{M}} x \in \pi_{C^{T-t_0}(x_0)}\hat{M}$, but there exists the point

$$x' \in \pi_{C^{T-t_0}(x_0)}\hat{M}_2 \bigcap w_\lambda,$$

where $w_\lambda = \lambda x + (1-\lambda)\pi_{\hat{M}} x$, $0 \leq \lambda \leq 1$, and using Lemma 9 for the set B representing a convex hull of the points x', $M_1, \dots, M_n$, we get that $\rho(x', M_i) < \rho(x, M_i)$ for all $i = 1, \dots, n$, i.e x' is the desired point. ■

Theorems 5–8 provide a constructive method for construction of the Pareto optimal set in the game $\Gamma_V^c(x_0, T - t_0)$ under various arrangements of "objective" points with respect to the reachability set. This implies the set of points corresponding to the Pareto optimal set of payoff vectors in the game $\Gamma_V^c(x_0, T - t_0)$, i.e. the set $P_x(x_0, T - t_0) = \{x \in C^{T-t_0}(x_0) \mid H(x) \in P(x_0, T - t_0)\}$ is obtained in all cases by employing the orthogonal projection of the convex hull of "objective" points onto the reachability set of system (7.3.2)–(7.3.3).

7.7 Existence of dynamic stable C–kernel and NM–solution in cooperative game of approaching

In this section, by the solution $W_V^c(x_0, T - t_0)$ of the game $\Gamma_V^c(x_0, T - t_0)$ is meant the c–kernel $C_V^c(x_0, T - t_0)$ and NM–solution $L_V^C(x_0, T - t_0)$.

Let $\bar{x}(\cdot)$ be such trajectory of system (7.3.2)–(7.3.3) that $\bar{x}(T) \in \pi_{C^{T-t_0}(x_0)}\hat{M}$, and $R_i(\bar{x}(t)) = R(M_i, r_i(\bar{x}(t)))$ be a closed sphere with center at the point M_i and of radius

$$r_i(\bar{x}(t)) = \max_{x \in \pi_{C^{T-t}(\bar{x}(t))}\hat{M}} \rho(x, M_i) \; (i = 1, \dots, n, \; t_0 \leq t \leq T).$$

For each coalition $S \subset N$ we construct the set

$$R_S(\bar{x}(t)) = \bigcap_{i \in S} R_i(\bar{x}(t)).$$

Theorem 13 *The Pareto–optimal set $P^c(x_0, T-t_0)$ in the game $\Gamma_v^c(x_0, T-t_0)$ is dynamic stable.*

Proof: Any trajectory $\bar{x}(\cdot)$ such that $\bar{x}(t_0) = x_0$, $\bar{x}(T) \in \pi_{C^{T-t_0}(x_0)}\hat{M}$ is conditionally–optimal. Because $\bar{x}(T) \in \pi_{C^{T-t_0}(x_0)}\hat{M}$, the point $\bar{x}(T)$ is the nearist in $C^{T-t_0}(x_0)$ to some point $z \in \hat{M}$, i.e.

$$\rho(z, \bar{x}(T)) = \min_{x \in C^{T-t_0}(x_0)} \rho(z, x).$$

But we have $C^{T-t}(\bar{x}(t)) \subset C^{T-t_0}(x_0)$, $\bar{x}(T) \in C^{T-t}(\bar{x}(t))$, $t_0 \leq t \leq T$, thus $\bar{x}(T)$ will remain the nearist point in the set $C^{T-t}(\bar{x}(t))$ to the point z for all $t_0 \leq t \leq T$, i.e.

$$\bar{x}(T) \in \pi_{C^{T-t}(\bar{x}(t))}\hat{M}$$

for all $t_0 \leq t \leq T$. But

$$P^c(\bar{x}(t), T-t) = \{H(x) \,:\, x \in \pi_{C^{T-t}(\bar{x}(t))}\hat{M}\},\ t_0 \leq t \leq T.$$

From $\bar{x}(t) \in \pi_{C^{T-t}(\bar{x}(t))}\hat{M}$ for all $t_0 \leq t \leq T$, it follows that $H(x(T)) \in P^c(\bar{x}(t), T-t)$, for all $t_0 \leq t \leq T$, i.e

$$H(\bar{x}(T)) \in \bigcap_{t_0 \leq t \leq T} P^c(\bar{x}(t), T-t).$$

The point $\bar{x}(t) \in \pi_{C^{T-t_0}(x_0)}\hat{M}$ was arbitrary, and thus the dynamic stability of the set $P^C(x_0, T-t_0)$ is proved. ■

Remark. The dynamic stability of the Pareto–optimal set may be easily proved also for the general n–person games.

Theorem 14 *For the existance of nonempty c–kernel coinciding with the Pareto optimal set $P^c(x_0, T-t_0)$ in the game $\Gamma_V^c(x_0, T-t_0)$, it is sufficient that the following condition be satisfied for each $S \subset N$ $(S \neq N)$:*

$$int\ R_S(x_0) \bigcap Y_S^{T-t_0}(x_0) = \emptyset, \tag{7.7.1}$$

where $int\ R_S(x_0)$ is the interior of the set $R_S(x_0)$, and

$$Y_S^{T-t_0}(x_0) = \{y \in C^{T-t_0}(x_0) \mid val\Gamma_S^y(x_0, T-t_0) = 0\}$$

(see 3 in this chapter).

Proof: Let the condition (7.7.1) be satisfied. We have to prove that the c–kernel $C_V^c(x_0, T-t_0)$ exists and coinsides with the set $P^c(x_0, T-t_0)$. To do this, it suffices to show that none of the vectors from $P^c(x_0, T-t_0)$ is dominated. Suppose the opposite true. Let the vector $H(x) \in P^c(x_0, T-t_0)$ be dominated by a vector $H(y) \in E_V^c(x_0, T-t_0)$. Then there exists such coalition $S \subset N$ that $H(y) \in V(S; x_0, T-t_0)$ and

$$H_i(y) > H_i(x) \text{ for all } i \in S. \tag{7.7.3}$$

Since $H(y) \in V(S; x_0, T-t_0)$, then $y \in Y_S^{T-t_0}(x_0)$ and the condition (7.7.1) implies $\rho(y, M_i) \geq r_i(x_0)$ at least for one $i \in S$. For all $x \in \pi_{C^{T-t_0}(x_0)}\hat{M}$ and $i = 1, \ldots, n$, however, $\rho(x, M_i) \leq r_i(x_0)$, so that for any $x \in \pi_{C^{T-t_0}(x_0)}\hat{M}$ the strict inequality $\rho(y, M_i) < \rho(x, M_i)$ is never satisfied for all $i \in S$. Therefore, (7.7.3) does not hold, i.e. $H(x)$ is nondominated imputation. Consequently, the c–kernel $C_V^c(x_0, T-t_0)$ exists and $P^t(x_0, T-t_0) \subset C_V^c(x_0, T-t_0)$. On the other hand, however,$C_V^c(x_0, T-t_0) \supset P^c(x_0, T-t_0)$. This completes the proof of the theorem. ■

It follows from Theorem 14 that under condition (7.7.1) each trajectory from the set

$$\bar{X} = \{\bar{x}(\cdot) \mid \bar{x}(t_0) = x_0,\ \bar{x}(T) \in \pi_{C^{T-t_0}(x_0)}\hat{M}\} \tag{7.7.4}$$

is a conditionally optimal (in the sense of c–kernel) trajectory in the game $\Gamma_V^c(x_0, T-t_0)$.

Theorem 15 *For a dynamic stable c–kernel to exist in the game $\Gamma_V^c(x_0, T-t_0)$, it is sufficient that along each trajectory $\bar{x}(\cdot) \in \bar{X}$ and for all $S \subset N$ $(S \neq N)$*

$$int\ R_S(\bar{x}(t)) \bigcap Y_S^{T-t}(\bar{x}(t)) = \emptyset,\ t - 0 \leq t < T. \tag{7.7.5}$$

In this case the dynamic stable c–kernel has the following structure: $C_V^c(x_0, T-t_0) = \{H(x) \mid x \in \pi_{C^{T-t_0}(x_0)}\hat{M}\}$. Each trajectory $\bar{x}(\cdot)$ is optimal.

Proof: Under condition (7.7.5), at each instant $t_0 \leq t \leq T$ along any trajectory $\bar{x}(\cdot) \in \bar{X}$ by Theorem 14 $C_V^c(\bar{x}(t), T-t) \subset P(\bar{x}(t), T-t)$, where $C_V^c(\bar{x}(t), T-t)$ and $P^c(\bar{x}(t), T-t)$ are c–cernel and Pareto–optimal sets in current subgame $\Gamma_v^c(\bar{x}(t), T-t)$. Therefore, the dynamic stability of the c–kernel $C_V^c(x_0, T-t_0)$ follows from that of the set $P^c(x_0, T-t_0)$. ■

Theorem 16 *For the NM–solution to exist in the game $\Gamma(x_0, T-t_0)$, it is sufficient that the following condition holds*

$$Y_{N\setminus\{i\}}^{T-t_0}(x_0) \bigcap \pi_{C^{T-t_0}(x_0)}\hat{M} = \emptyset,\ i = 1, \ldots, n. \tag{7.7.6}$$

In this case, the NM–solution coincides with the Pareto optimal set $P^c(x_0, T-t_0)$.

Proof: From condition (7.7.6) it follows, in particular, that

$$Y_S^{T-t_0}(x_0) \bigcap \pi_{C^{T-t_0}(x_0)} \hat{M} = \emptyset \text{ for all } S \neq N,$$

since for $S \neq N$ the superadditivity of V in S and $S \subset N \setminus \{i\}$ imply $Y_S^{T-t_0}(x_0) \subset Y_{N\setminus\{i\}}^{T-t_0}(x_0)$.

Let $H', H'' \in P^c(x_0, T-t_0)$ We will show that they can not dominate one another. From the definition of the set $P^c(x_0, T-t_0)$ it follows that the imputations H' and H'' do not dominate one another in the coalition N (in this respect, recall that in the games without side paiments the domination in the coalition N is possible). If, however, $H' \succ_S H''$, then by Definition 9 of domination $H'(y) \in V(S; x_0, T-t_0)$. But this is not possible, since $H'(y) \in V(S; x_0, T-t_0)$ implies $y \in Y_S^{T-t_0}(x_0)$ and $H'(y) \in P^c(x_0, T-t_0)$ implies $y \in \pi_{C^{T-t_0}(x_0)}$ with the result that

$$Y_S^{T-t_0}(x_0) \bigcap \pi_{C^{T-t_0}(x_0)} \hat{M} \neq \emptyset$$

for a particular $S \neq N$. The latter is inconsistent with the condition (7.7.6).

The second property of NM–solution is always satisfied for the set $P^c(x_0, T-t_0)$. This completes the proof of the theorem. ■

As follows from Theorem 16, under condition (7.7.6) each trajectory from the set (7.7.4) is a conditionally optimal trajectory (in the sense of HM–solution).

Theorem 17 *For a dynamic stable NM–solution to exist in the game $\Gamma_V^c(x_0, T-t_0)$, it is sufficient that along each trajectory $\bar{x}(\cdot) \in \bar{X}$ and for all $i = 1, \ldots, n$.*

$$Y_{N\setminus\{i\}}^{T-t}(\bar{x}(t)) \bigcap \pi_{C^{T-t_0}(\bar{x}(t))} \hat{M} = \emptyset, \; t_0 \leq t < T. \tag{7.7.7}$$

in this case the dynamic stable NM–solution has the following structure: $L_V^c(x_0, T-t_0) = \{H(x) \mid x \in \pi_{C^{T-t_0}(x_0)} \hat{M}\}$. Each trajectory from the set $\bar{X}$ is optimal where the set $\bar{X}$ is defined by the (7.7.4).

Under conditions (7.7.5), (7.7.7) the theorems 15 and 17 provide a constructive method for construction of a dynamic stable c–kernel and NM–solution.

Example. Suppose the game $\Gamma_V^c(x_0, T-t_0)$ is described as follows:

$$\dot{x} = u_1 + u_2 + u_3, \; x(0) = 0,$$

where $x \in R^2$, $u_i = (u_i^1, u_i^2)$, $|u_i| \leq 1$, $i = 1, 2, 3$, $T = 1$.

The reachability set $C^{T-t_0}(x_0)$ in this game is of the form: $(x^1)^2 + (x^2)^2 \leq 3$. For each $i = 1, 2, 3$ $V(\{i\}; x_0, T-t_0) = \emptyset$; for each of three coalitions $S : \{1,2\}, \{1,3\}, \{2,3\}$ $V(S; x_0, T-t_0) = \{H(x) \mid (x^1)^2 + (x^2)^2 \leq 1\}$.

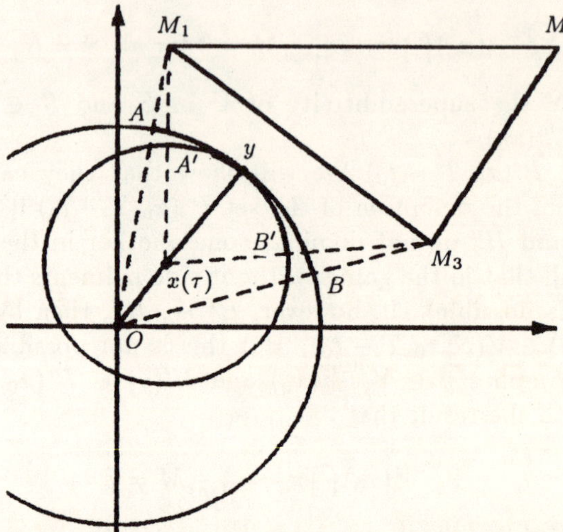

Fig. 41.

The set $\pi_{C^{T-t_0}(x_0)}\hat{M}$ coincides with the arc $\widehat{AB}$ (see Fig. 41.). Any trajectory $\bar{x}(\cdot) = Oy$, where $y \in \widehat{AB}$ is a conditionally optimal trajectory. For $0 \leq \tau \leq 1$ the set $C^{1-\tau}(\bar{x}(\tau))$ is a circle with center at the point $\bar{x}(\tau)$ and is tangent to the set $C^{T-t_0}(x_0)$ at the point y. The set $\pi_{C^{t-\tau}(\bar{x}(\tau))}\hat{M}$ coincides with the arc $\widehat{A'B'}$ that is the orthogonal projection of the triangle $\hat{M} = \Delta M_1 M_2 M_3$ onto the circle $C^{1-\tau}(\bar{x}(\tau))$. Since the condition of Theorem 17 is explicitly satisfied, i.e. for each $y \in \widehat{AB}$ the vector $H(y) = (-\rho(y, M_1), -\rho(y, M_2) - \rho(x, M_3))$ belongs to the set $P^c(\bar{x}(\tau), 1-\tau)$, $0 \leq \tau \leq 1$, then there exists a dinamic stable NM-solution $L^c_V(x_0, T-t_0) = \{H(y) \mid y \in \widehat{AB}\} = P^c(x_0, T-t_0)$.

The existence of the c–kernel, as is shown by condition (7.7.1) of Theorem 15, depends on the arrangement of "objective" points M_1, M_2, M_3. Let $M_1 = (2, 8)$, $M_2 = (3, 6)$, $M_3 = (6, 6)$. Then

$$\pi_{C^{T-t_0}(x_0)}\hat{M} = \overline{C^{T-t_0}(x_0)} \bigcup K,$$

where $\overline{C^{T-t_0}(x_0)}$ is the boundary of the region $C^{T-t_0}(x_0)$, and $K = \{x | 1/4x_2 \leq x_1 \leq x_2\}$. It is easily verified that for all these points the condition (7.7.5) of Theorem 15 is satisfied, and a dynamic stable c–kernel is of the form: $C^c_V(x_0, T-t_0) = \{H(y) \mid y \in \widehat{AB}\} = P^c(x_0, T-t_0)$.

Moreover, it can be easily shown that in the given game the c–kernel and HM–solution are not strongly–dynamically stable.

Chapter 8

Cooperative differential games with side payments

8.1 Definition of cooperative differential game in the characteristic function form

Evidently, the Nash equilibrium principle discussed in the previous chapter fails to refflect all the facets of optimal behavior in n–person game. This optimality principle in noncooperative games, takes no account of coalition formation. At the same time, this factor is a special feature of the majority of n–person decision problems and deserves comprehensive investigation.

This chapter considers a differential game $\Gamma(x_0, T - t_0)$ defined in 3 of Chapter 7 by (7.3.2.), assuming that any coalitions from the set N can be formed therein. By a coalition is meant a subset $S \subset N$ where players are in at least one of the following relations:

- between all members of coalition $S \subset N$ there is a full information exchange;
- members of coalition $S \subset N$ are capable of correlating their strategies, i.e. united in coalition S the players act cooperatively as one player with a set of strategies

$$\mathcal{D}_S = \prod_{i \in S} \mathcal{D}_i$$

where $\mathcal{D}_i$ is a set of admissible strategies for the player i;

- members of coalition S have common interests, i.e. the payoff function of coalition S to be maximized by joint efforts of its members can be expressed as

$$J_S = \sum_{i \in S} J_i$$

where J_i in the payoff function of the player $i \in S$.

These three types of relations between the members of one coalition we call levels of cooperation. In what follows, the levels of cooperations in all coalitions $S \subset N$ in the game $\Gamma(x_0, T - t_0)$ are assumed to be the same. The first level of cooperation can exist independently of two subsequent levels; without information exchange the strategies cannot be correlated; without strategy correlation, and hence without information exchange, maximization of the total payoff is senseless.

In connection with the community of interests (third level of cooperation), the possibility to transfer the payoff arises. When different players can compare their payoff, have an opportunity to sum up and devide them, and transfer part of the payoff to the other players (to effect side paiments), then such payoffs are called transferable. Otherwise we are dealing with untransferable payoffs. In contrast to the games with transferable payoffs, in the games with untransferable payoffs only the second level of cooperation (cooperative selection) of strategies can be reached. Since untransferable payoffs cannot be devided into parts and transferred to other players, the objective of selection of a joint strategy $u_s(\cdot) \in \mathcal{D}_s$ by the coalition S is a possible maximization of each component of the vector $J_S = (J_{i_1}, \ldots, J_{i_s})$, $i_k \in S$.

Definition 1. The game $\Gamma(x_0, T - t_0)$ is called a cooperative differential game with side payments if:

1. its rules provide for formation of any coalition;
2. the players' payoffs are transferable and the players are at the third level of cooperation (choosing joint strategies and effecting side payments);
3. the players' objective is to ensure a maximal total payoff and its equitable distribution under the agreement.

Under the most favorable conditions (in what follows they are specified), it is advantegeous for players to form one maximal coalition N, since in this case it is possible to ensure the greatest total payoff. As a result of such formation, the distribution of the total payoff among the players turns out to be the main problem in cooperative games. For this reason the prinsiples of optimal behavior of the players take in acount their advancing demands when the total payoff is shared. In this sense, cooperative games are the sharing games rather than the strategic games. Such an approach to cooperative games is based on the notion of characteristic function.

Definition 2. The function $v : 2^N \to R^1$, where 2^N in the set of all subsets of the set N, and R^1 is the space of real numbers, such that

1. $v(\emptyset) = 0$, where $\emptyset$ is an empty set,

2. $v(S \bigcup R) \geq v(S) + v(R)$ for any $S, R \subset N$ such that $S \bigcap R = \emptyset$ is called a characteristic function.

$v(S)$ is interpreted as a generated payoff of the coalition S, i.e. the payoff which can be obtained by the coalition S from the game independent of other players' actions. The second property of the characteristic function is superadditivity. This means that any two disjoint coalitions united in one coalition receive an amount at least not less than that they could have received without such union, i.e. without coordination of their actions.

We shall construct a characteristic function for the cooperative differential n–person games with side payments.

For each coalition $S \subset N$ we consider a zero–sum differential game $\Gamma_S(x_0, T - t_0)$ between coalition S and $N \setminus S = \{i \in N \mid i \notin S\}$ that is described by equations

$$\dot{x} = f(x, u_S, u_{N \setminus S}) \ , \ x(t_0) = x_0, \tag{8.1.1}$$

where

$$u_S \in \prod_{i \in S} V_i \ , \ u_{N \setminus S} \in \prod_{j \in N \setminus S} V_j \ ,$$

and the vectors x, f and the set U_i, $i = 1, \ldots, n$ are the same as in the game (7.3.2). The payoff for Player S (maximizer) in the game $\Gamma_S(x_0, T - t_0)$ in each situation $(u_S(\cdot), u_{N \setminus S}(\cdot))$, $u_s(\cdot) \in \mathcal{D}_S$, $u_{N \setminus S}(\cdot) \in \mathcal{D}_{N \setminus S}$,

$$\mathcal{D}_S = \prod_{i \in S} \mathcal{D}_i \ , \ \mathcal{D}_{N \setminus S} = \prod_{j \in N \setminus S} \mathcal{D}_j$$

is defined as

$$J_S(x_0, u_S(\cdot), u_{N \setminus S}(\cdot)) = \sum_{i \in S} J_i(x_0, u_S(\cdot), u_{N \setminus S}(\cdot)),$$

where the sets $\mathcal{D}_i$ and the functions J_i $(i = 1, \ldots, n)$ are defined just as in the game (7.3.2.). The payoff of Player $N \setminus S$ is assumed to be $-J_S(x_0, u_S(\cdot), u_{N \setminus S}(\cdot))$.

By existance theorems concerning zero–sum games, in the game $\Gamma_S(x_0, T - t_0)$ for any $\epsilon > 0$ there exist the ϵ–optimal piecewise open–loop strategies $u_S^\epsilon(\cdot)$, $u_{N \setminus S}^\epsilon(\cdot)$ and the value

$$val\Gamma_S(x_0, T - t_0) = \lim_{\epsilon \to 0} J_S(x_0, u_S^\epsilon(\cdot), u_{N \setminus S}^\epsilon(\cdot)).$$

For a fixed initial state $x_0 \in R^m$ the function $v(S; x_0, T - t_0)$, $S \in 2^N$ is defined as follows:

$$v(S; x_0, T - t_0) = \begin{cases} 0, & \text{if} \quad S = \emptyset, \\ val\Gamma_S(x_0, T - t_0), & \text{if} \quad S \subset N \ (\emptyset \neq S \neq N), \\ \max_{u_N(\cdot) \in \mathcal{D}_N} J_N(x_0, u_N(\cdot)) & \text{if} \quad S = N. \end{cases} \tag{8.1.2}$$

In what follows, for simplicity, we assume that a maximum in (8.1.2) is achieved. Otherwise all constructions become complicated, but can be made by employing sup operation instead of max and ϵ–optimal strategies instead of optimal strategies.

Lemma 1 *The function v defined by (8.1.2), is superadditive in S, i.e. for the fixed $x_0 \in R^m$ and any $S, R \subset N$, $S \bigcap R = \emptyset$ the following inequality holds*

$$v(S \bigcup R; x_0, T - t_0) \geq v(S; x_0, T - t_0) + v(R; x_0, T - t_0).$$

Proof: Let $K = S \bigcup R$. Using the property of inf operation, for all $u_K(\cdot) \in \mathcal{D}_K$

$$\inf_{u_{N\setminus K}(\cdot) \in \mathcal{D}_{N\setminus K}} J_K(x_0, u_K(\cdot), u_{N\setminus K}(\cdot)) \geq$$

$$\geq \inf_{u_{N\setminus K}(\cdot) \in \mathcal{D}_{N\setminus K}} J_S(x_0, u_K(\cdot), u_{N\setminus K}(\cdot)) + \inf_{u_{N\setminus K}(\cdot) \in \mathcal{D}_{N\setminus K}} J_R(x_0, u_K(\cdot), u_{N\setminus K}(\cdot)).$$

Evidently, with an increase in the left hand member, this inequality is preserved:

$$\sup_{u_K \in \mathcal{D}_K} \inf_{u_{N\setminus K}(\cdot) \in \mathcal{D}_{N\setminus K}} J_K(x_0, u_K(\cdot), u_{N\setminus K}(\cdot)) \geq$$

$$\geq \inf_{u_{N\setminus K}(\cdot) \in \mathcal{D}_{N\setminus K}} J_S(x_0, u_K(\cdot), u_{N\setminus K}(\cdot)) + \inf_{u_{N\setminus K}(\cdot) \in \mathcal{D}_{N\setminus K}} J_R(x_0, u_K(\cdot), u_{N\setminus K}(\cdot))$$

for all $u_K(\cdot) \in \mathcal{D}_K$. The wider is the set over which the infinum is taken the less is the infinum itself. Since $\mathcal{D}_{N\setminus K} \subset \mathcal{D}_{N\setminus S}$ and $\mathcal{D}_{N\setminus K} \subset \mathcal{D}_{N\setminus R}$, then

$$\inf_{u_{N\setminus K}(\cdot) \in \mathcal{D}_{N\setminus K}} J_S(x_0, u_K(\cdot), u_{N\setminus K}(\cdot)) + \inf_{u_{N\setminus K}(\cdot) \in \mathcal{D}_{N\setminus K}} J_R(x_0, u_K(\cdot), u_{N\setminus K}(\cdot)) \geq$$

$$\geq \inf_{u_{N\setminus S}(\cdot) \in \mathcal{D}_{N\setminus S}} J_S(x_0, u_S(\cdot), u_{N\setminus S}(\cdot)) + \inf_{u_{N\setminus R}(\cdot) \in \mathcal{D}_{N\setminus R}} J_R(x_0, u_R(\cdot), u_{N\setminus R}(\cdot)).$$

We get

$$v(S \bigcup R; x_0, T - t_0) = \sup_{u_K(\cdot) \in \mathcal{D}_K} \inf_{u_{N\setminus K}(\cdot) \in \mathcal{D}_{N\setminus K}} J_K(x_0, u_K(\cdot), u_{N\setminus K}(\cdot)) \geq$$

$$\geq \inf_{u_{N\setminus S}(\cdot) \in \mathcal{D}_{N\setminus S}} J_S(x_0, u_S(\cdot), u_{N\setminus S}(\cdot)) + \inf_{u_{N\setminus R}(\cdot) \in \mathcal{D}_{N\setminus R}} J_R(x_0, u_R(\cdot), u_{N\setminus R}(\cdot))$$

for all $u_S(\cdot) \in \mathcal{D}_S$ and $u_R(\cdot) \in \mathcal{D}_R$. Consequently, this inequality is preserved if in its right–hand member, we take

$$v(S; x_0, T - t_0) = \sup_{u_S \in \mathcal{D}_S} \inf_{u_{N\setminus S}(\cdot) \in \mathcal{D}_{N\setminus S}} J_S(x_0, u_S(\cdot), u_{N\setminus S}(\cdot)),$$

and instead of the addent, we take

$$v(R; x_0, T - t_0) = \sup_{u_R(\cdot) \in \mathcal{D}_R} \inf_{u_{N\setminus R}(\cdot) \in \mathcal{D}_{N\setminus R}} J_R(x_0, u_R(\cdot), u_{N\setminus R}(\cdot)).$$

Hence we have the inequality $v(S \bigcup R; x_0, T-t_0) \geq v(S; x_i, T-t_0)+v(R; x_0, T-t_0)$. This completes the proof of the lemma. ■

It follows from the superadditivity condition that it is advantageous for the players to form a maximal coalition N and obtain a maximal total payoff $v(N; x_0, T-t_0)$ that is possible in the game. Purposefully, the quantity $v(S; x_0, T-t_0)$ $(S \neq N)$ is equal to a garanteed payoff of the coalition S obtained irrespective of the behavior of other players, even though the latter form a coalition $N \setminus S$ against S.

Definition 3. The function $v : 2^N \times R^m \times R^1 \to R^1$ defined by (8.1.2), is called a characteristic function of the game $\Gamma(x_0, T-t_0)$.

Note that the positiveness of payoff fuunctions J_i, $i = 1, \ldots, n$ implies that of characteristic function. From the superadditivity of v it follows that $v(S'; x_0, T-t_0) \geq v(S; x_0, T-t_0)$ for any $S, S' \subset N$ such that $S \subset S'$, i.e. the superadditivity of the function v in S implies that this function is monotone in S.

Since the essence of cooperative game is the possibility to form coalitions and the main problem therein is distribution of the total payoff between players, then the subject of cooperative theory is characteristic function rather than strategy. In fact, the characteristic function displays the possibilities of coalitions in the best way and can form the basis for equitable distribution of the total payoff between players.

The pair $< N, v(S; x_0, T-t_0) >$, where $N$is the set of players, and v the characteristic function, defined by (8.1.2) is called the cooperative differential game in the form of characteristic function v. For short, it will be denotewd by $\Gamma_v(x_0, T-t_0)$.

Remark. The equality (8.1.2) is not the only way of defining the characteristic function and is used when the formation of counter–coalition is plausiblle in the formation of coalitions. Other methods may also be applied to provide estimation of the coalition strength.

Various methods for "equitable" distribution of the total profit between players are treated as optimality principles in cooperative games. The set of such distribution satisfying an optimality principle is called a solution to the cooperative game (in the sense of this optimality principle). We will now define solutions of the game $\Gamma_v(N; x_0, T-t_0)$.

Denote by ξ_i a share of the player $i \in N$ in the total gain $v(N; x_0, T-t_0)$.

Definition 4. The vector $\xi = (\xi_1, \ldots, \xi_n)$, whose components satisfy the conditions:

1. $\xi_i \geq v(\{i\}; x_0, T-t_0)$, $i \in N$;

2. $\sum_{i \in N} \xi_i = v(N; x_0, T-t_0)$,

is called an imputation in the game $\Gamma_v(x_0, T - t_0)$.

The equity of the distribution $\xi = (\xi_1, \ldots, \xi_n)$ representing an imputation is that each player receives at least his safe payoff and the entire maximal payoff is divided evenly without a remainder.

Theorem 1 *Suppose the function* $w : 2^N \times R^m \times R^1 \to R^1$ *is additive in* $S \in 2^n$, *i.e. for any* $S, R \in 2^N$, $S \bigcap R = \emptyset$ *we have* $w(S \bigcup R; x_0, T - t_0) = w(S; x_0, T - t_0) + w(R; x_0, T - t_0)$. *Then in the game* $\Gamma_w(x_0, T - t_0)$ *there is a unique imputation* $\xi = \{w(\{i\}; x_0, T - t_0),\ i = 1, \ldots, n\}$.

Proof: From the additivity of w we immediately obtain $w(N, x_0, T - t_0) = w(\{1\}; x_0, T - t_0) + \ldots + w(\{n\}; x_0, T - t_0)$, whence follows the statement of the theorem. ■

The game with additive characteristic function is called inessential. In the essential game $\Gamma_v(x_0, T - t_0)$ there is an infinite set of imputations. Indeed, any wector of the form

$$(v(\{1\}, x_0, T - t_0) + \alpha_1, \ldots, v(\{n\}, x_0, T - t_0) + \alpha_n),\ \alpha_i \geq 0,\ i \in N,$$

$$\sum_{i \in N} \alpha_i = v(N; x_0, T - t_0) - \sum_{i \in N} v(\{i\}; x_0, T - t_0) > 0, \tag{8.1.3}$$

is an imputation. The imputation set in the game $\Gamma_v(x_0, T - t_0)$ is denoted by $E_v(x_0, T - t_0)$.

Definition 5. We say that the imputation ξ dominates the imputation η in the coalition S ($\xi \succ_S \eta$) if

1. $\xi_i > \eta_i,\ i \in S$;
2. $\sum_{i \in S} \xi_i \leq v(S; x_0, T - t_0)$.

The imputation ξ is said to dominate the imputation η ($\xi \succ \eta$) if there is such coalition $S \subset N$ that $\xi \succ_S \eta$.

It follows from the definition of the imputation that domination in single–element coalition and coalition N, is not possible.

Definition 6. The set of nondominated imputations is called the c–kernel of the game $\Gamma_v(x_0, T - t_0)$ and it denoted by $C_v(x_0, T - t_0)$.

The equity of the imputation belonging to the c–kernel is that none of the coalitions can offer a reasonable alternative against this imputation.

Definition 7. The set $L_v(x_0, T - t_0) \subset E_v(x_0, T - t_0)$ is called the Neumann–Morgenstern solution (the NM–solution) of the game $\Gamma_v(x_0, T - t_0)$ if:

1. $\xi, \eta \in L_v(x_0, T - t_0)$ implies $\xi \not\succ \eta$ (ξ does not dominate η),
2. for $\eta \notin L_v(x_0, T - t_0)$ there is such $\xi \in L_v(x_0, T - t_0)$ that $\xi \succ \eta$.

As is seen from Definition 7, the conditions placed on the imputations from the NM–solution are weaker than those on the imputation from the c–kernel and, as a result, the NM–solution alwais contains the c–cernel. Unlike the c–kernel and NM–solution, the Shapley value representing an optimal distribution principle of the total gain $v(N; x_0, T - t_0)$ is defined without using the concept of domination.

Definition 8. The vector $\Phi^v(x_0, T - t_0) = \{\Phi_i^v(x_0, T - t_0),\ i = 1, \dots, n\}$ is called the Shapley value if it satisfies the following conditions:

1. if v, w are two chararcteristic functions, then $\Phi^v(x_0, T - t_0) + \Phi^w(x_0, T - t_0) = \Phi^{v+w}(x_0, T - t_0)$;
2. $\pi\Phi^v(x_0, T - t_0) = \Phi^{\pi v}(x, T - t_0)$, where π is any permutation of players, $\pi\Phi^v(x_0, T - t_0) = \{\Phi_{\pi(i)}^v(x_0, T - t_0),\ i = 1, \dots, n\}$, where πv is characteristic function such that for any coalition $S = \{i_1, \dots, i_s\}$ $\pi v(\{\pi(i_1), \dots, \pi(i_s)\}; x_0, T - t_0) = v(S; x_0, T - t_0)$;
3. $\sum_{i \in N} \Phi_i^v(x_0, T - t_0) = v(N, x_0, T - t_0)$;
4. if $v(S; x_0, T - t_0) - v(S \setminus i; x_0, T - t_0) = 0$ for all $S \subset N$ $(S \ni i)$, then $\Phi_i^v(x_0, T - t_0) = 0$.

It is well known that there exists a unique vector $\Phi^v(x_0, T - t_0)$ satisfying these four conditions, and its components are computed by the formulas

$$\Phi_i^v(x_0, T - t_0) =$$

$$= \sum_{S \subset N\ (S \ni i)} \frac{(n-s)!(s-1)!}{n!} [v(S; x_0, T - t_0) - v(S \setminus i; x_0, T - t_0)], \qquad (8.1.4)$$

$$i = 1, \dots, n.$$

The components of the Shapley value have the meaning of the players' expected shares in the total gain. Also, it may be shown that the Shaplay value is an imputation.

8.2 Principle of dynamic stability (time–consistancy)

Formalization of the notion of optimal behavior constitutes one of the fundamental problems in the theory of n-person game. At present, quite for the various classes of games different optimality principels are constructed. Some of them are stated in the previous section. Recall that the players' behavior (strategies in noncooperative games or imputation in cooperatives games) satisfying some given optimality principle is called a solution of the game in the sence of this principle and must posses two properties. On the one hand, it must adequately reflect the conceptual notion of optimality providing special features of the class of games for which it is defined. On the one hand, it must be feasible under conditions of the game where it is applied. This property reduces to the existence of a the solution of the game generated by a specified principle of optimality.

In dynamic games, one more requirement is naturally added to the above–mentioned requirements, viz. the purposefulness and feasibility of an optimality principle are to be preserved throughout the game. This requirement is called the dynamic stability of a solution of the game (time consistency).

The dynamic stability of a solution of the differential game is the property that, when the game proceedes along an "optimal" trajectory, at each instant of time the players are to be guided by the same optimality principle, and hence do not have any ground for deviation from the previously adopted "optimal" behavior throughout the game. When the dynamic stability is betrayed, at some instant of time there are conditions under which the continuation of the initial behavior becomes non–optimal and hence the initially chosen principle of optimality proves to be unfeasible.

In what follows the concept of dynamic stability is specified for each class of games.

Some principles of optimality in noncooperative differential games (and, in particular, in zero–sum two–person games) are dynamic stable. We will show, for example, that the Nash equilibria are always dynamic stable.

Let $u^\epsilon(\cdot) = (u_1^\epsilon(\cdot), \dots, u_n^\epsilon(\cdot))$ be the Nash ϵ–equilibrium in the noncooperative game $\Gamma(x_0, T-t_0)$ (see Definition 4 in 2, chapter 1). The trajectory of system (7.3.2.)–(7.3.3.) corresponding to the equilibrium $u^\epsilon(\cdot)$ is denoted by $x^\epsilon(\cdot)$. With the game $\Gamma(x_0, T-t_0)$, we consider the current games $\Gamma(x^\epsilon(t), T-t)$, $t_0 \le t \le T$ determined along the trajectory $x^\epsilon(\cdot)$. The game $\Gamma(x^\epsilon(t), T-t)$ is determined along the trajectory $x^\epsilon(\cdot)$. The game $\Gamma(x^\epsilon(t), T-t)$ is described by the equation $\dot{x} = f(x, u_1, \dots, u_n)$ with the initial state $x^\epsilon(t)$, is of duration $T-t$, and the payoff function of players are of the form

$$\int_t^T h_i(x(\tau))d\tau + H_i(x(T)),\ i = 1, \dots, n.$$

Denote $\Delta_i[t_k^{\Delta_i}, T] = \{t_k^{\Delta_i}, t_{k+1}^{\Delta_i}, \dots, t_{q_i}^{\Delta_i} = T\}$, $a_i[t_k^{\Delta_i}, T] = (a_i^k, \dots, a_i^{q_i})$. Let $t \in [t_k^{\Delta_i}, t_{k+1}^{\Delta_i})$, and $\bar{u}_i(t) \in \mathcal{U}_i[t_k^{\Delta_i}, t_{k+1}^{\Delta_i})$ be the image of the map $a_i^k(t_k^{\Delta_i}, x(t_k^{\Delta_i}))$. We construct a truncated map $\bar{a}_i^t(t, x(t))$ of the map $a_i^k(t_k^{\Delta_i}, x(t_k^{\Delta_i}))$, setting for any $t \in [t_k^{\Delta_i}, t_{k+1}^{\Delta_i})$ the image of the map $\bar{a}_i^t(t, x(t))$ equal to $\bar{u}_i(\tau)$, $\tau \in [t, t_{k+1}^{\Delta_i})$ or, in the other words, $\bar{a}_i^t(t, x(t))(\tau) = a_i^k(t_k^{\Delta_i}, x(t_k^{\Delta_i}))(\tau)$, $\tau \in [t, t_{k+1}^{\Delta_i})$.

Now let $t \in [t_0, T]$, and $u_i(\cdot)[t,T]$ be a strategy truncation on the time interval $[t,T]$, i.e.

$$u_i(\cdot)[t,T] = \begin{cases} (\Delta_i[t_k^{\Delta_i}, T], a_i[t_k^{\Delta_i}, T]) & \text{if } t = t_k^{\Delta_i}, \\ (\Delta_i'[t,T], a_i'[t,T]), & \text{if } t \in (t_k^{\Delta_i}, t_{k+1}^{\Delta_i}), \end{cases}$$

$k = 0, 1, \ldots, q_i$, where $\Delta_i'[t,T] = \{t, t_{k+1}^{\Delta_i}, \ldots, t_{q_i}^{\Delta_i} = T\}$, $a_i'[t,T] = (\bar{a}_i^t, a_i^{k+1}, \ldots, a_i^{q_i})$, $\bar{a}_i^t \; : \; \bar{a}_i^t(t, x(t))(\tau) \in \mathcal{U}_i[t, t_{k+1}^{\Delta_i})$, $\tau \in [t, t_{k+1}^{\Delta_i})$. Then, for any equilibrium $u(\cdot)$ and its corresponding trajectory,

$$J_i(x(t), u(\cdot)[t,T]) = \int_t^T h_i(x(\tau))d\tau + H_i(x(T)),$$

where $x(t) = x(t, x_0, u(\cdot))$.

The equilibrium $u^\epsilon(\cdot)$ is called dinamic stable in the noncooperative game $\Gamma(x_0, T - t_0)$ if for any $t \in [t_0, T]$, $u_i(\cdot)[t,T] \in \mathcal{D}_i[t,T]$ and $i = 1, \ldots, n$

$$J_i(x^\epsilon(t), u^\epsilon(\cdot)[t,T]) \geq J_i(x^\epsilon(t), u^\epsilon(\cdot)[t,T] \| u_i(\cdot)[t,T]) - \epsilon,$$

where $(u^\epsilon(\cdot)[t,T] \| u_i(\cdot)[t,T]) = (u_1^\epsilon(\cdot)[t,T], \ldots, u_{i-1}^\epsilon(\cdot)[t,T], u_i(\cdot)[t,T], u_{i+1}^\epsilon(\cdot)$ $[t,T], \ldots, u_n^\epsilon(\cdot)[t,T])$ i.e. from the fact that the situation $u^\epsilon(\cdot)$ is ϵ–equilibrium in the game $\Gamma(x_0, T - t_0)$ it follows that the truncated situation $u^\epsilon(\cdot)[t,T]$ is ϵ–equilibrium in the current game $\Gamma(x^\epsilon(t), T - t)$ $(t_0 \leq t \leq T)$.

To prove the dynamic stability of the ϵ–equilibrium situation u^ϵ in the noncooperative game $\Gamma(x_0, T - t_0)$, we construct the strategy

$$v_i(\cdot) = \begin{cases} u_i^\epsilon(\cdot)[t_0, t] & \text{on } [t_0, t], \\ u_i(\cdot)[t,T] & \text{on } [t,T], \end{cases}$$

where $u_i(\cdot)[t,T] \in \mathcal{D}_i[t,T]$, and $\mathcal{D}_i[t,T]$ is a restriction of $\mathcal{D}_i$ on $[t,T]$. From the definition of the ϵ–equilibrium situation $u^\epsilon(\cdot)$ for all $u_i(\cdot)[t,T] \in \mathcal{D}_i[t,T]$ we get

$$J_i(x_0, u^\epsilon(\cdot)) \geq J_i(x_0, u^\epsilon(\cdot) \| u_i(\cdot)) - \epsilon, \; i = 1, \ldots, n,$$

or

$$\int_{t_0}^t h_i(x^\epsilon(\tau))d\tau + J_i(x^\epsilon(t), u^\epsilon(\cdot)[t,T]) \geq \int_{t_0}^t h_i(x^\epsilon(\tau))d\tau +$$
$$+ J_i(x^\epsilon(t), u^\epsilon(\cdot)[t,T] \| u_i(\cdot)[t,T]) - \epsilon, \; i = 1, \ldots, n,$$

Hence for all $u_i(\cdot)[t,T] \in \mathcal{D}_i[t,T]$

$$J_i(x^\epsilon(t), u^\epsilon(\cdot)[t,T]) \geq J_i(x^\epsilon(t), u^\epsilon(\cdot)[t,T] \| u_i(\cdot)[t,T]) - \epsilon, \; i = 1, \ldots, n,$$

This shows that the situation $u^\epsilon(\cdot)[t,T]$ is the ϵ–equilibrium in the sence of Nash in the current game $\Gamma(x^\epsilon(t), T - t)$. Consequently, the Nash ϵ–equilibrium in the game $\Gamma(x_0, T - t_0)$ is dynamic stable.

The dynamic stability of the Nash equilibrium in the game $\Gamma(x_0, T - t_0)$ may be shown in much the same way.

Thus, the employment of an equilibrium in noncooperative (and in particular, zero–sum) games does not involve any dynamic stability problem. The dynamic stability of the equilibrium is ensured by the very fact of its existence. This validates the consideration of noncooperatives games in normal form which reduces the game to a unique choice of strategies at the initial instant of the game.

The situation is different in cooperative differential games.

Assume that at the start of the game the players adopt an optimality principle and construct a solution based on it (a imputation set satisfying the chosen principle of optimality, say the c–kernel, NM–solution, etc.). From the definition of cooperative game it follows that the evolution of the game is to be along the trajectory providing a maximal total payoff for the players. When moving, the players arrive at the subgames featuring current initial states and current duration. In due course, not only the conditions of the game and the players opportunities, but even the players' interests may change. Therefore, at an instant t the initially optimal solution of the current game may not exist or satisfy the players now. Then, at the instant t, the players will have no ground to keep to the initially chosen trajectory. The latter exactly means the dynamic instability of the chosen optimality principle and, as a result, the dynamic instability of the motion itself.

We now focus our attention on dynamic stable solutions in the cooperative differential games with side payments.

Let an optimality principle be chosen in the game $\Gamma_v(x_0, T - t_0)$. The solution of this game constructed in the initial state $x(t_0) = x_0$ based on the chosen principle of optimality is denoted by $W_v(x_0, T - t_0)$. The set $W_v(x_0, T - t_0)$ is a subset of the imputation set $E_v(x_0, T - t_0)$ in the game $\Gamma_v(x_0, T - t_0)$. Assume that $W_v(x_0, T - t_0) \neq \emptyset$.

Definition 9. Any trajectory $\bar{x}(\cdot)$ of the system (7.3.2.)–(7.3.3.) such that

$$J_N(\bar{x}(\cdot)) = \sum_{i \in N} J_i(\bar{x}(\cdot)) = v(N; x_0, T - t_0)$$

is called a conditionally optimal trajectory in the game $\Gamma_v(x_0, T - t_0)$.

The definition suggests that along the conditionally optimal trajectory the players obtain the largest total payoff. For simplicity, we assume henceforth that such a trajectory exists. In the absence of the conditionally optimal trajectory we may introduce the notion of "ϵ–conditionally optimal trajectory" and carry out the necessary constructions with an accuracy ϵ.

We will now consider the behavior of the set $W_v(x_0, T - t_0)$ along the conditionally optimal trajectory. Towards this end, in each current state $\bar{x}(t)$ the current subgame $\Gamma_v(\bar{x}(t), T-t)$ is defined as follows. In the state $\bar{x}(t)$, we define the characteristic function

$$v(S; \bar{x}(t), T-t) =$$
$$\begin{cases} 0 & \text{if } S = \emptyset, \\ val\Gamma_S(\bar{x}(t), T-t) & \text{if } S \subset N \ (\emptyset \neq S \neq N), \\ \max_{u_N(\cdot)[t,T] \in \mathcal{D}_N[t,T]} J_N(\bar{x}(t), u_N(\cdot)[t,T]) & \text{if } S = N. \end{cases}$$

Here $J_N(\bar{x}(t), u_N(\cdot)[t,T])$ is the remaining total payoff of the players from the initial state $\bar{x}(t)$ on the conditionally optimal trajectory, i.e.

$$J_N(\bar{x}(t), u_N(\cdot)[t,T]) = \sum_{i \in N} \left[\int_t^T h_i(\bar{x}(\tau))d\tau + H_i(\bar{x}(T)) \right];$$

$val\Gamma_S(\bar{x}(t), T-t)$ is the value of the zero–sum differential game $\Gamma_S(\bar{x}(t), T-t)$ between coalitions S and $N \setminus S$ which is described by the equation $\dot{x} = f(x, u_S, u_{N \setminus S})$ and from the initial state $\bar{x}(t)$ and duration $T-t$. Here the payoff of Player S (maximizer) in each situation $(u_S(\cdot)[t,T], u_{N \setminus S}(\cdot)[t,T])$, $u_S(\cdot)[t,T] \in \mathcal{D}_S[t,T]$, $u_{N \setminus S}(\cdot)[t,T] \in \mathcal{D}_{N \setminus S}[t,T]$ equals to

$$J_S(\bar{x}(t), u_S(\cdot)[t,T], u_{N \setminus S}(\cdot)[t,T]) = \sum_{i \in S} J_i(\bar{x}(t), u_S(\cdot)[t,T], u_{N \setminus S}(\cdot)[t,T]).$$

The payoff of player $N \setminus S$ is set equal to $-J_S$. The truncation of the admissible strategy $u_S(\cdot)$ and the admissible strategy $u_S(\cdot)$ and the admissible strategy set $\mathcal{D}_S$ of the coalition S over the time interval $[t,T]$ are denoted by $u_S(\cdot)[t,T]$ and $\mathcal{D}_S[t,T]$ respectively.

The current subgame $\Gamma_v(\bar{x}(t), T-t)$ is defined as $< N, v(S, \bar{x}(t), T-t) >$. The imputation set in the game $\Gamma_v(\bar{x}(t), T-t)$ is of the form:

$$E_v(\bar{x}(t), T-t) = \Big\{ \xi \in R^n \mid \xi_i \geq v(\{i\}; \bar{x}(t), T-t), \ i = 1, \ldots, n;$$
$$\sum_{i \in N} \xi_i = v(N; \bar{x}(t), T-t) \Big\},$$

where

$$v(N; \bar{x}(t), T-t) = v(N; x_0, T-t_0) - \int_{t_0}^t \sum_{i \in N} h_i(\bar{x}(\tau))d\tau.$$

The quantity

$$\int_{t_0}^t \sum_{i \in N} h_i(\bar{x}(\tau))d\tau$$

is interpreted as the total gain of the players on the time interval $[t_0, t]$ when the motion is caried out along the trajectory $\bar{x}(\cdot)$.

Consider the family of current games

$$\{\Gamma_v(\bar{x}(t), T-t) =< N, v(S; \bar{x}(t), T-t) >, t_0 \leq t \leq T\},$$

determined along the conditionally optimal trajectory $\bar{x}(\cdot)$ and their solutions $W_v(\bar{x}(t), T-t) \subset E_v(\bar{x}(t), T-t)$ generated by the same principle of optimality as the initially solution $W_v(x_0, T-t_0)$.

Lemma 2 *The set $W_v(\bar{x}(T), 0)$ is a solution of the current game $\Gamma_v(\bar{x}(T), 0)$ and is composed of the only imputation $H(\bar{x}(T)) = \{H_i(\bar{x}(T)),\ i = 1, \ldots, n\}$, where $H_i(\bar{x}(T))$ is the terminal part of the player i's payoff along the trajectory $\bar{x}(\cdot)$.*

Proof: Since the game $\Gamma_v(\bar{x}(T), 0)$ is of zero–duration, then for all $i \in N$ $v(\{i\}; \bar{x}(T), 0) = H_i(\bar{x}(T))$. Hence

$$\sum_{i \in N} v(\{i\}; \bar{x}(T), 0) = \sum_{i \in N} H_i(\bar{x}(T)) = v(N; \bar{x}(T), 0),$$

i.e. the characteristic function of the game $\Gamma_v(\bar{x}(T), 0)$ is additive for S and, by Theorem 1,

$$E_v(\bar{x}(T), 0) = H(\bar{x}(T)) = W_v(\bar{x}(T), 0)$$

This completes the proof of the lemma. ■

Dynamic stability of solution. Let the conditionally optimal trajectory $\bar{x}(\cdot)$ be such that $W_v(\bar{x}(t), T-t) \neq \emptyset$, $t_0 \leq t \leq T$. If this condition is not satisfied, it is impossible for players to adhere to the chosen principle of optimality, since at the very first instant t, when $W_v(\bar{x}(t), T-t) = \emptyset$, the players have no possibility to follow this principle. Assume that in the initial state x_0 the players agree upon the imputation $\xi^0 \in W_v(x_0, T-t_0)$. This means that in the state x_0 the players agree upon such imputation of the gain $v(N; x_0, T-t_0)$ that (when the game terminates at the instant T) the share of the i–th player is equal to ξ_i^0, i.e. the i–th component of the imputation ξ^0. Suppose the player i's payoff (his share) on the time interval $[t_0, t]$ is $\xi_i(\bar{x}(t))$. Then, on the remaining time interval $[t, T]$ according to the ξ^0 he is to receive the gain $\eta_i^t = \xi_i^0 - \xi_i(\bar{x}(t))$. For the original agreement (the imputation ξ^0) to remain in force at the instant t, it is essential that the vector $\eta^t = (\eta_1^t, \ldots, \eta_n^t)$ belong to the set $W_v(\bar{x}(t), T-t)$, i.e. a solution of the current game $\Gamma_v(\bar{x}(t), T-t)$. If such a condition is satisfied at each instant of time $t \in [t_0, T]$ along the trajectory $\bar{x}(\cdot)$, then the imputation ξ^0 is realized. Such is the conceptual meaning of the dynamic stability of the sharing.

Along the trajectory $\bar{x}(\cdot)$ on the time interval $[t, T]$, $t_0 \leq t \leq T$, the coalition N obtains the payoff

$$v(N; \bar{x}(t), T-t) = \sum_{i \in N} \left[\int_t^T h_i(\bar{x}(\tau)) d\tau + H_i(\bar{x}(T)) \right].$$

Then the difference

$$v(N;x_0,T-t_0)-v(N;\bar{x}(t),T-t)=\int_{t_0}^{t}\sum_{i\in N}h_i(\bar{x}(\tau))d\tau$$

is equal to the payoff the coalition N obtains on the time interval $[t_0,t]$. The share of the i-th player in this payoff, considering the transferability of payoffs, may be represented as

$$\gamma_i(t)=\int_{t_0}^{t}\beta_i(\tau)\sum_{i=1}^{n}h_i(\bar{x}(\tau))d\tau=\gamma_i(\bar{x}(t),\beta), \tag{8.2.1}$$

where $\beta_i(\tau)$ is the $[t_0,T]$ integrable function satisfying the condition

$$\sum_{i=1}^{n}\beta_i(\tau)=1,\ \beta_i(t)\geq 0,\ t_0\leq\tau\leq T,\ (i=1,\dots,n).$$

From (8.2.1) we have

$$\frac{d\gamma_i}{dt}=\beta_i(t)\sum_{i\in N}h_i(\bar{x}(t)).$$

This quantity may be interpreted as an instantaneous gain of the player i at the moment t. Hence it is clear the vector $\beta(t)=(\beta_1(t),\dots,\beta_n(t))$ references to a distribution of the total gain among the members of the coalition N. By properly choosing $\beta(t)$, the players can ensure the dessirable outcome, i.e. to regulate the players' gain receipt with respect to time, so that et each instant $t\in[t_0,T]$ there will be no objection against realization of the original agreement (the imputation ξ^0).

Definition 10. The imputation $\xi^0\in W_v(x_0,T-t_0)$ is called dynamic stable in the game $\Gamma_v(x_0,T-t_0)$ if the following conditions are satisfied:

1. there exists a conditionally optimal trajectory $\bar{x}(\cdot)$ along which $W_v(\bar{x}(t),T-t)\neq\emptyset$, $t_0\leq t\leq T$,
2. there exists such $[t_0,T]$ integrable function $\beta(t)=(\beta_1(t),\dots,\beta_n(t))$ that for each $t_0\leq t\leq T$ $\beta_i(t)\geq 0$, $\sum_{i=1}^{n}\beta_i(t)=1$ and

$$\xi^0\in\bigcap_{t_0\leq t\leq T}[\gamma(\bar{x}(t),\beta)\oplus W_v(\bar{x}(t),T-t)], \tag{8.2.2}$$

where $\gamma(\bar{x}(t),\beta)=(\gamma_1(\bar{x}(t),\beta),\dots,\gamma_n(\bar{x}(t),\beta))$, and $W_v(\bar{x}(t),T-t)$ is a solution of the current game $\Gamma_v(\bar{x}(t),T-t)$.

The cooperative differential game $\Gamma_v(x_0,T-t_0)$ with side payments has a dymamic stable solution $W_v(x_0,T-t_0)$ if all of the imputations $\xi\in W_v(x_0,T-t_0)$ are dynamic stable.

The conditionally optimal trajectory along which there exists a dynamic stable solution of the game $\Gamma_v(x_0, T-t_0)$ is called an optimal trajectory.

If there exists at least one dynamic stable imputation $\xi^0 \in W_v(x_0, T-t_0)$, but not all of the imputations from the set $W_v(x_0, T-t_0)$ have such a property, then we may discuss a partial dynamic stability of the solution $W_v(x_0, T-t_0)$ of the game $\Gamma_v(x_0, T-t_0)$

The sum $\oplus$ in above Definition 10 has the following meaning: for $\eta \in R^n$ and $A \subset R^n$ $\eta \oplus A = \{\eta + a \mid a \in A\}$.

From the definition of the dynamic stability at the instant $t = T$ we have $\xi^0 \in \gamma(\bar{x}(T), \beta) \oplus W_v(\bar{x}(T), 0)$, where $W_v(\bar{x}(T), 0)$ is a solution of the current game $\Gamma_v(\bar{x}(T), 0)$ is made up of the only imputation $\xi^T = H(\bar{x}(T))$, the sharing ξ^0 may represented as $\xi^0 = \gamma(\bar{x}(T), \beta) + H(\bar{x}(T))$ or

$$\xi^0 = \int_{t_0}^{T} \beta(\tau) \sum_{i \in N} h_i(\bar{x}(\tau)) d\tau + H(\bar{x}(T)).$$

The dynamic stable imputation $\xi^0 \in W_v(x_0, T-t_0)$ may be realized as follows. From (8.2.2) at any instant $t_0 \leq t \leq T$ we have

$$\xi^0 \in [\gamma(\bar{x}(t), \beta) \oplus W_v(\bar{x}(t), T-t)], \tag{8.2.3}$$

where

$$\gamma(\bar{x}(t), \beta) = \int_{t_0}^{t} \beta(\tau) \sum_{i \in N} h_i(\bar{x}(\tau)) d\tau$$

is the payoff vector on the time interval $[t_0, t]$, the player i's share in the gain on the same interval being

$$\gamma_i(\bar{x}(t), \beta) = \int_{t_0}^{t} \beta_i(\tau) \sum_{i \in N} h_i(\bar{x}(\tau)) d\tau.$$

When the game proceeds along the optimal trajectory, the players on each time interval $[t_0, t]$ share the total gain

$$\int_{t_0}^{t} \sum_{i \in N} h_i(\bar{x}(\tau)) d\tau$$

among themselves

$$\eta_0 - \gamma(\bar{x}(t), \beta) \in W_y(\bar{x}(t), T-t) \tag{8.2.4}$$

so that the inclusion (8.2.3) is satisfied. Furthemore, (8.2.3) implies the existence of such vector $\xi^t \in W_v(\bar{x}(t), T-t)$ that $\xi^0 = \gamma(\bar{x}(t), \beta) + \xi^t$. That is, in the description of the above method of choosing $\beta(\tau)$, the vector of the gains to be obtained by the players at the remaining stage of the game

$$\xi^t = \xi^0 - \gamma(\bar{x}(t), \beta) = \int_t^T h(\bar{x}(\tau))d\tau + H(\bar{x}(T))$$

belongs to the set $W_v(\bar{x}(t), T-t)$. Geometrically, this means that by varying the vector $\gamma(\bar{x}(t), \beta) = (\gamma_1(\bar{x}(t), \beta), \ldots, \gamma_n(\bar{x}(t), \beta))$ resticted by the only condition

$$\sum_{i \in N} \gamma_i(\bar{x}(t), \beta) = \int_{t_0}^t \sum_{i \in N} h_i(\bar{x}(\tau))d\tau,$$

the players ensure diplacement of the set $\gamma(\bar{x}(t), \beta) \oplus W_v(\bar{x}(t), T-t)$ in such a way that the inclusion (8.2.3) is satisfied.

In general, it is fairly easy to see that there may exist an infinite number of vectors $\beta(\tau)$ satisfying conditions (8.2.3),(8.2.4). Therefore the sharing method proposed here seems to lack true uniqueness. However, for any vector $\beta(\tau)$ satisfying conditions (8.2.3)–(8.2.4) at each time instant $t_0 \le t \le T$ the players are guided by the imputation $\xi^t \in W_v(\bar{x}(t), T-t)$ and the same optimality principle throughout the game, and hence have no reason to violate the previously concluded agreement.

Let us make the following additional assumptions

a) the set $W_v(\bar{x}(t), T-t)$ is continuously dependent on $\bar{x}(t), t$ in Hausdorf metric;

b) the vector $\xi^t \in W_v(\bar{x}(t), T-t)$ may be chosen as a continuously differentiable monotone nonincreasing function of the argument t.

Show that by properly choosing $\beta(t)$ we may always ensure dynamic stability of the sharing $\xi^0 \in W_v(x_0, T-t_0)$ under assumptions a),b) and the first condition of Definition 10 (i.e. along the conditionally optimal trajectory at each time instant $t_0 \le t \le T$ $W_v(\bar{x}(t), T-t) \neq \emptyset$).

We choose $\xi^t \in W_v(\bar{x}(t), T-t)$ to be a continuously differentiable function of t, $t_0 \le t \le T$. Construct the difference $\xi^0 - \xi^t = \alpha(t)$ then we get $\xi^t + \alpha(t) \in W_v(x_0, T-t_0)$. Let $\beta(t) = (\beta_1(t), \ldots, \beta_n(t))$ be the $[t_0, T]$ integrable vector function satisfying condition (8.2.4). Solve the equation (with respect to $\beta(t)$)

$$\int_{t_0}^t \beta(\tau) \sum_{i \in N} h_i(\bar{x}(\tau))d\tau = \alpha(t):$$

$$\beta(t) = \frac{1}{\sum_{i \in N} h_i(\bar{x}(t))} \cdot \frac{d\alpha(t)}{dt} = -\frac{1}{\sum_{i \in N} h_i(\bar{x}(t))} \cdot \frac{d\xi^t}{dt}. \tag{8.2.5}$$

Make sure that for such $\beta(t)$ the condition (8.2.4) is satisfied. Indeed,

$$\sum_{i\in N}\beta_i(t) = -\frac{\sum_{i\in N}\frac{d\xi_i^t}{dt}}{\sum_{i\in N}h_i(\bar{x}(t))} = -\frac{\frac{d}{dt}v(N;\bar{x}(t),T-t)}{\sum_{i\in N}h_i(\bar{x}(t))} =$$

$$= -\frac{\frac{d}{dt}\left[\sum_{i\in N}\left(\int_t^T h_i(\bar{x}(\tau))d\tau + H_i(\bar{x}(T))\right)\right]}{\sum_{i\in N}h_i(\bar{x}(t))} = \frac{\sum_{i\in N}h_i(\bar{x}(t))}{\sum_{i\in N}h_i(\bar{x}(t))} = 1$$

(since $\sum_{i\in N}\xi_i^t = v(N;\bar{x}(t),T-t)$).

From condition (7.3.13.) we have $h_i, H_i > 0$, $i \in N$, and since $\frac{d\xi^t}{dt} \leq 0$, then $\beta_i(\tau) \geq 0$.

Thus, if along the conditionally optimal trajectory all current games have nonempty solutions possessing conditions a), b) then the original game $\Gamma_v(x_0, T-t_0)$ has a dynamic stable solution. Theoretically, the main problem is to study conditions imposed on the vector function $\beta(t)$ in order to ensure dynamic stability of specific forms of solutions $W_v(x_0, T-t_0)$ in various classes of games.

8.3 Classification of dynamic stable solutions

In this section we will consider the concept of strongly dynamic stability and define dynamic stable solutions for cooperative games with terminal payoffs.

Strongly–dynamic stable solution. For the dynamic stable imputation $\xi^0 \in W_v(x_0, T-t_0)$, as follows from Definition 10, for $t_0 \leq t \leq T$ there exist such $[t_0, T]$ integrable vector function $\beta(t)$ and imputation ξ^t (generally nonunique) from the solution $W_v(\bar{x}(t), T-t)$ of the current game $\Gamma_v(\bar{x}(t), T-t)$ that $\xi^0 = \gamma(\bar{x}(t), \beta) + \xi^t$. The conditions of dynamic stability do not affect the imputation from the set $W_v(\bar{x}(t), T-t)$ which fail to satisfy this equation. Furthermore, of interest is the case where any imputation from the current solution $W_v(\bar{x}(t), T-t)$ may provide a "good" continuation for the original agreement, i.e. for a dynamic stable imputation $\xi^0 \in W_v(x_0, T-t_0)$ at any instant $t_0 \leq t \leq T$ and for every $\xi^t \in W_v(\bar{x}(t), T-t)$ the condition $\gamma(\bar{x}(t), \beta) + \xi^t \in W_v(x_0, T-t_0)$, where $\gamma(\bar{x}(T), \beta) + H(\bar{x}(T)) = \xi^0$, be satisfied. By slightly strengthening this requirement, we obtain a qualitatively new dynamic stability concept of the solution $W_v(x_0, T-t_0)$ of the game $\Gamma_v(x_0, T-t_0)$ and call it a strongly dynamic stability.

Definition 11 The imputation $\xi^0 \in W_v(x_0, T-t_0)$ is called strongly–dynamic stable in the game $\Gamma_v(x_0, T-t_0)$ if the following conditions are satisfied:

1. the imputation ξ^0 is dynamic stable;

2. for any $t_0 \le t_1 \le t_2 \le T$ and $\beta^0(t)$ corresponding to the imputation ξ^0,

$$\gamma(\bar{x}(t_2, \beta^0) \oplus W_v(\bar{x}(t_2), T - t_2)) \subset \gamma(\bar{x}(t_1, \beta^0) \oplus W_v(\bar{x}(t_1), T - t_1)) \quad (8.3.1)$$

The cooperative differential game $\Gamma_v(x_0, T - t_0)$ with side payments has a strongly–dynamic stable solution $W_v(x_0, T - t_0)$ if all the imputation from $W_v(x_0, T - t_0)$ are strongly–dynamic stable.

The conditionally optimal trajectory along which there exist a strongly dynamic stable solution of the game $\Gamma_v(x_0, T - t_0)$ is called a strongly optimal trajectory.

If there exists at least one strongly–dynamic stable, imputation $\xi \in W_v(x_0, T - t_0)$, but not all of the imputation from the set $\xi \in W_v(x_0, T - t_0)$ have such a property, then we are dealing with a partial straongly dinamic stability of the solution $W_v(x_0, T - t_0)$ of the game $\Gamma_v(x_0, T - t_0)$.

The dynamic instability of the solution of the cooperative differential game leads to abandonment of the optimality principle generating this solution, since none of the imputation from the set $W_v(x_0, T - t_0)$ remains optimal until the game terminates. Therefore, the set $W_v(x_0, T - t_0)$ is generally to be called a solution to the game $\Gamma_v(x_0, T - t_0)$ only if it is dinamic stable. Otherwise the game $\Gamma_v(x_0, T - t_0)$ is assumed to have no solution.

Terminal payoffs. In (7.3.2.), let $h_i \equiv 0$, $i = 1, \ldots, n$.. The cooperative differential game with terminal payoffs is denoted by the same symbol $\Gamma_v(x_0, T - t_0)$. In such games the payoffs are made when the game terminates.

Denote by $C^{T-t_0}(x_0)$ the set of points $y \in R^m$ for which there exist an open–loop control $(u_1(t), \ldots, u_n(t))$, $u_i(t) \in \mathcal{U}_i$, $i = 1, \ldots, n$ transferring (because of system (7.3.2.)), a phase point from the initial state x_0 to the point y in time $T - t_0$. The set $C^{T-t_0}(x_0)$ is called the reachability set in the game $\Gamma_v(x_0, T - t_0)$.

It is naturally assumed that in the game with terminal payoffs.

$$v(N; x_0, T - t_0) = \max_{x \in C^{T-t_0}(x_0)} H_N(x) = H_N(x^*), \quad (8.3.2)$$

where

$$H_N(x) = \sum_{i \in N} H_i(x)$$

(for simplicity, assume that a maximum in (8.3.2) is achieved, otherwise constructions become somewhat complicated and we have to deal with the ϵ–dynamic stability).

Definition 12. Any trajectory $\bar{x}(\cdot)$ of system (7.3.2.)–(7.3.3.) such that $\bar{x}(T) = \bar{x}^*$ is called a conditionally optimal trajectory in the cooperative differential game $\Gamma_v(x_0, T - t_0)$ with terminal payoffs.

Definition of a dynamic stable imputation from the solution $W_v(x_0, T-t_0)$ is obtained as a special case of Definition 10. Since the games with terminal payoffs are frequently encountered in this book, this definition is provided separately.

Consider the current games $\Gamma_v(\bar{x}(t)), T-t)$, $t_0 \le t \le T$ along a conditionally optimal trajectory $\bar{x}(\cdot)$. As before, their solutions are denoted by $W_v(\bar{x}(t), T-t) \subset E_v(\bar{x}(t), T-t)$, $t_0 \le t \le T$. The game $\Gamma_v(\bar{x}(t), T-t)$ is of duration $T-t$, has the initial state $\bar{x}(t)$, and the payoff functions therein are defined just as in the game $\Gamma_v(x_0, T-t_0)$. Note that, with the motion along the conditionally optimal trajectory, at each time instant $t_0 \le t \le T$ the point $\bar{x}^*$ remains in the reachability region

$$C^{T-t}(\bar{x}(t)) =$$

$$= \{y \in R^m \mid x(T; \bar{x}(t), u_1(t), \ldots, u_n(t)) = y,\ u_i(t) \in \mathcal{U}_i,\ i = 1, \ldots, n\},$$

and at the instant $t = T$, $C^0(\bar{x}(T)) = \bar{x}^*$.

Definition 13. The imputation $\xi^0 \in W_v(x_0, T-t_0)$ is called dynamic stable in the game with terminal payoffs if:

1. there is a conditionally optimal trajectory $\bar{x}(\cdot)$ along which $W_v(\bar{x}(t), T-t) \ne \emptyset$, $t_0 \le t \le T$;
2. $\xi^0 \in \bigcap_{t_0 \le t \le T} W_v(\bar{x}(t), T-t)$.

The conditionally optimal trajectory along which there exists a dynamic stable imputation $\xi^0 \in W_v(x_0, T-t_0)$ is called an optimal trajectory.

Theorem 2 *In the coopeyrative differential game $\Gamma_v(x_0, T-t_0)$ with terminal payoffs $H_i(x(T))$, $i = 1, \ldots, n$, only the vector $H(x^*) = \{H_i(x^*),\ i = 1, \ldots, n\}$ whose components are equal to the players payoffs at the end point of the conditionally optimal trajectory may be dynamic stable.*

Proof: It follows from the dynamic stability of the imputation $\xi^0 \in W_v(x_0, T-t_0)$ that

$$\xi^0 \in \bigcap_{t_0 \le t \le T} W_v(\bar{x}(t), T-t).$$

But since the current game $\Gamma_v(\bar{x}(T), 0)$ is of zero duration, then therein $E_v(\bar{x}(T), 0) = W_v(\bar{x}(T), 0) = H(\bar{x}(T)) = H(x^*)$. Hence

$$\bigcap_{t_0 \le t \le T} W_v(\bar{x}(t), T-t) = H(x^*),$$

i.e. $\xi^0 = H(x^*)$ and there are no other imputations. ∎

Theorem 3 *For the existence of the dynamic stable solution in the game with terminal payoff it is necessary and sufficient that for all* $t_0 \leq t \leq T$

$$H(x^*) \in W_v(\bar{x}(t), T - t),$$

where $H(x^*)$ *is the players payoff vector at the end point of the conditionally optimal trajectory* $\bar{x}(\cdot)$, *with* $W_v(\bar{x}(t), T - t)$, $t_0 \leq t \leq T$ *being the solutions of the current games along the conditionally optimal trajectory generated by the chosen principle of optimality.*

This theorem is a corollary of Theorem 2.

Thus, if in the game with terminal payoffs there is a dynamic stable imputation, then the players in the initial state x_0 have to agree upon realization of the vector (imputation) $H(x^*) \in W_v(x_0, T - t_0)$ and, with the motion along the optimal trajectory $\bar{x}(\cdot)$, at each time instant $t_0 \leq t \leq T$ this imputation $H(x^*)$ belongs to the solution of the current games $\Gamma_v(\bar{x}(t), T - t)$.

As Theorem 2. shows, in the game with terminal payoffs only a unique imputation from the set $W_v(x_0, T - t_0)$ may be dynamic stable. Therefore, in such games there is no point in discussing both the dynamic stability of the solution $W_v(x_0, T - t_0)$ as a whole and its strong dynamic stability.

Chapter 9

New optimality principles in n–person differential games

9.1 Integral optimality principles

Most of the optimality principles (OP) in n–person differential game theory are taken from the classical (static) theory and are dynamic instable (time inconsistent), or strongly dynamic (unstable strongly time inconsistent). In this chapter for the cooperative case we try to purpose the family of strongly time consistent optimality principles (STCOP) constructed on the base of integration of the locally optimal behavior in classical sense. Using the regularizing procedure we get new STCOP's based on core, Shapley value, NM–solution.

Consider n–person differential game $\Gamma(x_0, T - t_0)$

$$\dot{x} = f(x, u_1, \dots, u_n),\ u_i \in U_i \subset CompR^\ell,$$

with integral payoffs

$$K_i(x_0, T - t_0; u_1, \dots, u_n) = \int_{t_0}^{T} h_i(x(t))dt,$$

$$h_i > 0,\ i = 1, \dots, n.$$

Cooperative form of $\Gamma(x_0, T - t_0)$. In this formalization we suppose that the players before starting the game agree to play $u_1^*, \dots, u_n^*$ such that the corresponding trajectory maximizes the sum of the payoffs

$$\max \sum_{i=1}^{n} K_i(x_0, T - t_0; u_1, \dots, u_n) =$$

$$= \sum_{i=1}^{n} K_i(x_0, T - t_0; u_1^*, \dots, u_n^*) = \tag{9.1.1}$$

$$= \sum_{i=1}^{n} \int_{t_0}^{T} h_i(x^*(t))dt = V(N; x_0, T - t_0),$$

where N is the set of all players in $\Gamma(x_0, T - t_0)$. The trajectory $x^*(t)$ is called optimal. Let $V(S; x_0, T - t_0)$ be the characteristic function ($S \subset N$) and $C(x_0, T - t_0)$ the core. Consider the family of subgames $\Gamma(x^*(t), T - t)$ along $x^*(t)$, $t \in [t_0, T]$, corresponding cores $C(x^*(t), T - t)$ (which are supposed to be nonvoid) and c.f. $V(S; x^*(t), T - t)$. The core $C(x_0, T - t_0)$ is strongly time inconsistent and more over in all nontrivial cases even time inconsistent. But using the c.f. $V(S; x^*(t), T - t)$ and $C(x^*(t), T - t)$, $t \in [t_0, T]$ we shall construct a new c.f. and based on it a new strongly dynamic stable time consistent (STC) optimality principle (OP).

Let us introduce the following function

$$\bar{V}(S; x_0, T - t_0) =$$

$$\frac{1}{T - t_0} \int_{t_0}^{T} \left[\sum_{i \in S} \int_{t_0}^{t} h_i(x^*(\tau))d\tau + V(S; x^*(t), T - t) \right] dt, \tag{9.1.2}$$

where $S \subset N$, $x^*(\tau)$ is the optimal trajectory from (9.1.1), $V(S; x^*(t), T - t)$, $t \in [t_0, T]$ c.f. in the subgame $\Gamma(x^*(t), T - t)$. We suppose $V(S; x^*(t), T - t)$ integrable on $[t_0, T]$.

We have for $S = N$

$$\bar{V}(N; x_0, T - t_0) =$$

$$= \frac{1}{T - t_0} \int_{t_0}^{T} \left[\sum_{i \in N} \int_{t_0}^{t} h_i(x^*(\tau))d\tau + V(N; x^*(t), T - t) \right] dt = \tag{9.1.3}$$

$$= \frac{1}{T - t_0} \int_{t_0}^{T} V(N; x_0, T - t_0)dt = V(N; x_0, T - t_0),$$

because along the optimal trajectory $x^*(\tau)$, $\tau \in [t_0, T]$ the Bellmann's optimality principle holds for the function $V(N; x^*(t), T - t)$, i.e.

$$V(N; x_0, T - t_0) = \sum_{i \in N} \int_{t_0}^{t} h_i(x^*(\tau))d\tau +$$

$$+ V(N; x^*(t), T - t), \quad t \in [t_0, T).$$

It is easily seen that

$$\bar{V}(S_1 \bigcup S_2; x_0, T - t_0) \geq$$

$$\geq \bar{V}(S_1, x_0, T - t_0) + \bar{V}(S_2; x_0, T - t_0), \tag{9.1.4}$$

for $S_1 \bigcap S_2 = \emptyset$, $S_1 \subset N$, $S_2 \subset N$. From the superadditivity of c.f. we have for all $t \in [t_0, T]$

$$V(S_1; x^*(t), T-t) + V(S_2; x^*(t), T-t) \leq$$
$$\leq V(S_1 \bigcup S_2; x^*(t), T-t). \quad (9.1.5)$$

Adding to both sites of (9.1.5)

$$\sum_{i \in S_1} \int_{t_0}^{t} h_i(x^*(\tau))d\tau + \sum_{i \in S_2} \int_{t_0}^{t} h_i(x^*(\tau))d\tau$$

and integrating on $[t_0, T]$ we get (9.1.4).

Thus from (9.1.3) and (9.1.4) it follows that $\bar{V}(S; x_0, T-t_0)$, $S \subset N$ is c.f. in the game $\Gamma(x_0, T-t_0)$. Define now the analogue of $\bar{V}(S; x_0, T-t_0)$, $S \subset N$ for the subgames $\Gamma(x^*(\Theta), T-\Theta)$, $\Theta \in [t_0, T]$. Let

$$\bar{V}(N; x^*(\Theta), T-\Theta) =$$
$$= \frac{1}{T-t_0} \int_{\Theta}^{T} \left[\sum_{i \in N} \int_{\Theta}^{t} h_i(x^*(\tau))d\tau + V(N; x^*(t), T-t) \right] dt \quad (9.1.6)$$

and

$$\bar{V}(S; x^*(\Theta), T-\Theta) =$$
$$= \frac{1}{T-t_0} \int_{\Theta}^{T} \left[\sum_{i \in S} \int_{\Theta}^{t} h_i(x^*(\tau))d\tau + V(S; x^*(t), T-t) \right] dt, \quad (9.1.7)$$
$$S \subset N$$

From (9.1.6) we get that

$$\bar{V}(N; x^*(\Theta), T-\Theta) = \frac{T-\Theta}{T-t_0} V(N; x^*(\Theta), T-\Theta). \quad (9.1.8)$$

We see that $\bar{V}$ is not a c.f. in a common sense in the subgame $\Gamma(x^*(\Theta), T-\Theta)$, $t_0 \leq \Theta \leq T$, because $\bar{V}(N; x^*(\Theta), T-\Theta)$ is not equal to the maximal sum of the payoffs of all the players in this subgame.

In the way we have done it for $\bar{V}(S; x_0, T-t_0)$ we show that $\bar{V}(S; x^*(\Theta), T-\Theta)$, $\Theta \in [t_0, T]$, $S \subset N$ is a superadditive function of S.

Let $C(x_0, T-t_0)$ and $C(x^*(t), T-t)$ be the nonvoid cores in the game $\Gamma(x_0, T-t_0)$ and $\Gamma(x^*(t), T-t)$, $t \in [t_0, T]$ respectively. Let

$$\xi(t) = \{\xi_1(t), \ldots, \xi_i(t), \ldots, \xi_n(t)\} \in$$
$$\in C(x^*(t), T-t), \; t \in [t_0, T]$$

be an integrable selector which is an imputation from the core of the subgame $\Gamma(x^*(t), T-t)$ at each instant t. Consider the quantities

$$\bar{\xi}_i^{t_0} = \frac{1}{T-t_0}\left\{\int_{t_0}^{T}\left[\int_{t_0}^{t} h_i(x^*(\tau))d\tau + \hat{\xi}_i(t)\right] dt\right\},$$

$$\bar{\xi}_i^{\Theta} = \frac{1}{T-t_0}\left\{\int_{\Theta}^{T}\left[\int_{\Theta}^{t} h_i(x^*(\tau))d\tau + \xi_i(t)\right] dt\right\}, \tag{9.1.9}$$

$$i = 1, \ldots, n.$$

Let $\bar{C}(x_0, T-t)$, $\bar{C}(x^*(\Theta, T-\Theta)$, be the sets of all possible vectors $\bar{\xi}^{t_0} = \{\bar{\xi}_i^{t_0}\}$ and correspondingly $\bar{\xi}^{\Theta} = \{\xi_i^{\Theta}\}$ defined by (9.1.9) for all possible integrable selectors

$$\xi(t) \in C(x^*(t), T-t), \quad t \in [t_0, T],$$

from the cores of the subgames $\Gamma(x^*(t), T-t)$. Then using the set integration we may write,

$$\bar{C}(x_0, T-t_0) = \frac{1}{T-t_0}\int_{t_0}^{T} C(x^*(t), T-t)dt,$$

$$\bar{C}(x^*(\Theta), T-\Theta) = \frac{1}{T-t_0}\int_{\Theta}^{T} C(x^*(t), T-t)dt. \tag{9.1.10}$$

The necessary and sufficient condition for an imputation $\xi(t)$ belong to the core $C(x^*(t), T-t)$ is

$$\sum_{i\in S} \xi_i(t) \geq V(S; x^*(t), T-t), \quad S \subset N \tag{9.1.11}$$

adding to both sides of (9.1.11) $\sum_{i\in S}\int_{\Theta}^{t} h_i(x^*(\tau))d\tau$ and integrating cover $[\Theta, T]$ we get

$$\sum_{i\in S}\frac{1}{T-t_0}\left\{\int_{\Theta}^{T}\left[\sum_{i\in S}\int_{\Theta}^{t} h_i(x^*(\tau))d\tau + \xi_i(t)\right] dt\right\} \geq$$

$$\geq \frac{1}{T-t_0}\left\{\int_{\Theta}^{T}\left[\sum_{i\in S}\int_{\Theta}^{t} h_i(x^*(\tau))d\tau + V(S; x^*(t), T-t)\right] dt\right\}. \tag{9.1.12}$$

Using (9.1.7) and (9.1.9) we get

$$\sum_{i\in S} \bar{\xi}_i^{\Theta} \geq \bar{V}(S; x^*(\Theta), T-\Theta), \quad S \subset N$$

i.e. that any vector $\bar{\xi}^{\Theta} \in \bar{C}(x^*(\Theta), T-\Theta)$, $\Theta \in [t_0, T]$ belongs also to the core of the subgame $\Gamma(x^*(\Theta), T-\Theta)$ defined by the c.f. $\bar{V}(S; x^*(\Theta), T-\Theta)$.

Theorem 1 *The set* $\bar{C}(x^*(\Theta), T-\Theta)$ *belongs to the core of the subgame* $\Gamma(x^*(\Theta, T-\Theta)$ *with the c.f.* $\bar{V}(S; x^*(\Theta), T-\Theta)$, $S \subset N$.

Now we have the intuitive background to introduce $\bar{C}(x^*(\Theta), T-\Theta)$ as an OP in the subgame $\Gamma(x^*(\Theta), T-\Theta)$, $\Theta \in [t_0, T]$ (in the case $\Theta = t_0$ we have an OP for the original game $\Gamma(x_0, T-t_0)$). Define now a natural procedure of distribution of the imputation on the time interval $[t_0, T]$ which leads to the STCOP.

Let $\bar{\xi} \in C(x_0, T-t_0)$ and a function $\beta_i(t)$, $i = 1, \ldots, n$, $t \in [t_0, T]$ satisfies the condition

$$\int_{t_0}^{T} \beta_i(t)dt = \bar{\xi}_i, \ \beta_i(t) \geq 0.$$

The function $\beta(t) = \{\beta_i(t)\}$ we shall call the imputation distribution procedure (IDP). Define

$$\int_{t_0}^{\Theta} \beta_i(t)dt = \bar{\xi}_i(\Theta), \ i = 1, \ldots, n.$$

Definition 1. The OP $\bar{C}(x_0, T-t_0)$ is called STC if there exist such an IDP $\beta(t) = \{\beta_i(t)\}$, that

$$\bar{\xi}(\Theta) + \bar{C}(x^*(\Theta), T-\Theta) \subset \bar{C}(x_0, T-t_0), \tag{9.1.14}$$

for all $\Theta \in [t_0, T]$.

The definition of STC is applicable for larger class o OP's (such that Shaply value, NM–solution, Kalai–Smorodinsky solution, etc.)

The STC of the OP means that if an imputation $\bar{\xi} \in \bar{C}(x_0, T-t_0)$ and an IDP $\beta(t) = \{\beta_i(t)\}$ of $\bar{\xi}$ are selected, then after getting by the players, on the time interval $[t_0, \Theta]$, the amount

$$\bar{\xi}_i(\Theta) = \int_{t_0}^{\Theta} \beta_i(t)dt, \ i = 1, \ldots, n,$$

the optimal income (in the sense of the OP $\bar{C}(x^*(\Theta), T-\Theta)$) on the time interval $[\Theta, T]$ in the subgame $\Gamma(x^*(\Theta), T-\Theta)$ together with $\bar{\xi}(\Theta)$ constitutes the imputation belonging to the OP in the original game $\Gamma(x_0, T-t_0)$. This condition is stronger than time consistency, where we have only

$$\bar{\xi} - \bar{\xi}(\Theta) \subset \bar{C}(x^*, T-\Theta),$$

which means that the part of the previously considered "optimal" imputation belongs to the OP in the corresponding current subgame $\Gamma(x^*(\Theta), T-\Theta)$.

Theorem 2 *OP* $\bar{C}(x_0, T-T_0)$ *is STC in* $\Gamma(x_0, T-t_0)$.

Proof: Define $\beta_i(t)$

$$\beta_i(t) = \frac{\hat{\xi}_i(t)}{T - t_0} + \frac{T - t}{T - t_0} \cdot h_i(x^*(t)), \tag{9.1.15}$$

where $\hat{\xi}(t) \in C(x^*(t), T - t)$, $t \in [t_0, T]$. Consider the set $\bar{\xi}(\Theta) + \bar{C}(x^*(\Theta), T - \Theta)$, where

$$\bar{\xi}(\Theta) = \int_{t_0}^{\Theta} \beta(t)dt. \tag{9.1.16}$$

From (9.1.16) we get

$$\bar{\xi}(\Theta) = \frac{1}{T - t_0} \cdot \int_{t_0}^{\Theta} \hat{\xi}(t)dt + \frac{1}{T - t_0} \int_{t_0}^{\Theta} (T - t)h(x^*(t))dt =$$

$$= \frac{1}{T - t_0} \int_{t_0}^{\Theta} \hat{\xi}(t)dt + \frac{1}{T - t_0} \int_{t_0}^{\Theta} \left[\int_{t_0}^{t} h(x^*(\tau))d\tau\right] dt +$$

$$+ \frac{1}{T - t_0} \int_{t_0}^{\Theta} (T - t)h(x^*(t))dt - \frac{1}{T - t_0} \int_{t_0}^{\Theta} \left[\int_{t_0}^{t} h(x^*(\tau))d\tau\right] dt = \tag{9.1.17}$$

$$= \frac{1}{T - t_0} \int_{t_0}^{\Theta} \left[\int_{t_0}^{t} h(x^*(\tau))d\tau + \hat{\xi}(t)\right] dt +$$

$$+ \frac{1}{T - t_0} \int_{t_0}^{\Theta} (T - t)h(x^*(t))dt - \frac{1}{T - t_0} \int_{t_0}^{\Theta} \left[\int_{t_0}^{t} h(x^*(\tau))d\tau\right] dt =$$

$$= \frac{1}{T - t_0} \int_{t_0}^{\Theta} \left[\int_{t_0}^{t} h(x^*(\tau))d\tau + \hat{\xi}(t)\right] dt + \frac{T - \Theta}{T - t_0} \int_{t_0}^{\Theta} h(x^*(\tau))d\tau,$$

by using the formula

$$\Theta \int_{t_0}^{\Theta} h(x^*(\tau))d\tau =$$

$$= \int_{t_0}^{T} th(x^*(t))dt + \int_{t_0}^{\Theta} \left[\int_{t_0}^{t} h(x^*(\tau))d\tau\right] dt.$$

From (9.1.17) and (9.1.16) we get

$$\bar{\xi}(\Theta) + \bar{C}(x^*(\Theta), T - \Theta) =$$

$$= \int_{t_0}^{\Theta} \frac{1}{T - t_0} \left[\int_{t_0}^{t} h(x^*(\tau))d\tau + \hat{\xi}(t)\right] dt +$$

$$+ \frac{T - \Theta}{T - t_0} \int_{t_0}^{\Theta} h(x^*(\tau))d\tau + \bar{C}(x^*(\Theta), T - \Theta) =$$

$$= \frac{1}{T - t_0} \int_{t_0}^{\Theta} \left[\int_{t_0}^{t} h(x^*(\tau))d\tau + \hat{\xi}(t)\right] dt +$$

$$+\frac{T-\Theta}{T-t_0}\int_{t_0}^{\Theta} h(x^*(\tau))d\tau+$$

$$+\Big\{\bar{\xi}^{\Theta} : \bar{\xi}^{\Theta} = \frac{1}{T-t_0}\int_{\Theta}^{T}\Big[\int_{\Theta}^{t} h(x^*(\tau))d\tau + \xi'(t)\Big]\,dt,$$

$$\text{for all } \xi'(t) \subset C(x^*(t), T-t),\ t \in [\Theta, T]\Big\}.$$

We see that every element $\bar{\xi}$ of the set $\bar{\xi}(\Theta) + \bar{C}(x^*(\Theta), T-\Theta)$ is represented in the form

$$\bar{\xi} = \frac{1}{T-t_0}\int_{t_0}^{\Theta}\Big[\int_{t_0}^{t} h(x^*(\tau))d\tau + \hat{\xi}(t)\Big]\,dt+$$

$$+\frac{T-\Theta}{T-t_0}\int_{t_0}^{\Theta} h(x^*(\tau))d\tau + \frac{1}{T-t_0}\int_{\Theta}^{T}\Big[\int_{\Theta}^{t} h(x^*(\tau))d\tau + \xi'(t)\Big]\,dt,$$

where $\xi'(t)$ is some imputation from the core $C(x^*(t), T-t)$, $t \in [\Theta, T]$ in the subgame $\Gamma(x^*(t), T-t)$. But the function

$$\tilde{\xi}(t) = \begin{cases} \hat{\xi}(t), & t \in [t_0, \Theta), \\ \xi'(t), & t \in [\Theta, T] \end{cases}$$

also is a selector from $C(x^*(t), T-t)$, $t \in [t_0, T]$, thus

$$\bar{\xi} = \frac{1}{T-t_0}\int_{t_0}^{\Theta}\Big[\int_{t_0}^{t} h(x^*(\tau))d\tau + \hat{\xi}(t)\Big]\,dt+$$

$$+\frac{T-\Theta}{T-t_0}\int_{t_0}^{\Theta} h(x^*(\tau))d\tau + \frac{1}{T-t_0}\int_{\Theta}^{T}\Big[\int_{\Theta}^{t} h(x^*(\tau))d\tau + \xi'(t)\Big]\,dt =$$

$$= \frac{1}{T-t_0}\int_{t_0}^{\Theta}\Big[\int_{t_0}^{t} h(x^*(\tau))d\tau + \hat{\xi}(t)\Big]\,dt+$$

$$+\frac{1}{T-t_0}\int_{\Theta}^{T}\int_{t_0}^{\Theta} h(x^*(\tau))d\tau dt+$$

$$+\frac{1}{T-t_0}\int_{\Theta}^{T}\Big[\int_{\Theta}^{t} h(x^*(\tau))d\tau + \xi'(t)\Big]\,dt =$$

$$= \frac{1}{T-t_0}\int_{t_0}^{\Theta}\Big[\int_{t_0}^{t} h(x^*(\tau))d\tau + \hat{\xi}(t)\Big]\,dt+$$

$$+\frac{1}{T-t_0}\int_{\Theta}^{T}\Big[\int_{t_0}^{t} h(x^*(\tau))d\tau + \xi'(t)\Big]\,dt =$$

$$= \frac{1}{T-t_0}\int_{t_0}^{T}\Big[\int_{t_0}^{t} h(x^*(\tau) + \tilde{\xi}(t)\Big]\,dt \subset \bar{C}(x_0, T-t_0)$$

and we have

$$\bar{\xi}(\Theta) + \bar{C}(x^*(\Theta), T-\Theta) \subset \bar{C}(x_0, T-t_0),$$

for all $\Theta \in [t_0, T]$. The theorem is proved. ■

It may be easily seen that in the case of terminal payoff, when

$$K_i(x_0, T - t_0; u_1, \ldots, u_n) = H_i(x(T)),\ i = 1, \ldots, n$$

all the results remain valid if we put

$$\xi = \frac{1}{T - t_0} \int_{t_0}^{T} \xi(t) dt,$$

where $\xi(t) \in C(x^*(t), T - t)$,

$$\bar{\xi}(\Theta) = \frac{1}{T - t_0} \int_{t_0}^{\Theta} \xi(t) dt,$$

$$\beta(t) = \frac{\xi(t)}{T - t_0},$$

$$\bar{C}(x^*(\Theta), T - \Theta) = \frac{1}{T - t_0} \int_{\Theta}^{T} C(x^*(t), T - t) dt,$$

$$\Theta \in [t_0, T].$$

Using the c.f. $\bar{V}(S; x_0, T - t_0)$ we may compute the Shapley value $\bar{Sh} = \{\bar{Sh}_i\}$ in the game $\Gamma(x_0, T - t_0)$. We get the following formula

$$\bar{Sh}(x_0, T - t_0) = \frac{1}{T - t_0} \int_{t_0}^{T} Sh(x^*(t), T - t) dt,$$

where $Sh(x^*(t), T - t)$ is a Shapley value for the subgame $\Gamma(x^*(t), T - t)$ with c.f. $V(S; x^*(t), T - t)$, $S \subset N$. The following equality holds

$$\bar{Sh}(x_0, T - t_0) = \frac{1}{T - t_0} \int_{t_0}^{\Theta} Sh(x^*(t), T - t) dt +$$

$$+ \bar{Sh}(x^*(\Theta), T - \Theta),\ \Theta \in [t_0, T],$$

where $\bar{Sh}(x^*(\Theta), T - \Theta)$ is a Shapley value in the subgame $\Gamma(x^*(t), T - t)$ with c.f. $\bar{V}(S; x^*(t), T - t)$. As in the case of core we may construct the IDP with guaranties the STC (which in this case coincides with TC) of the $\bar{Sh}$. For this reason it is enough to put

$$\beta(t) = \frac{1}{T - t_0} Sh(x^*(t), T - t).$$

The construction of the STC NM-solution precedes in much the same way.

9.2 Differential strongly time consistent optimality principles

In 1 we have introduced a family of strongly time consistent optimality principles (SPCOP) based upon the integration of optimality principles (OP) in current subproblems occurring along the optimal path. When trying to apply this new OP's to the optimization of complicated economic systems one faces the problem of the redistribution of the investments on the time interval under consideration. The latter may require additional investments in intermediate time intervals. This tightens the applicability of the purposed integral STCOP. To overcome the difficulty we purpose here another approach suitable for the optimization of developing economic systems and close to one considered by S. Chistjakov.

Let n–person dynamic game $\Gamma(x_0, T-t_0)$ with prescribed deraction $T-t_0$, from the initial position x_0.

$$\dot{x} = f(x, u_1, \ldots, u_n),\ x \in R^n,\ u_i \in U_i \subset CompR^\ell, \tag{9.2.1}$$

with integral payoffs

$$K_i(x_0, T-t_0; u_1, \ldots, u_n) = \int_{t_0}^{T} h_i(x(t))dt$$

$$h_i > 0,\ i = 1, \ldots, n.$$

be given.

Consider the cooperative form of the game $\Gamma(x_0, T-t_0)$. In this formalization as before we suppose that the players before starting the game agree to play $u_1^*, \ldots, u_n^*$ such that the corresponding trajectory maximizes the sum of the payoffs

$$\max \sum_{i=1}^{n} K_i(x_0, T-t_0; u_1, \ldots, u_n) =$$

$$= \sum_{i=1}^{n} K_i(x_0, T-t_0; u_1^*, \ldots, u_n^*) =$$

$$= \sum_{i=1}^{n} \int_{t_0}^{T} h_i(x^*(t))dt = V(N; x_0, T-t_0),$$

where N is the set of all players in $\Gamma(x_0, T-t_0)$. The trajectory $x^*(t)$ is called optimal. Let $S \subset N$, and $V(S; x_0, T-t_0)$ be a characteristic function. Denote by $L(x_0, T-t_0)$ the set of all imputations in $\Gamma(x_0, T-t_0)$, i.e.,

$$L(x_0, T-t_0) = \{\xi = \{\xi_i\}\ :\ \sum_{i=1}^{n} \xi_i = V(N; x_0, T-t_0),$$

$$\xi_i \geq V(\{i\}; x_0, T - t_0),\ i = 1, \dots, n\}.$$

Let $C^{t-t_0}(x_0)$ ($t \in [t_0, T]$) be a reachable set of the system. I.e., the set of all points in R^n which could be reached at instant $t \in [t_0, T]$ from the initial position $x_0 = x(t_0)$ according to (9.2.1) with the help of some admissible open–loop control $u(\tau)$, $\tau \in [t_0, t]$. For each $y \in C^{t-t_0}(x_0)$ consider a subgame $\Gamma(y, T-t)$ of the game $\Gamma(x, T-t_0)$ with corresponding characteristic function $V(S; y, T-t)$ and set of imputations $L(y, T-t)$.

Definition 2. A point to set mapping

$$C(y, T-t) \subset L(y, T-t)$$

defined for all $y \in C^{t-t_0}(x_0)$, $t \in [t_0, T]$ is called optimality principle (OP) in the family of subgames $\Gamma(y, T-t)$.

In special cases $C(y, T-t)$ may be a core, NM–solution, Shaply–value e.t.c.

Consider the family of subgames $\Gamma(x^*(t), T-t)$, along the optimal trajectory $x^*(t)$, $t \in [t_0, T]$, with corresponding characteristic functions $V(S; x_0^*(t), T-t)$, and sets of imputations $L(x^*(t), T-t)$.

Define now a natural procedure of distribution of the imputation on the time interval $[t_0, T]$ which leads to the differential STCOP.

Let $\bar{\xi} \in C(x_0, T-t_0)$ and a function $\beta_i(t)$, $i = 1, \dots, n$, $t \in [t_0, T]$ satisfies the condition

$$\int_{t_0}^{T} \beta_i(t)dt = \bar{\xi}_i,\ \beta_i(t) \geq 0.$$

The function $\beta(t) = \{\beta_i(t)\}$ we shall call the imputation distribution procedure (IDP). Define

$$\int_{t_0}^{\Theta} \beta_i(t)dt = \bar{\xi}_i(\Theta),\ i = 1, \dots, n.$$

Definition 3. The OP $\bar{C}(x^*(t), T-t)$, $t \in [t_0, T]$ is called dynamic stable TC (time consistent) if there exist such an IDP $\beta(t) = \{\beta_i(t)\}$, that

$$\bar{\xi} - \bar{\xi}(\Theta) \subset \bar{C}(x^*(\Theta), T-\Theta) \tag{9.2.2}$$

for all $\Theta \in [t, T]$ (see Ch. 8).

Definition 4. The OP $\bar{C}(x^*(t), T-t)$, $t \in [t_0, T]$ is called strongly dynamic stable STC (strongly time consistent) if there exist such an IDP $\beta(t) = \{\beta_i(t)\}$, that

$$\bar{\xi} + \bar{C}(x^*(\Theta), T-\Theta) \subset \bar{C}(x_0, T-t_0) \tag{9.2.3}$$

for all $\Theta \in [t_0, T]$. Here $\bar{\xi} + \bar{C}(x^*(\Theta), T-\Theta)$ means the set of all possible vectors $\bar{\xi} + \eta$, for all $\eta \in \bar{C}(x^*(\Theta), T-\Theta)$.

The STC of the OP means that if an imputation $\bar{\xi} \in \bar{C}(x_0, T-t_0)$ and IDP $\beta(t) = \{\beta_i(t)\}$ of $\bar{\xi}$ are selected, then after getting by the players on the time interval $[t_0, \Theta]$ of the amount

$$\bar{\xi}_i(\Theta) = \int_{t_0}^{\Theta} \beta_i(t)dt, \; i = 1, \ldots, n$$

any optimal income (in the sense of the OP $\bar{C}(x^*(\Theta), T-\Theta)$) on the time interval $[\Theta, T]$ in the subgame $\Gamma(x^*, T-\Theta)$ together with $\bar{\xi}_i(\Theta)$ constitutes the imputation belonging to the OP in the original game $\Gamma(x_0, T-t_0)$. This condition is stronger than time–consistency.

Suppose $t_0 = \Theta_0 < \Theta_1 < \ldots < \Theta_n < \Theta_{n+1} < \ldots < \Theta_m = T$ is a partition of the time interval $[t_0, T]$ such that $\Theta_{n+1} - \Theta_n = \delta$, $k = 0, 1, \ldots, m-1$.

If (9.2.2) holds only in the points Θ_k, $k = 0, 1, \ldots, m-1$, i.e.

$$\bar{\xi} - \bar{\xi}(\Theta_k) \subset \bar{C}(x^*(\Theta_k), T - \Theta_k) \tag{9.2.4}$$

we call OP $\bar{C}(x^*(t), T-t)$ δTC (δ time–consistent). If (9.2.3) holds only in the points Θ_k, $k = 0, 1, \ldots, m-1$, i.e.

$$\bar{\xi}(\Theta_k) + \bar{C}(x^*(\Theta_k), T - \Theta_k) \subset \bar{C}(x_0, T - t_0) \tag{9.2.5}$$

we call OP $\bar{C}(x^*(t), T-t)$ δSTC (δ strongly time–consistent).

Now we have everything necessary to construct differential STCOP. Introduce the following functions.

$$\beta_i^1(\tau) = \frac{\xi_i^1 \sum_{i=1}^n h_i(x^*(\tau))}{V(N; x_0, T)}$$

$\tau \in [\Theta_0, \Theta_1)$, where $\xi^1 \in C(x^*(t_0), T-t)$, $x_0 = x(\Theta_0) = x(t_0)$.

$$\beta_i^2(\tau) = \frac{\xi_i^2 \sum_{i=1}^n h_i(x^*(\tau))}{V(N; x^*(\Theta_1), T - \Theta_1)} \tag{9.2.6}$$

$\tau \in [\Theta_1, \Theta_2)$, where $\xi^2 \in C(x^*(\Theta_1), T - \Theta_1)$.

$$\beta_i^k(\tau) = \frac{\xi_i^k \sum_{i=1}^n h_i(x^*(\tau))}{V(N; x^*(\Theta_{k-1}), T - \Theta_{k-1})}$$

$\tau \in [\Theta_{k-1}, \Theta_k)$, where $\xi^k \in C(x^*(\Theta_{k-1}), T - \Theta_{k-1})$.

$$\beta_i^m(\tau) = \frac{\xi_i^m \sum_{i=1}^n h_i(x^*(\tau))}{V(N; x^*(\Theta_{m-1}), T - \Theta_{m-1})}$$

$\tau \in [\Theta_{m-1}, \Theta_m = T]$, where $\xi^m \in C(x^*(\Theta_{m-1}), T - \Theta_{m-1})$.

Define the IDP $\beta(\tau)$ by formula

$$\bar{\beta}(\tau) = \beta^k(\tau),\ \tau \in [\Theta_{k-1}, \Theta_k) \tag{9.2.7}$$

$k = 1, \ldots, n-1$. It is easily seen that $\bar{\beta} > 0$.

Consider the formulas (9.2.6). For different imputations $\xi^k \in C(x^*(\Theta_{k-1}), T-\Theta_{k-1})$, $k = 1, \ldots, n-1$ we get different functions $\beta^k(\tau)$, $\tau \in [\Theta_k, \Theta_{k+1})$, and thus different functions $\beta(\tau)$ defined by (9.2.7). Let B be the set of all such IDP $\beta(\tau)$, $\tau \in [t_0, T]$ for all possible $\xi^k \in C(x^*(\Theta_{k-1}), T-\Theta_{k-1})$, $k = 1, \ldots, n-1$.Consider the set

$$\bar{\bar{C}}(x_0, T-t_0) = \left\{\bar{\xi}\ :\ \bar{\xi} = \int_{t_0}^{T} \beta(\tau)d\tau,\ \beta(\tau) \in B\right\}.$$

and the sets

$$\bar{\bar{C}}(x^*(\Theta_k), T-\Theta_k) = \left\{\bar{\xi}^k\ :\ \bar{\xi}^k = \int_{\Theta_k}^{T} \beta(\tau)d\tau,\ \beta(\tau) \in B\right\}.$$

The set $\bar{\bar{C}}(x_0, T-t_0)$ is called the regularized OP $\bar{C}(x_0, T-t_0)$, and correspondingly $\bar{\bar{C}}(x^*(\Theta_k), T-\Theta_k)$ is a regularized OP $\bar{C}(x^*(\Theta_k), T-\Theta_k)$.

We consider $\bar{\bar{C}}(x_0, T-t_0)$ as a new optimality principle in the game $\Gamma(x_0, T-t_0)$.

Theorem 3 *If the IDP $\beta(\tau)$, $\tau \in [t_0, T]$ is defined by (9.2.7) then always*

$$\bar{\xi}(\Theta_k) + \bar{\bar{C}}(x^*(\Theta_k), T-\Theta_k) \subset \bar{\bar{C}}(x_0, T-t_0)$$

i.e. the OP $\bar{\bar{C}}(x^*(\tau), T-t)$ is δ STCOP.

Proof: Suppose $\bar{\xi} = \bar{\xi}(\Theta_k) + \bar{\bar{C}}(x^*(\Theta_k), T-\Theta_k)$. Then $\bar{\xi} = \bar{\xi}(\Theta_k) + \int_{\Theta_k}^{T} \beta(\tau)d\tau$ for some $\beta(\tau) \in B$. But $\bar{\xi}(\Theta_k) = \int_{t_0}^{\Theta_k} \beta'(\tau)d\tau$ for some $\beta'(\tau) \in B$. Consider

$$\beta''(\tau) = \begin{cases} \beta'(\tau), & \tau \in [t_0, \Theta_k), \\ \beta(\tau), & \tau \in [\Theta_k, T], \end{cases}$$

then $\beta''(\tau) \in B$, and

$$\bar{\xi} = \int_{t_0}^{T} \beta''(\tau)d\tau$$

and thus $\bar{\xi} \in \bar{\bar{C}}(x_0, T-t_0)$. The theorem is proved. ■

The defined IDP has the advantage (compared with integral one defined in 1 of this chapter)

$$\sum_{i=1}^{n} \beta_i(\tau) = 1,\ \tau \in [t_0, T],$$

and thus

$$\sum_{i=1}^{n} \bar{\xi}_i(\Theta) = \sum_{i=1}^{n} \int_{t_0}^{\Theta} h_i(x_i(\tau))d\tau \tag{9.2.8}$$

which is the actual amount to be divided between the players on the time interval $[t_0, \Theta]$ and which is as it is seen by the formula (9.2.8), exactly equal to the amount earned by them on this time interval. Thus for the realization of the purposed IDP no additional investments are needed ((9.2.8) may not hold for integral OP's).

If δ tends to zero we may get STCOP by introducing the IDP $\beta(\tau)$, $\tau \in [t_0, T]$, by the formula

$$\beta_i(\tau) = \frac{\xi_i(\tau) \sum_{i=1}^{n} h_i(x^*(\tau))}{V(N; x^*(\tau), T - \tau)}$$

where $\xi_i(\tau) \in C(x^*(\tau), T - \tau)$ is an integrable selector.

9.3 Strongly time consistent optimality principles for the games with discount payoffs

The problem of dynamic stability (time consistency) for the n–person differential games with discount payoffs was first mentioned in [68], where it was proved that even the Pareto optimal solutions may be time inconsistent in this case. The reason is that in a discount payoff case the payoffs of the players in subgames acquiring along an optimal path essentially change their structure implieing the time inconsistency of the chosen optimality principle (OP). Till the last time no attempts have been made for the regularizing of the OP's in discount payoff case. We refer to the report [69] on V International Symposium on Dynamic Games where this question was once more stated. Here we try to use the approach from 2 of this chapter to construct a family of strongly dynamic stable optimality principles in the case under discussion. Here we shall consider core as OP in the game, but all the results remain valid for any other subset of imputations, considered as optimality principle.

Consider n–person differential game $\Gamma(x_0)$

$$\dot{x} = F(x, u_1, \ldots, u_n),\ u_i \in u \subset CompR^{\ell}$$

$$x(t_0) = x_0$$

with payoffs

$$K_i(x_0, u_1, \ldots, u_n) = \int_{t_0}^{\infty} e^{\lambda_i (t - t_0)} h_i(x(t))dt,\ h_i > 0,\ \lambda_i > 0 \tag{9.3.1}$$

Cooperative form of $\Gamma(x_0)$.

Consider the n-tuple of open loop controls $u_1^*, \ldots, u_n^*$ such that the corresponding trajectory maximizes the sum of the payoffs (utilities)

$$\max_u \sum_{i=1}^n K_i(x_0; u_1, \ldots, u_n) = \sum_{i=1}^n K_i(x_0; u_1^*, \ldots, u_n^*) =$$

$$= \sum_{i=1}^n \int_{t_0}^\infty e^{-\lambda_i(t-t_0)} h_i(x^{1*}(t))dt = V(N; x_0),$$

where N is the set of all players in $\Gamma(x_0)$. The trajectory $x^{1*}(t)$ is called conditionally optimal. Let $V^1(S; x_0)$ be the characteristic function ($S \subset N$) and $C^1(x_0)$ the core. Consider the family of subgames $\Gamma(x^{1*}(t))$, along $x^{1*}(t)$, $t \in [t_0, \infty)$, corresponding cores and c.f. $V^1(S; x^{1*}(t))$. The payoff functions in the subgames have the form

$$K_i^t(x^*(t), u_i, \ldots, u_n) = \int_t^\infty e^{-\lambda_i(\tau-t)} h_i(x^{1*}(\tau))d\tau,$$

and differs by multiplier $e^{\lambda_i t}$ from the payoff functions in the subgames defined for the games without discount factor. This essentially changes the relative weights of payoff functions of different players, when the game develops, and thus the hole game itself.

Consider the partition of the time interval $[t_0, \infty)$ by the points $\Theta\ t_0 = \Theta_0 < \Theta_1 < \ldots < \Theta_k < \Theta_{k+1} < \ldots$, where $\Theta_{k+1} - \Theta_k = \delta > 0$ does not depend upon k, the subgame $\Gamma(x^{1*}(\Theta_1))$,c.f. $V^2(S; x^{1*}(\Theta_1))$ and core $C^2(x^{1*}(\Theta_1))$. Let $x^{2*}(t)$, $t \geq \Theta_1$ be the conditionally optimal trajectory in the subgame $\Gamma(x^{1*}(\Theta_1))$, i.e. such that

$$\max_n \sum_{i=1}^n K_i^\Theta(x^{1*}(\Theta_1), u_1, \ldots, u_n) =$$

$$= \sum_{i=1}^n K_i^\Theta(x^{1*}(\Theta_1), u_i^*, \ldots, u_*^n) =$$

$$= \sum_{i=1}^n \int_{\Theta_1}^\infty e^{-\lambda_i(\tau-\Theta_1)} h_i(x^{2*}(\tau))d\tau.$$

Then consider the subgame $\Gamma(x^{2*}(\Theta_2))$, c.f. $V^3(S; x^{2*}(\Theta_2))$, core $C^3(x^{2*}(\Theta_2))$. Continuing in the same manner we get the sequence of subgames $\Gamma(x^{k*}(\Theta_k))$, c.f. $V^{k+1}(S; x^{k*}(\Theta_n))$, cores $C^{k+1}(x^{k*}(\Theta_k))$ and conditionally optimal trajectories $x^{(k+1)*}(t)$, $t \geq \Theta_k$.

Definition 5. The trajectory $x^* = x^{k*}(t)$, $t \in [\Theta_{k-1}, \Theta_k)$, $k = 1, 2, \ldots$ is called optimal in the game $\Gamma(x_0)$.

In this formalization of the cooperative game we suppose that the players starting the game agree to use the optimal trajectory $x^*(t)$, $t \geq \Theta_0$, $t_0 = \Theta_0$. Our formalization depends upon $\delta > 0$, and we denote the cooperative form of $\Gamma(x_0)$ by $\Gamma^\delta(x_0)$.

The vector function $\beta_1(\tau), \ldots, \beta_i(\tau), \ldots, \beta_n(\tau)$ is called the utily distribution procedure (UDP) in $\Gamma^\delta(x_0)$ if

$$\sum_{i=1}^{n} \int_{\Theta_k}^{\Theta_{k+1}} \beta_i(t)dt =$$

$$= \sum_{i=1}^{n} \int_{\Theta_k}^{\Theta_{k+1}} e^{-\lambda_i(t-\Theta_k)} h_i(x^{(k+1)*}(t))dt =$$
$$= V^{k+1}(N; x^{k+1*}(\Theta_k)) - V^{k+1}(N; x^{k+1*}(\Theta_{k+1})),$$
$$\beta_i(t) \geq 0,\ i = 1, \ldots, n,\ k = 0, 1, \ldots.$$

Let $\xi^k \in C^{k+1}(x^{k*}(\Theta_k))$, $k = 0, 1, \ldots$ be any imputation in the subgame $\Gamma(x^{k*}(\Theta_k))$ belonging to the core of this subgame. Define the function $\beta_i(t)$, $t \in [\Theta_k, \Theta_{k+1})$ by the formula

$$\beta_i(t) = \frac{\xi_i^k \sum_n^{i=1} \int_{\Theta_k}^{\Theta_{k+1}} e^{-\lambda_i(t-\Theta_k)} h_i(x^{(k+1)*}(t))dt}{V^{k+1}(N; x^{k*}(\Theta_k))(\Theta_{k+1} - \Theta_k)} =$$

$$= \frac{\xi_i^k [V^{k+1}(N; x^{k+1*}(\Theta_k)) - V^{k+1}(N; x^{k+1*}(\Theta_{k+1}))]}{V^{k+1}(N; x^{k*}(\Theta_k))\delta} \geq 0.$$

$i = 1, \ldots, n$, $k = 0, 1, \ldots$. The functions $\{\beta_i(t)\}$, $t \geq t_0$ constitute for each $\xi^k \in C^{k+1}(x^{k*}(\Theta_k))$, $k = 0, 1, \ldots$ a UDP in $\Gamma^\delta(x_0)$. Let $\bar{\xi} = \int_{t_0}^{\infty} \beta(t)dt$. Denote by $\bar{C}(x_0)$ the set of all such vectors $\bar{\xi}$, for all possible UDP's $\beta(t)$, for different $\xi^k \in C^{k+1}(x^{k*}(\Theta_k))$, $k = 0, 1, \ldots$. Consider $\bar{C}(x_0)$ as optimality principal (OP) in $\Gamma^\delta(x_0))$, and call it the regularized core (RC). Define $\bar{C}(x^*(\Theta_k))$ for subgame $\Gamma^\delta(x^*(\Theta_k))$.

Definition 6. The $\bar{C}(x_0)$ is called STC if for all $\beta \in \bar{C}(x_0)$ we have

$$\left\{\int_{\Theta_0}^{\Theta_k} \beta_i(t)dt\right\} + \bar{C}(x^*(\Theta_k)) \subset \bar{C}(x_0) = \bar{C}(x^*(\Theta_0)).$$

Theorem 4 *The (RC) $\bar{C}(x_0)$ is a STC optimality principle in $\Gamma^\delta(x_0)$.*

Bibliography

[1] R. Isaacs, *Differential games* , (Kryger Publishing Company, Huntington, New–York, 1975), p. 474.

[2] N. N. Vorobyev, *Advances in Math. Sc.*, vol. 14, no. 4, (1959).

[3] N. N. Vorobyev and I. V. Romanovsky, *Len. Univ. Bulletin*, no. 7, part 2, (1959).

[4] D. Geil, *The Theory of Linear Economic Models*, (Mc–Hill, New–York, 1960), p. 410.

[5] M. I. Zelikin and N. T. Tiniansky, *Advances in Math. Sc.*, vol. 20, no. 4, (1965).

[6] V. I. Zubov, *USSR Ac. of Sc., Reports*, vol 190, no. 4, (1970).

[7] D. Isbell, *Positional Games*, (M., Nauka , Moskow, 1967).

[8] N. Karlin, *Mathematical Methods in Game Theory, Programming and Economics* , (Pergamon Press, London, 1959), p. 838.

[9] D. Z. Kelendzheridze, *USSR Ac. of Sc., Reports*, vol. 133, no. 3, (1961).

[10] N. N. Krasovskii, *Game Theoretic Problems of Motion Control* , (M., Nauka , Moscow, 1970), p. 420.

[11] N. N. Krasovskii and A. I. Subbotin, *Positional Differential Games*, (M., 1967), p. 455.

[12] N. N. Krasovskii, *Appl. Math. and Mech.*, vol. 27, no. 2, (1963).

[13] G. U. Kuhn, *Positional games*, (M., 1967).

[14] V. N. Lagunov, *Theory of games*, (Yerevan, 1973).

[15] D. I. Littlewood, *Mathematician's Miscellany*, (Methuen & Co, London, 1960).

[16] O. A. Malafeyev, *Len. Univ. Bulletin*, no. 19, part 4, (1972).

[17] O. A. Malafeyev, *Controlled Systems*, no. 4–5, (1970).

[18] E. F. Mishchenko, *Automatics and Telemechanics*, no. 9, (1972).

[19] E. F. Mishchenko and L. S. Pontryagin, *USSR Ac. of Sc., Reports*, vol. 174, no. 1, (1967).

[20] M. S. Nikolsky, *USSR Ac. of Sc., Reports*, vol. 205, no. 4, (1972).

[21] M. S. Nikolsky, *Cybernetics*, no. 2, (1973).

[22] J. S. Osipov, *USSR Ac. of Sc., Reports*, vol. 203, no. 1, (1972).

[23] N. M. Petrov, *USSR Ac. of Sc., Reports*, vol. 190, no. 6, (1970).

[24] L. A. Petrosjan, *USSR Ac. of Sc., Reports*, vol. 161, no. 2, (1965).

[25] L. A. Petrosjan, *USSR Ac. of Sc., Reports*, vol. 195, no. 3, (1970).

[26] L. A. Petrosjan, *Advances in Theory of Games*, (Vilnius, 1973).

[27] L. A. Petrosjan, *Armenian Ac. of Sc., Reports*, vol. 44, no. 1, (1967).

[28] L. A. Petrosjan, *Len. Univ. Bulletin*, no. 19, (1972).

[29] L. A. Petrosjan, *Armenian Ac. of Sc., Bulletin*, vol. 8, no. 2, (1973).

[30] L. A. Petrosjan, *Armenian Ac. of Sc., Bulletin*, vol. 3, no. 5, (1968).

[31] L. A. Petrosjan, *USSR Ac. of Sc., Reports*, vol. 161, no. 1, (1965).

[32] L. A. Petrosjan, *Problems of Cybernetics*, no. 27, (1973).

[33] L. A. Petrosjan, *Armenian Ac. of Sc., Reports*, vol. 40, no. 4, (1965).

[34] L. A. Petrosjan and J. G. Dutkevich, *Len.Univ. Bulletin*, no. 13, part 3, (1969).

[35] L. A. Petrosjan and N. V. Murzov, *USSR Ac. of Sc., Reports*, vol. 172, no. 6, (1967).

[36] L. S. Pontryagin and E. F. Mishenko, *Differential Equations*, vol. 7, no. 3, (1971).

[37] L. S. Pontryagin, *USSR Ac. of Sc., Reports*, vol. 191, no. 2, (1970).

[38] L. S. Pontryagin, *USSR Ac. of Sc., Reports*, vol. 174, no. 6, (1967).

[39] L. S. Pontryagin, *USSR Ac. of Sc., Reports*, vol. 175, no. 4, (1967).

[40] L. S. Pontryagin, *USSR Ac. of Sc., Reports*, vol. 156, no. 4, (1967).

[41] L. S. Pontryagin and E. F. Mischenko, *USSR Ac. of Sc., Reports*, vol. 7, no. 3, (1971).

[42] B. N. Pshenichny, *Cybernetics* , no. 1, (1968).

[43] B. N. Pshenichny, *Mathematical Research Methods and System Optimization*, Kiev, no. 3, (1969).

[44] A. A. Savelov, *Plane curves*, (Fizmatgiz, Moscow, 1960).

[45] N. Y. Satimov, *Differential Equations*, vol 9, no. 10, (1973).

[46] N. Y. Satimov, *USSR Ac. of Sc., Engineering Cybernetics*, no. 6, (1972).

[47] H. E. Scarf and L. S. Shapley, *Applications of Game Theory in Military Affairs*, (Moscow, 1961).

[48] N. T. Tiniansky, *USSR Acad of Sc., Reports*, vol. 207, no. 1, (1972).

[49] I. P. Jachauskas, *Lith. Math. Articles*, no. 1, (1967).

[50] L. D. Berkovitz, *Ann. Math. Study*, vol. 52, (Princeton, 1964).

[51] L. D. Berkovitz, *Math. Theory of Control*, ed. by Balakrishnan and Newstadt, (Academic press, 1967).

[52] W. H. Fleming, *J. Math. and Appl.* , vol. 3, (1961).

[53] W. H. Fleming, *Advances in Game Theory*, (Princeton, 1964).

[54] A. Friedman, *Differential Equations*, vol. 12, no. 1, (1972).

[55] A. Friedman, *Differential Equations*, vol. 12, no. 2, (1972).

[56] A. Friedman, *Differential Equations*, vol. 12, no. 3, (1972).

[57] J. C. Ho, *SIAM J. on Control*, vol. 4, no. 3, (1966).

[58] J. C. Ho, A. E. Bryson and S. Barson, *IEEE Trans. Aut. Control*, vol. 10, no. 4, (1965).

[59] G. Leitman, *Intern. J. of Non Linear Mechanics*, vol. 3, (1969).

[60] C. Ryll Nardzenski, *Ann. Math. Studies*, no. 3, (1957).

[61] H. E. Scarf, *Ann. Math Studies*, no. 3, (1957).

[62] H. Steinhaus, *Mysl. Acad.*, vol. 1, no. 1, (1925).

[63] P. P. Varaiya, *SIAM J. on Control*, vol. 5, (1967).

[64] P. P. Varaiya and J. Lin, *SIAM J. on Control*, vol. 7, (1969).

[65] A. Zieba, *Studia Mat.*, vol. 22, (1962).

[66] L. A. Petrosjan, *Vestnik of the Len State Univ.*, no. 13, (1977).

[67] L. A. Petrosjan, *Dynamic control*, (Sverdlovsk, 1979).

[68] R. H. Strotz, *Review of Economic Studies XXIII*, (1955–56).

[69] V. Kaitala and M. Pohjola, *Reports of the Fifth International Symposium of Dynamic Games and Applications*, (Jeneve, 1992).

[70] T. Basar and G. J. Olsder, *Dynamic Noncooperative Game Theory*, (Academic Press, London, 1982), p. 255.

[71] A. Friedman, *Differential Games*, (John Wiley, New York, 1971), p. 280.

[72] J. Nash, *Annals of Mathematics*, vol. 54, no. 2, (1951).

[73] J. Von Neumann and O. Morgenstern, *Theory of Games and Economic Behavior*, (Princeton University Press, Princeton, NJ, 1947), p. 850.

[74] G. Owen, *Game Theory*, (Sauders, Philadelphia, 1968), p. 320.

[75] Arunabha Bagchi, *Stackelberg Differential Games in Economic Models*, in *Lecture Notes in Control and Information Sciences*, ed. A. V. Balakrishnan and M. Thoma, no. 64, (Springer–Verlag, Berlin, 1984), p. 203.

[76] *Differential Games in Economic Analysis*, ed. R. P. Hamalainen, H. K. Ehtamo, in *Lecture Notes in Control and Information Sciences*, ed. M. Thoma and A. Winer, no. 157, (Springer–Verlag, Berlin, 1992), p. 312.

[77] *Differential Games–Developments in Modelling and Computation*, in *Lecture Notes in Control and Information Sciences*, ed. M. Thoma and A. Winer, no. 156, (Springer–Verlag, Berlin, 1992), p. 294.

[78] A. I. Subbotin and A. Y. Chentsov, *Optimization of Guarantee in Control Problems*, (Nauka, Moskow, 1981), p. 264.

[79] F. L. Chernousko and A. A. Melikjan, *Game-theoretic Problems of Control and Search*, (Nauka, Moskow, 1978), p. 268.

[80] N. L. Grigorenko, *USSR Ac. of Sc., Reports*, vol. 258, no. 2, (1988).

[81] M. I. Zelikin, *USSR Ac. of Sc., Reports*, vol. 202, no. 5, (1972).

[82] A. F. Kononenko, *USSR Ac. of Sc., Reports*, vol. 231, no. 2, (1976).

[83] E. F. Mishenko, *USSR Ac. of Sc., Engineering Cybernetics*, no. 5, (1971).

[84] M. S. Nikolsky, *Math. Proceedings*, vol. 116, no. 1, (1981).

[85] N. Y. Satimov, *Math. Notes*, vol. 21, no. 3, (1977).

[86] F. L. Chernousko, *Appl. Math. and Mech*, vol. 40, no. 1, (1976).

[87] A. A. Chikry, *Appl. Math. and Mech*, vol. 39, no. 5, (1975).

[88] A. N. Krasovskii, *USSR Ac of Sc., Reports*, vol. 253, no. 6, (1980).

[89] A. I. Subbotin, *USSR Ac. of Sc., Reports*, vol. 254, no. 2, (1980).

[90] A. I. Subbotin and N. N. Subbotina, *Appl. Math. and Mech*, vol. 46, no. 2, (1982).

[91] B. B. Rihsiev, *Differential Games with Simple Motions*, (Fan, Tashkent, 1989), p. 232.

[92] L. A. Petrosjan and N. N. Danilov, *Cooperative Differential Games and their Applications*, (Izd. Tomskogo Univ., Tomsk, 1985), p. 276.

[93] N. N. Krasovskii, *Dynamic Systems Control*, (Nauka, Moscow, 1985), p. 518.

[94] L. A. Petrosjan and G. V. Tomskii, *Geometry of Simple Pursuit*, (Nauka, Novosibirsk, 1983), p. 142.

[95] L. A. Petrosjan and G. V. Tomskii, *Dynamic Games and Their Applications*, (Izd. Len Univ., Leningrad, 1982).

[96] L. A. Petrosjan and G. V. Tomskii, *Differential Games with Incomplete Information*, (Izd. Irk. Univ., Irkutsk, 1984), p. 188.

[97] L. A. Petrosjan and B. B. Rihsiev, *Pursuit on the Plane*, (Nauka, Moskow, 1991), p. 96.

[98] L. A. Petrosjan, *Len. Univ. Bulletin*, Ser. 1, no. 8(2), (1992).

[99] L. A. Petrosjan, in *IFIP Technical Conference Reports*, (Springer–Verlag, Berlin, 1974).

[100] L. A. Petrosjan and T. I. Kuzmina, *Non–zero Sum Differential Games*, (Jakutsk, 1989), p. 147.

[101] L. A. Petrosjan, in *Reports of VII conference of Applied Mathematics in University of Osijck*, ed. R. Skitovsky, (Osijck, 1990).

[102] L. A. Petrosjan and V. V. Zakharov, *Introduction to the Mathematical Ecology*, (Izd. Len. Univ.,Leningrad 1986), p. 220.

[103] L. A. Petrosjan, in *Proceedings of the International Conference of Differential Equations and Applications*, hold in Ruse, Bulgaria, Aug. 13–19, 1989. (Ruse, 1991).

[104] N. A. Zenkevich, *Len. Univ. Bulletin*, no. 19 (1981).

[105] V. V. Zakharov, *Len. Univ. Bulletin*, no. 1 (1988).

[106] G. V. Tomskii, *Appl. Math. and Mech.*, vol. 45 no. 2 (1981).

[107] V. I. Zhukovskii and N. T. Tiniansky, *Equilibrium Controls in Multicriterial Dynamic Systems*, (Moskow Univ., Moskow, 1984), p. 224.

[108] V. I. Zhukovskii, in *Operation Reseach*, (Bulgar, Acad. Sci., Sofia, 1885).

[109] N. M. Sloboshanin, *Len. Univ. Bulletin*, vol. 3, no. 13 (1981).

Index